Generalized Functionals of Brownian Motion and Their Applications

Nonlinear Functionals of Fundamental Stochastic Processes

Generalized Functionals of Brownian Motion and Their Applications

Nonlinear Functionals of Fundamental Stochastic Processes

N U Ahmed

University of Ottawa, Canada

World Scientific

NEW JERSEY · LONDON · SINGAPORE · BEIJING · SHANGHAI · HONG KONG · TAIPEI · CHENNAI

Published by

World Scientific Publishing Co. Pte. Ltd.

5 Toh Tuck Link, Singapore 596224

USA office: 27 Warren Street, Suite 401-402, Hackensack, NJ 07601

UK office: 57 Shelton Street, Covent Garden, London WC2H 9HE

British Library Cataloguing-in-Publication Data
A catalogue record for this book is available from the British Library.

GENERALIZED FUNCTIONALS OF BROWNIAN MOTION AND THEIR APPLICATIONS

ISBN-13 978-981-4366-36-6
ISBN-10 981-4366-36-6

Printed in Singapore.

In memory of my Mother, Father,
Uncles and my wife Feroza and my teacher Professor
G.S.Glinski who gave so much.

Dedicated to my children Jordan, Schockley, Pamela, Rebeka,
and my grandchildren Reynah-Sofia, Maximus, Achilles, Eliza,
Pearl,
Austin, Rio, Kira, Jazzmine.

Preface

In this monograph we present some recent developments in the theory of generalized functionals of Brownian motion and their applications to stochastic differential equations, inverse problems (identification), nonlinear filtering, Navier-Stokes equations, optimization, and optimal control. First, we give a brief summary of the Wiener-Itô multiple integrals, construct an abstract L_2 space over the Wiener measure space, and present a Fourier analysis on this space leading to the Riesz-Fischer theorem. We discuss some of the basic properties of these multiple integrals and related functional analysis on Wiener measure space. Subsequently these results are extended by constructing Hilbert spaces of generalized functionals which contain the abstract L_2 space as a dense subspace. This leads to a Gelfand triple of generalized functionals. We consider some important topics on compactness and weak compactness, including their characterization on such spaces. This is used to study optimization and inverse (identification) problems related to stochastic differential equations in finite and infinite dimensional spaces. Similar results on functionals of fractional Brownian motion and Lévy process, in particular the centered Poisson process, are briefly presented. These results, a significant part of which are collected from the work of the author, have not been presented in book form previously.

We then discuss the pioneering work of Hida who constructed, for the first time, a class of generalized functionals of Brownian motion. These functionals of Hida are based on L_2-Sobolev spaces admitting $H^s, s \in R$, valued kernels in the multiple stochastic integrals. These functionals are more general than the classical Wiener-Itô class. More recent developments by the author allows us to introduce an even broader class of generalized functionals which are based on $L_p, 1 \leq p < \infty$, spaces and L_p-Sobolev spaces admitting kernels from the spaces $\mathcal{W}^{p,s}$, $s \in R$. This permits analysis of a very broad class of nonlinear functionals of Brownian motion, which cannot be handled by either the Wiener-Itô class or the Hida class. For $s \leq 0$, they represent generalized functionals on the Wiener measure space like Schwartz distributions on finite dimensional spaces. Also, we introduce some further generalizations and construct locally convex topological vector spaces of generalized functionals and discuss their applications. For the first time in the literature, we introduce vector measures with values in the space of generalized functionals of Brownian motion and consider their application to nonlinear filtering with solutions which are stochastic measure valued processes.

There are 9 chapters in this monograph. In Chapter 1 we introduce some background materials necessary for application of generalized functionals. Chapters 2-7 present functionals of Brownian motion, including fractional Brownian motion and Lévy processes, in the order of increasing generality. Chapter 2 presents regular functionals of Brownian motion (in finite and infinite dimensional Hilbert spaces), Gaussian random fields, fractional Brownian motion and Lévy process. In Chapter 3, we present a class of generalized functionals covering the classical Wiener-Itô class that leads to a Gelfand triple possessing Hilbertian structure. In Chapter 4 we consider some important aspects of functional analysis on such spaces and their application to optimization and inverse (identification) problems. In Chapter 5 we present the class of generalized functionals due to Hida. In Chapter 6 these results are further extended by constructing several classes of generalized functionals based on $L_p, 1 \leq p < \infty$, spaces. Also studied in this chapter are vector measures with values in the space of generalized functionals of Brownian motion and Gaussian random fields. These are new developments, and are used there in the study of inverse problems and nonlinear filtering. In Chapter 7 we construct generalized functionals based on the duals of Sobolev spaces $W^{m,p}$. This is further extended by constructing abstract generalized functionals using the so-called Gelfand triple. Here also, for the first time in the literature, we introduce and study the properties of vector measures with values in the space of Wiener-Itô distributions. These results are then applied to Nonlinear Filtering problems and Stochastic Navier-Stokes equations, including their optimal control. The natural space for solutions of such systems is the space of measure valued processes, or equivalently vector measures with values in the space of Wiener-Itô distributions as mentioned above. In Chapter 8, we describe some fundamental elements of the Malliavin calculus and their applications. We construct Sobolev spaces on Wiener measure space and utilize them in the study of abstract evolution equations on Fock spaces. The Malliavin calculus is also used here in the study of stochastic differential equations. In particular, we consider the questions of existence and smoothness of densities of measure valued functions induced by the solutions of stochastic differential equations. Further, using Malliavin calculus we generalize the central limit theorem due to Nualart from finite to infinite dimensional Hilbert spaces. In Chapter 9, we consider evolution equations on Fock spaces. We start with the Malliavin and Ornstein-Uhlenbeck operators and consider linear and semilinear evolution equations on abstract Fock spaces and Wiener-Sobolev spaces. We prove existence, uniqueness and regularity properties of solutions for these equations.

Target Audience: The book is intended for an audience that includes research workers in the field, mathematicians, and mathematically inclined scientists who want to apply stochastic analysis to physical problems. The book should also be useful to graduate students who may wish to apply or advance the theory presented here. For this purpose, problems for exercise are added at the end of each chapter. The recommended prerequisites for smooth reading are measure theory and functional analysis.

The author would like to thank Professor T. Hida for his encouragement throughout the writing of this monograph. Also thanks are due to Dr. Swee Cheng Lim of the World Scientific Publishers for his many valuable technical suggestions. I would also like to thank Ms. Lai Fun Kwong of WSPC for her incredible patience in proofreading the entire manuscript.

N. U. Ahmed

Contents

Chapter 1

Background Material

1.1 Introduction

In this chapter we present some background materials required for study of the rest of the monograph. Further, we include some areas of applications involving stochastic differential equations to illustrate the usefulness of generalized functionals of Brownian motion in the following chapters. In section 1.2, we present basic properties of Wiener process and define the classical Wiener measure space. In the following section, we present some results on existence, uniqueness and regularity properties of solutions of stochastic differential equations in finite dimensional spaces. In section 1.4, this is extended to stochastic differential equations on infinite dimensional Hilbert spaces. This section also includes briefly some results on measure valued solutions. In section 1.5, we consider filtering problems in both finite and infinite dimensional spaces and present the main results used later in the monograph. In section 1.6, we include some basic results from the theory of vector measures which are useful for the rest of this book.

1.2 Wiener Process and Wiener Measure

Let $(\Omega, \mathcal{F}, P)$ be a complete probability space and $\{\mathcal{F}_t, t \geq 0\}$ be an increasing family of complete subsigma algebras of the sigma algebra $\mathcal{F}$. The family $\{\mathcal{F}_t, t \geq 0\}$ is assumed to be right continuous having left limits. By right continuity one means that $\mathcal{F}_t = \mathcal{F}_{t+} \equiv \cap_{s>t}\mathcal{F}_s$, and by left limit, it is understood that $\lim_{s\uparrow t} \mathcal{F}_s = \mathcal{F}_{t-} \equiv \sigma(\cup_{s<t}\mathcal{F}_s))$ where $\sigma(\cup_{s<t}\mathcal{F}_s))$ denotes

1

the sigma field generated by the union. We call $(\Omega, \mathcal{F} \supset \mathcal{F}_t, t \geq 0, P)$ the filtered probability space or a probability space with a filtration. For any $\mathcal{F}$-measurable random variable Z, we let

$$m(Z) \equiv \mathbb{E}(Z) \equiv \int_\Omega Z(\omega) P(d\omega)$$

denote its expectation (mean) and

$$V(Z) \equiv \mathbb{E}\{(Z - \mathbb{E}(Z))^2\}$$

its variance.

Definition 1.2.1 *A real valued process* $\{W(t), t \in I \equiv [0, \infty)\}$ *is called a standard Brownian motion (or Wiener process) if it satisfies the following two properties:*

(i): $P\{ W(0) = 0\} = 1$,

(ii): Over disjoint intervals $\{(t_i, t_{i+1}] \subset I\}$, *the increments of W given by* $\{W(t_{i+1}) - W(t_i)\}$ *are independent Gaussian random variables with mean zero and variance* $\{(t_{i+1} - t_i)\}$. *Hence for any Borel set* $\Gamma \subset R$

$$P\{W(t_{i+1}) - W(t_i) \in \Gamma\}$$
$$= \int_\Gamma (1/\sqrt{2\pi(t_{i+1} - t_i)}) \exp\{-(1/2)(x^2/(t_{i+1} - t_i))\} dx. \quad (1.1)$$

From these properties one can easily deduce the following important properties of Brownian motion.

(P1): $\mathbb{E}(W(t)W(s)) = t \wedge s$.

(P2): $\mathbb{E}(W(t) - W(s))^2 = |t - s|$, $\mathbb{E}(W(t) - W(s))^4 = 3(t - s)^2$.

(P3): For any $\varepsilon > 0$, $\lim_{s \to t} P\{|W(t) - W(s)| > \varepsilon\} = 0$.

(P4): $\lim_{s \to t} P\{|\frac{W(t) - W(s)}{t - s}| < r\} = 0$ for any $r \in [0, \infty)$.

(P5): $P\left\{\sup_{\pi_n} \sum_{i=0}^{n-1} |W(t_{i+1}) - W(t_i)| < \infty\right\} = 0$, where

$$\pi_n \equiv \{0 = t_0 < t_1 < t_2 \cdots < t_{n-1} < t_n = T < \infty\}$$

is any partition of the interval $I \equiv [0, T]$ with the supremum taken over all such partitions.

(P6): $\lim_{s \to t} P\left\{\frac{|W(t) - W(s)|}{|t - s|^\alpha} > \varepsilon\right\} = 0, \forall \varepsilon > 0$, and $\alpha \in [0, 1/2)$.

It follows from the property (P2) and the well-known Kolmogorov's criterion for continuity of stochastic processes that the Brownian motion W

is continuous with probability one. Even better, it follows from the property (P6) that Brownian motion is actually Hölder continuous with exponent $\alpha \in (0, 1/2)$. According to (P4), Brownian motion is nowhere differentiable (in the classical sense) and according to (P5), it has unbounded variation on any interval with probability one. Hence the length of the path traced by a Brownian particle over any finite time interval is $+\infty$ with probability one. It is very clear from these properties that Brownian motion is quite a wild process.

An n-dimensional standard Brownian motion defined on the probability space $(\Omega, \mathcal{F}, P)$ is nothing but n independent copies of the scalar Brownian motion and it has exactly similar properties as stated above except the change necessary for dimensionality as indicated below. For any Borel set $\Gamma \subset R^n$, and any $s, t \in I$,

$$P\{W(t) - W(s) \in \Gamma\}$$
$$= \int_\Gamma (1/(2\pi|t - s|)^{n/2}) \exp\{(-1/2)|x|^2_{R^n}/|t - s|\} dx. \quad (1.2)$$

From this expression, it follows that for any $\xi, \eta \in R^n$ and $t, s \in I$, we have

$$\mathbb{E}(W(t), \xi) = 0, \quad \mathbb{E}\{(W(t), \xi)^2\} = t|\xi|^2_{R^n},$$

and $\mathbb{E}\{(W(t), \xi)(W(s), \eta)\} = t \wedge s(\xi, \eta)$.

The reader can also verify that the characteristic function of the random variable $W(t)$ is given by

$$\mathbb{E}\{e^{i(W(t), \xi)}\} = \exp\{(-1/2)t|\xi|^2\}$$

for every $t \in I$ and $\xi \in R^n$.

In view of the above, we see that the sample paths of the Brownian motion are elements of the space of continuous functions starting from the origin. We denote these by $C_0(I, R)$ for the scalar case, and $C_0(I, R^n)$ for the n-vector case. These are Banach spaces with respect to the sup-norm topology. We can view the Brownian motion W as a random variable $W : \Omega \ni \omega \longrightarrow W(\omega)$ with values in $C_0(I, R^n)$. In other words $W(\omega) \equiv \{\omega(t), t \in I\}$ is a particular realization corresponding to the elementary event ω. Using an increasing family of finite dimensional distributions defined on cylinder sets of the form

$$
\left\{ C_k \equiv \{x \in C_0(I, R^n) : x(t_1) \in \Gamma_1, \cdots, x(t_k) \in \Gamma_k\}, \right.
$$

$$
\left. \{t_i\} \in I, \{\Gamma_i\} \in \mathcal{B}_n, k \in N \right\},
$$

and using the finite dimensional Gaussian distributions such as (1.2) induced by the Brownian motion one can construct a sequence of countably additive measures on an increasing family of cylinder sets. Then by use of the well-known Kolmogorov extension theorem, one can construct a countably additive measure μ^W on the sigma algebra $\sigma(\mathcal{C})$ generated by the class of cylinder sets $\mathcal{C}$ of the Banach space $\Omega_0 \equiv C_0(I, R^n)$.

This measure is then extended to the class of Carathéodory measurable sets of the set Ω_0 by use of the outer measure constructed from the probability measure μ^W. We denote this extension also by μ^W. This is the classical Wiener measure space which we denote by $(\Omega_0, \mathcal{B}_0, \mu^W)$. It was Norbert Wiener [Wiener (1938)] who first constructed this measure and hence it is usually known as the Wiener measure. A detailed account of the method of construction of the Wiener measure as discussed above is given by Yeh [Yeh (1973)]. Note that $\mu^W = PW^{-1}$ where W is the map $W : \Omega \to C_0(I, R^n)$ as introduced above. Conversely, for any measurable map $X : \Omega \longrightarrow C_0(I, R^n)$, let $\mu^X \equiv PX^{-1}$ denote the corresponding measure induced on the path space $C_0(I, R^n)$. If, under this measure, the elements of $C_0(I, R^n)$ satisfy component wise the properties (i) and (ii) of Definition 1.2.1, then we have the Wiener measure $\mu^W = \mu^X$.

In the sequel, we consider both measurable functions and generalized functions on the Wiener measure space.

It is well known that for decades Physicists and Engineers have been using Gaussian white noise which is basically the time derivative of the Wiener process. We have already noted that Brownian motion is nowhere differentiable in the classical sense. But by virtue of the well-known Schwartz distribution theory [Adams (1975)] Brownian motion has derivatives of all orders in the distribution sense. For example, for any $\xi \in C_0^\infty(I, R^n)$,

C^∞-functions with compact supports, and any integer $m \geq 1$,

$$(d^m/dt^m W(\cdot), \xi(\cdot)) = \int_I (W^{(m)}(t), \xi(t))dt = (-1)^m \int_I (W(t), \xi^{(m)}(t))dt,$$

and it is a well defined Gaussian random variable with mean zero and variance

$$V = \int_I |\xi^{(m-1)}(t)|^2_{R^n} dt < \infty.$$

For $m = 0$, the reader can easily verify that

$$\mathbb{E}\left((W(\cdot), \xi(\cdot))\right)^2 = \mathbb{E}\left(\int_I (W(t), \xi(t))dt\right)^2 = \int_{I \times I} t \wedge s \ (\xi(t), \xi(s))dsdt.$$

Thus, the m-th derivative of Brownian motion denoted by $W^{(m)}$ is a random distribution (in the sense of Schwartz), not a function, and, for any pair $\{\xi, \eta\} \in C_0^\infty(I, R^n)$, it is easy to verify that

$$\mathbb{E}\{(W^{(m)}, \xi)(W^{(m)}, \eta)\} = \int_{I \times I} t \wedge s \ (\xi^{(m)}(t), \eta^{(m)}(s))_{R^n} dsdt$$

$$= \int_I (\xi^{(m-1)}(t), \eta^{(m-1)}(t))_{R^n} dt.$$

More on Wiener measure space and the measure space constructed by Hida on the dual of a suitable nuclear space (white noise processes) see the work of Hida [Hida (1980)] [Hida (2008)].

1.3 Stochastic Differential Equations in R^d

Before we consider stochastic differential equations, it may be useful to present a brief motivation for the first time readers. In the study of dynamical systems, differential equations are used as models for the physical system under consideration. This is given by an ordinary differential equation on R^d as follows

$$\dot{x}(t) = b(t, x(t)), x(0) = x_0, t \in I \equiv [0, T], \tag{1.3}$$

where $b : [0, \infty) \times R^d \longrightarrow R^d$, and $x_0 \in R^d$ is the initial state and $T \in (0, \infty)$. Under suitable assumptions on the vector field b, such as continuity on $I \times R^d$, one can prove the existence of a solution locally in time (possibly with finite blow up time). Under the assumptions of measurability in $t \in I$

and local Lipschitz and (at most) linear growth property in the argument $x \in R^d$, one can prove the existence of a unique solution $x(t) = x(t, x_0), t \in I$, continuously dependent on the initial state x_0. Given the vector field b, the evolution of the process x is entirely determined by the initial state. Often in the real world, the vector field b may not be precisely known. For example, the contact friction of the car tire on the roadbed, which determines the dynamics of motion of the car on the road, is not often precisely known. Thus, if one uses an imprecise b, one may find significant discrepancy between the natural process and the process determined by the model b. Also there are situations where the initial state x_0 is not precisely known. Under these conditions, one is forced to consider a model that can capture such discrepancies and indeterminacies in a probabilistic fashion. This can be done by introducing an additional perturbation ξ to the dynamics given by

$$\dot{x}(t) = b(t, x(t)) + \xi(t), t \geq 0, \tag{1.4}$$

where the most popular model for the process ξ is the white noise of compatible dimension. As seen in the previous section, white noise is the formal derivative of Brownian motion and so, as it stands, the equation (1.4) has no classical meaning. The correct way is to write it as an integral equation which is symbolically written as

$$dx(t) = b(t, x(t))dt + dW(t), t \geq 0. \tag{1.5}$$

However, this model also is not very satisfactory since the fluctuation here is purely additive and independent of state. The intensity of random fluctuation may also depend on time and the current state, and hence a more satisfactory model is obtained by considering the following system,

$$dx(t) = b(t, x(t))dt + \sigma(t, x(t))dW(t), t \geq 0, \tag{1.6}$$

where σ is a matrix valued function of time and the current state with compatible dimension. Given the current sate $x(t) \equiv \eta$, an intuitive interpretation of this model can be given by writing it as an incremental equation as follows

$$x(t + \Delta t) \approx \eta + b(t, \eta)\Delta t + \sigma(t, \eta)\Delta W \tag{1.7}$$

where $\Delta W(t) \equiv W(t + \Delta t) - W(t)$ denotes the increment of the Brownian motion over the time interval $(t, t + \Delta t]$. Clearly, given the current state $x(t) = \eta$, the conditional expectation of the random variable $x(t + \Delta t)$ is given by $m(\eta) \equiv \eta + b(t, \eta)\Delta t$ and its covariance (matrix) is given by $V(\eta) \equiv \Delta t \, \sigma(t, \eta)\sigma(t, \eta)^*$ provided W is a standard Brownian motion. In case of nonstandard Brownian motion, the incremental covariance matrix of W is a positive square matrix Q. In this case $V = \Delta t \, \sigma(t, \eta)Q\sigma(t, \eta)^*$. It is clear from the expression (1.7) that $x(t + \Delta t)$ is conditionally Gaussian and hence one may consider the process x to be locally Gaussian, though globally it is far from it. Geometrically, in the state space R^d, given that $x(t) = \eta$, one may picture $x(t + \Delta t)$ as a random vector describing a ball of radius $r(\eta) \equiv \sqrt{TrV(\eta)}$ centered at the point $m(\eta)$. Clearly, for Δt sufficiently small and V nonsingular, the conditional probability law of the random variable $x(t + \Delta t)$, given $x(t) = \eta$, is approximately Gaussian and for any Borel set $\Gamma \subset R^d$, we have

$$P\{x(t + \Delta t) \in \Gamma | x(t) = \eta\}$$
$$= \left(1/(2\pi)^{d/2}|V|^{1/2}\right) \int_\Gamma \exp -(1/2)\{(\xi - m(\eta))V^{-1}(\xi - m(\eta))\}d\xi,$$

where $|V|$ is the determinant of the matrix V. Taking the ball $B_{r(\eta)}(m(\eta))$ for Γ one can verify that most of the probability mass is concentrated in this ball and hence it is reasonable to call this the ball of uncertainty of the position of the random vector $x(t + \Delta t)$ issuing from $m(\eta)$. Clearly, as $\Delta t \to 0$, the ball shrinks to the point η.

In order to study the system (1.6), or equivalently the associated integral equation

$$x(t) = x_0 + \int_0^t b(s, x(s))ds + \int_0^t \sigma(s, x(s))dW(s), t \in I \equiv (0, T], \quad (1.8)$$

it is necessary to understand the notion of stochastic (Itô) integral to justify the last term in the above equation. This can be found in any text book on stochastic differential equations [Skorohod (1965)], [Ahmed (1998)]. We present a brief outline. Let $L_2^a(I, \mathcal{L}(R^n, R^d))$ denote the class of matrix valued functions $\{F\}$ (or functions with values in the space of linear operators $\mathcal{L}(R^n, R^d)$) defined on I which are $\mathcal{F}_t$ adapted and satisfy

$$\mathbb{E} \int_I \| F(t) \|_{HS}^2 \, dt \equiv \mathbb{E} \int_I Tr(F^*(t)F(t))dt < \infty. \quad (1.9)$$

Then, let $\mathcal{S}^a(I, \mathcal{L}(R^n, R^d)) \subset L_2^a(I, \mathcal{L}(R^n, R^d))$ denote the class of simple functions in the sense that, for every $F \in \mathcal{S}^a(I, \mathcal{L}(R^n, R^d))$, there exists a finite disjoint partition of the interval I giving $I \equiv \cup_{i=0}^{m-1}(t_i, t_{i+1}]$, and random matrices $\{M_i, 0 \le i \le m-1\} \in \mathcal{L}(R^n, R^d)$ with M_i being $\mathcal{F}_{t_i}$ measurable, so that

$$F(t) = \sum_{i=0}^{m-1} \chi_{(t_i, t_{i+1}]}(t) M_i.$$

For any such F the stochastic integral is given by

$$\mathcal{I}(F) = \int_I F(t) dW(t) = \sum_{i=0}^{m-1} M_i[W(t_{i+1}) - W(t_i)].$$

This is a well defined R^d valued $\mathcal{F}$-measurable random variable. By using conditional expectation and their properties, it is easy to verify that $\mathbb{E}\mathcal{I}(F) = 0$, and that

$$\mathbb{E}|\mathcal{I}(F)|^2_{R^d} = \sum \mathbb{E} \parallel M_i \parallel^2_{HS} (t_{i+1} - t_i) = \mathbb{E} \int_I \parallel F(t) \parallel^2_{HS} dt.$$

Now we use the fact that simple functions are dense in $L_2^a(I, \mathcal{L}(R^n, R^d))$. Thus for $F \in L_2^a(I, \mathcal{L}(R^n, R^d))$ there exists a sequence $F_k \in \mathcal{S}^a(I, \mathcal{L}(R^n, R^d))$ such that $F_k \xrightarrow{s} F$ in $L_2^a(I, \mathcal{L}(R^n, R^d))$. The Itô integral of F with respect to the Brownian motion W is then given by

$$\mathcal{I}(F) = \lim_{k \to \infty} \int_I F_k(t) dW(t).$$

This gives a brief outline of Itô integral. For more on this topic see [Skorohod (1965)], [Hida (1980)], [Ahmed (1998)]. Now we present some basic results on the question of existence and regularity properties of solutions of stochastic differential equations driven by standard Brownian motion. Recall that a measurable stochastic process $x \equiv \{x(t), t \ge 0\}$ is said to be $\mathcal{F}_t$-adapted if for each $t \ge 0$, the random variable $x(t)$ is measurable with respect to the sigma algebra $\mathcal{F}_t$.

Definition 1.3.1 *Let $B_\infty^a(I, L_2(\Omega, R^d))$ denote the class of $\mathcal{F}_t$ adapted R^d valued random processes having uniformly bounded second moments given by*

$$\sup\{\mathbb{E}|x(t)|^2_{R^d}, t \in I\} < \infty.$$

Endowed with the norm topology, $\| x \| \equiv \sqrt{\sup\{\mathbb{E}|x(t)|^2_{R^d}, t \in I\}}$ for $x \in B^a_\infty(I, L_2(\Omega, R^d))$, this is a Banach space.

Theorem 1.3.2 *Consider the system (1.6) in the integral form (1.8) and suppose the initial state x_0 has finite second moment and there exists a $K \in L_2(I)$ such that*

$$|b(t,x)|^2_{R^d} + \| \sigma(t,x) \|^2_{HS} \leq K^2(t)\{1 + |x|^2_{R^d}\} \tag{1.10}$$

$$|b(t,x) - b(t,y)|^2_{R^d} + \| \sigma(t,x) - \sigma(t,y) \|^2_{HS} \leq K^2(t)\{|x - y|^2_{R^d}\}. \tag{1.11}$$

Then, the integral equation has a unique solution $x \in B^a_\infty(I, L_2(\Omega, R^d))$ and further, with probability one, $x \in C(I, R^d)$.

Proof Detailed proof can be found in any of the books [Skorohod (1965)], [Hida (1980)], [Ahmed (1998)]. We present a brief outline. Define the operator G by

$$(Gx)(t) \equiv x_0 + \int_0^t b(s, x(s))ds + \int_0^t \sigma(s, x(s))dW(s), t \in I. \tag{1.12}$$

Under the growth assumption (1.10), it is easy to verify that G maps $B^a_\infty(I, L_2(\Omega, R^d))$ to itself. Further, under the Lipschitz assumption (1.11), for sufficiently large m, the m-th iterate G^m of G is a contraction in the Banach space $B^a_\infty(I, L_2(\Omega, R^d))$. Hence, by Banach fixed point theorem, G^m and consequently, G itself has one and the same fixed point in $B^a_\infty(I, L_2(\Omega, R^d))$. To prove that $x \in C(I, R^d)$ almost surely, one uses Borel-Cantelli Lemma [25]. •

Itô Differential: In the study of stochastic differential equations, Itô differential plays a significant role. First consider the ordinary differential equation

$$\dot{x} = b(t,x), x(0) = x_0, t \in I,$$

and let $\varphi \in C^{1,2}(I \times R^d)$, and consider the approximation

$$\Delta x \approx x(t + \Delta t) - x(t) \approx b(t, x(t))\Delta t.$$

Then

$$\varphi(t + \Delta t, x(t + \Delta t)) \approx \varphi(t, x(t)) + (\partial\varphi/\partial t)\Delta t + (D\varphi, b(t, x(t)))\Delta t$$
$$+ (1/2)(D^2(\varphi)b, b)(\Delta t)^2 + o(\Delta t) \tag{1.13}$$

where $\partial\varphi/\partial t$ denotes the partial of φ with respect to t, $D\varphi = \{\partial_i\varphi\}$ denotes the gradient of φ with respect to x and $D^2\varphi = \{\partial_i\partial_j\varphi\}$ the matrix of second partials with respect to x. Thus the differential increment of φ is given by

$$\Delta\varphi \approx (\partial\varphi/\partial t)\Delta t + (D\varphi, b)\Delta t + o(\Delta t).$$

Neglecting all terms of (small) order $o(\Delta t)$, we have

$$d\varphi(t, x(t)) = (\partial\varphi/\partial t) \, dt + (D\varphi(t, x(t)), b(t, x(t))dt. \qquad (1.14)$$

This is valid for all $\varphi \in C^{1,1}(I \times R^d)$. Now consider the stochastic differential equation (1.6) and let $\varphi \in C^{1,2}(I \times R^d)$ and compute the differential increment of the function $\varphi(t, x(t))$ along the solution trajectory of equation (1.6). Using similar approximation as (1.13), it is easy to verify that the differential increment is given by

$$\Delta\varphi \approx (\partial\varphi/\partial t)\Delta t + (D\varphi, b)\Delta t + (\sigma^* D\varphi, \Delta W)$$
$$+ (1/2) < (\sigma^* D^2\varphi\sigma)\Delta W, \Delta W > + o(\Delta t),$$
$$(1.15)$$

where $\Delta W = W(t + \Delta t) - W(t)$. Note that the last two terms of the expression (1.13) are of order $o(\Delta t)$ for ordinary differential equations and hence has not made any contribution to the final result (1.14). In contrast, for stochastic differential equations, the infinitesimal increment of φ given by (1.15) contains terms such as $(\sigma^* D\varphi, \Delta W)$ and the quadratic term $(1/2) < (\sigma^* D^2\varphi\sigma)\Delta W, \Delta W >$ which are not of small order $o(\Delta t)$ and hence cannot be omitted. Since the components of W are mutually independent standard Brownian motions, it can be verified that

$$(1/2) < (\sigma^* D^2\varphi\sigma)\Delta W, \Delta W > \longrightarrow (1/2)tr(D^2\varphi\sigma\sigma^*)dt$$

with probability one as $\Delta t \longrightarrow dt$. The rules of thumb are $dtdW = 0, dtdt = 0, (dW_i dW_j) = \delta_{i,j}dt$. Thus, in the limit, it follows from the expression (1.15) that the Itô differential of φ is given by

$$d\varphi = L\varphi dt + (\sigma^* D\varphi, dW), \qquad (1.16)$$

where the operator L is a second order partial differential operator given by

$$L\varphi \equiv (\partial\varphi/\partial t) + (b, D\varphi) + (1/2)tr(D^2\varphi\sigma\sigma^*). \qquad (1.17)$$

For rigorous and detailed proof see Ikeda-Watanabe [Ikeda (1989)].

Some simple examples: For illustration we present few simple examples.

(E1) Consider computing the stochastic integral

$$J \equiv \int_0^T w(t)dw(t)$$

where w is the scalar Brownian motion. For φ, take $\varphi(w) = w^2$. Using the
Itô differential rule given by (1.16) it is easy to see that

$$d\varphi = 2wdw + dt.$$

Thus

$$w^2(T) = 2\int_0^T w(t)dw(t) + T,$$

and hence

$$\int_0^T wdw = (1/2)\{w^2(T) - T\}.$$

On the other hand, if we consider this as an ordinary Stieltjes integral, then
we will have

$$\int_0^T wdw = (1/2)w^2(T).$$

Since w has unbounded variation, this integral cannot be defined as Stieltjes
integral. However, if the integral is interpreted in the sense of Stratonovich
[[Ahmed (1998)], p23] the result is correct. Unlike Lebesgue and Lebesgue-
Stieltjes integrals, the stochastic integral is quite illusive. Consider the
partition $0 = t_0 \le t_1 \le t_2 \cdots t_i \le t_{i+1} \le t_n = T$; and suppose we wish
to evaluate the integral $\mathbb{E}J \equiv \mathbb{E}\{\int_0^T w(t)dw(t)\}$ by taking the limit of
the sum $\mathbb{E}\{\sum_{i=0}^{n-1} w(\tau_i)(w(t_{i+1}) - w(t_i))\}$, $\tau_i \in [t_i, t_{i+1}]$. For existence of a
unique limit one must specify the exact value of τ_i from the set $[t_i, t_{i+1}]$. In
general, for an arbitrary $\tau_i \in [t_i, t_{i+1}]$, the reader can easily verify (using
the properties of iterated conditional expectations) that

$$EJ = \lim_{n\to\infty} \sum_{i=0}^{n-1}(\tau_i - t_i).$$

Clearly if we choose $\tau_i = t_i$ we have the Itô integral, giving $EJ = 0$. If
we choose the mid point, $\tau_i = (1/2)(t_{i+1} + t_i)$, we obtain the Stratonovich

integral giving $EJ = T/2$. Since τ_i can be any point from the interval $[t_i, t_{i+1}]$, this integral has infinitely many values and the range is the closed interval $[0, T]$.

(E2) Consider the Scalar equation

$$d\xi = \alpha\xi dt + \beta\xi dw,$$

where α and β are constants. This model is widely popular in finance where it is used to model the stock price $\{\xi(t), t \geq 0\}$. This model was first proposed by Fischer Black and Myron Scholes for stock price dynamics and hence it is known as the Black-Scholes model. To solve this, take $\varphi = log\xi$. Then using the Itô differential of φ, it is easily seen that

$$d\varphi = (D\varphi)d\xi + (1/2)(D^2\varphi) < d\xi, d\xi > = (\alpha - (1/2)\beta^2)dt + \beta dw.$$

Hence

$$\xi(t) = \xi(0)\exp\left\{\int_0^t (\alpha - (1/2)\beta^2)ds + \int_0^t \beta dw\right\}, t \geq 0.$$

If one uses the classical calculus, the solution is given by the same expression but without the β^2 term. Again this is not justified since the stochastic integral with respect to Brownian motion cannot be computed as the Stieltjes integral.

Before closing this section we present a result on continuous dependence of solutions. The reader will have no difficulty proving this result. Let $B_\infty^a(I, L_2(\Omega, R^d))$ denote the space of bounded measurable stochastic processes adapted to $\mathcal{F}_t, t \geq 0$, having finite second moments. Equipped with norm topology,

$$\sup\{(\mathbb{E}|x(t)|_{R^d}^2)^{1/2}, t \in I\},$$

it is a Banach space. The class of $\mathcal{F}_t$-adapted R^d-valued stochastic processes which are continuous with probability one having finite second moments is denoted by $L_2^a(\Omega, C(I, R^d))$. Endowed with the norm topology,

$$\mathbb{E}(|x|_{C(I,R^d)}^2)^{1/2},$$

this is also a Banach space and it is clear that $L_2^a(\Omega, C(I, R^d)) \subset B_\infty^a(I, L_2(\Omega, R^d))$.

Theorem 1.3.3 *Let $\mathcal{I}_{ad} \equiv \{(b, \sigma)\}$ denote the set of all admissible drifts and diffusion coefficients satisfying the Lipschitz and growth properties (1.10) and (1.11) uniformly for a fixed $K \in L_2(I)$. Let $(b_n, \sigma_n) \in \mathcal{I}_{ad}$ satisfying $b_n \longrightarrow b_o$ and $\sigma_n \longrightarrow \sigma_o$ for almost all $t \in I$ and each $x \in R^d$, and let $x_{0,n} \longrightarrow x_0$ in the mean square sense. Then the solutions of equation (1.6), corresponding to the sequence $\{(b_n, \sigma_n)\}$ and denoted by $x^n = \mathcal{I}(b_n, \sigma_n)$, converge uniformly on I with probability one to the solution $x^o = \mathcal{I}(b_o, \sigma_o)$ corresponding to the pair $\{b_0, \sigma_0\}$, and further $x^o \in L_2^a(\Omega, C(I, R^d)) \subset B_\infty^a(I, L_2(\Omega, R^d))$.*

Proof. The proof is based on the Doob's martingale inequality, the Borel-Cantelli Lemma, the Lebesgue dominated convergence theorem, and the generalized Gronwall Lemma. For detailed proof see [[Ahmed (1988)], Theorem 7.4.7, p360], and [[Skorohod (1965)], Theorem 1, p83] where more general processes with jumps are considered. •

Remark 1.3.4 Consider the path space $C(I, R^d)$ furnished with the usual sup-norm topology and the associated metric giving a complete separable metric space $X_\rho \equiv (C(I, R^d), \rho)$. Let $\mathcal{M}(X_\rho)$ denote the space of probability measures on X_ρ. It is well known [[Parthasarthy (1967)], Theorem 6.7, p47] that the necessary and sufficient condition for weak compactness of any family of probability measures $\{\mu^n\}$ on a complete separable metric spaces X_ρ is tightness. That is, for each $\varepsilon > 0$, there exists a compact set $K_\varepsilon \subset X_\rho$ such that $\mu^n(X_\rho \setminus K_\varepsilon) < \varepsilon$ for all $n \in N$. Since the solutions of the system equation (1.3.6) belong to $L_2^a(\Omega, C(I, R^d))$, in view of Theorem 1.3.3, $x^n \longrightarrow x^o$ in the metric topology ρ with probability one. Thus the corresponding sequence of measures $\mu^n \in \mathcal{M}(X_\rho)$ converges weakly to the measure $\mu^o \in \mathcal{M}(X_\rho)$ corresponding to x^o. This implies that the sequence $\{\mu^n\} \subset \mathcal{M}(X_\rho)$ is uniformly tight. This is useful in the study of Markovian feedback control problems

$$dx = b(t, x, u(t, x))dt + \sigma(t, x, u(t, x))dW, x(0) = x_0, t \in I, u \in \mathcal{U}_{ad}$$

where $\mathcal{U}_{ad}$ is a compact metric space and the cost functional is given by

$$J(u) \equiv \int_{X_\rho} \varphi(x, u)\mu^u(dx)$$

with $\varphi : X_\rho \times \mathcal{U}_{ad} \longrightarrow [0, \infty]$. The problem is to find a control from the admissible set $\mathcal{U}_{ad}$ that minimizes J.

Remark 1.3.5 Let $\mathcal{M}(R^d)$ denote the space of probability measures defined on the sigma algebra of Borel sets $\mathcal{B}_d$ of R^d. In view of Theorem 1.3.3, it is clear that for each $t \in I$, $x^n(t)$ is a well defined R^d valued random variable and hence for each $D \in \mathcal{B}_d$, $\mu_t^n(D) \equiv Prob\{x^n(t) \in (D)\}$ is well defined. Let $\mu_t^o, t \in I$, denote the measure induced by $x^o(t), t \in I$. In view of the preceding result, it is clear that for each $t \in I$, the sequence of measures $\{\mu_t^n\}_n \subset \mathcal{M}(R^d)$ is tight and converges weakly to μ_t^o. Again, this result is very useful in control theory where the drift and diffusion parameters of the system (1.6) are given by $\{b_n(t, x), \sigma_n(t, x)\} = \{b(t, x, u^n), \sigma(t, x, u^n)\}$ with $\{u^n\}$ taking values (possibly) from a compact metric space U. In general, for $u \in \mathcal{U}_{ad}$, let μ_t^u denote the measure induced by the solution $x^u(t)$ for $t \in I$. Then, for each $t \in I$, the reachable set of measures induced by the controlled SDE is given by

$$\mathcal{R}(t) \equiv \{\nu \in \mathcal{M}(R^d) : \nu = \mu_t^u \equiv \mathcal{L}(x^u(t)), u \in U\}.$$

Under some additional assumptions it can be shown that $\mathcal{R}(t) \subset \mathcal{M}(R^d)$ is relatively weakly compact for each $t \in I$. This fact is useful in the study of optimal control. For example, one may wish to find a feedback control policy that minimizes the probability of hitting an insecure zone $\mathcal{O}$, an open subset of R^d. The appropriate cost functional for this problem is given by $J(u) \equiv \sup\{\mu_t^u(\mathcal{O}), t \in I\}$. The problem is to find a control $u^o \in \mathcal{U}_{ad}$ such that $J(u^o) \leq J(u)$ for all $u \in \mathcal{U}_{ad}$.

1.4 Stochastic Differential Equations in H

In this section we wish to present a brief outline of stochastic differential equations on infinite dimensional Hilbert spaces. Let H and U be two separable Hilbert spaces each with a complete orthonormal basis. Let $(\Omega, \mathcal{F}, \mathcal{F}_t, P)$ be a filtered probability space with $\mathcal{F}_t$ being an increasing family of subsigma algebras of the sigma algebra $\mathcal{F}$ and $W \equiv \{W(t), t \in I\}$, an $\mathcal{F}_t$ measurable cylindrical Brownian motion taking values from U. A Brownian motion W taking values from U is said to be cylindrical if it has the representation

$$W(t) = \sum_{i=1}^{\infty} \beta_i(t) u_i$$

where $\{u_i\}$ is a complete orthonormal basis of U and $\{\beta_i\}$ are mutually independent standard Brownian motions. Clearly, the covariance operator for this process is given by tI where I is the identity operator in U. In other words, the characteristic function of this process is given by $\varphi(t, \xi) \equiv \mathbb{E}\{e^{i(W(t),\xi)}\} = e^{(1/2)|\xi|_U^2}$. The corresponding Gaussian measure is known as the cylindrical Wiener measure and it follows from the well-known Minlos-Sazanov theorem [Gihman (1971)] that it is not supported on U since the identity operator is not nuclear.

In general, a nonlinear (semilinear) stochastic differential equation on Hilbert space H is given by

$$dx = Ax\,dt + B(x)\,dt + C(x)\,dW, x(0) = x_0, t \in I, \tag{1.18}$$

where A (generally unbounded) is the infinitesimal generator of a C_0-semigroup $S(t), t \geq 0$, of bounded linear operators in H. We use standard notations to denote the resolvent set $\rho(A)$ of the operator A and $R(\lambda, A) \equiv (\lambda I - A)^{-1}$ for the corresponding resolvent. The maps $B : H \longrightarrow H$ and $C : H \longrightarrow \mathcal{L}(U, H)$ are bounded nonlinear operators where $\mathcal{L}(U, H)$ is the space of bounded linear operators from U to H. In infinite dimensional spaces, for deterministic systems, such as (1.18) with $C(x) \equiv 0$ and x_0 a fixed element of H, there are several notions of solutions such as classical, strong, mild and weak [Ahmed (1991)]. In the study of stochastic systems there are two major notions of solutions known as strong and weak (= martingale) [Da Prato (1992)]. This applies to both finite and infinite dimensional spaces. For the stochastic system (1.18), we present a formal definition of a solution.

Definition 1.4.1 *An $\mathcal{F}_t$ adapted process $x \equiv \{x(t), t \in I\}$, with values in H is said to be a mild solution (considered as strong solution in the stochastic sense) if it satisfies the following stochastic integral equation*

$$x(t) = S(t)x_0 + \int_0^t S(t - r)B(x(r))dr + \int_0^t S(t - r)C(x(r))dW(r),$$

$$\tag{1.19}$$

for $t \in I$, P-almost surely.

For proof of existence and regularity properties of solutions we need some basic assumptions.

Basic Assumptions:

(A): The operator A with domain and range in H is the infinitesimal generator of a C_0-semigroup of bounded linear operators in H satisfying $\sup\{\| S(t) \|_{\mathcal{L}(H)}, t \in I\} \leq M$.

(B): The operator $B : H \longrightarrow H$ is continuous and there exists a constant $b > 0$ such that

$$|B(x)|_H^2 \leq b(1 + |x|_H^2), \quad \text{and} \quad |B(x) - B(y)|_H^2 \leq b|x - y|_H^2$$

for all $x, y \in H$.

(C): The operator $C : H \longrightarrow \mathcal{L}_2(U, H)$ (the space of Hilbert-Schmidt operators) is continuous and there exists a constant $c > 0$ such that

$$\| C(x) \|_{\mathcal{L}_2(U,H)}^2 \leq c(1 + |x|_H^2), \quad \text{and} \quad \| C(x) - C(y) \|_{\mathcal{L}_2(U,H)}^2 \leq c|x - y|_H^2.$$

Theorem 1.4.2 *Consider the system (1.18) and suppose the assumptions (A),(B) and (C) hold. Let U be a separable Hilbert space and W a cylindrical Brownian motion with incremental covariance being an identity operator in U. Then, for every $\mathcal{F}_0$-measurable initial state $x_0 \in L_2(\Omega, H)$, the system (1.18) has a unique mild solution $x \in B_\infty^a(I, L_2(\Omega, H))$. Further, the solution has continuous modification in the sense that there exists a continuous process $\tilde{x}$ such that $Prob.\{x(t) \neq \tilde{x}(t), t \in I\} = 0$.*

Proof As in the finite dimensional case, define the operator G by

$$(Gx)(t) \equiv S(t)x_0 + \int_0^t S(t - r)B(x(r))dr + \int_0^t S(t - r)C(x(r))dW(r),$$

$$(1.20)$$

$t \in I$. We show that G is a bounded operator in $B_\infty^a(I, L_2(\Omega, H))$. Clearly, for $x \in B_\infty^a(I, L_2(\Omega, H))$ $x(t)$ is $\mathcal{F}_t$ adapted and hence $(Gx)(t)$ is also $\mathcal{F}_t$ adapted. Using our assumptions (A), (B), (C) we will verify that

$$\mathbb{E}|(Gx)(t)|_H^2 \leq M^2\mathbb{E}|x_0|_H^2 + tM_b^2 \int_0^t (1 + \mathbb{E}|x(r)|_H^2)dr$$

$$+ M_c^2 \int_0^t (1 + \mathbb{E}|x(r)|_H^2)dr. \quad (1.21)$$

The first two terms on the right hand side are standard. Considering the first term, we have

$$\mathbb{E}|S(t)x_0|_H^2 \leq M^2 Tr(P_0) = M^2 \mathbb{E}|x_0|_H^2 < \infty. \tag{1.22}$$

Considering the drift term (second term) in (1.20) and denoting it by $z_1(t)$ it is easy to verify that

$$\mathbb{E}|z_1(t)|_H^2 \leq t M_b^2 \int_0^t (1 + \mathbb{E}|x(r)|_H^2)dr, \tag{1.23}$$

where $M_b \equiv M\sqrt{b}$. We prove the validity of the last term. Define

$$z_2(t) \equiv \int_0^t S(t-r)C(x(r))dW(r).$$

Let $\{e_n\}$ be any complete orthonormal basis of H and $\{u_n\}$ any complete orthonormal basis of the Hilbert space U. Since W is a cylindrical Brownian motion, it has the representation

$$W(t) = \sum_{i=1}^{\infty} u_i \beta_i(t)$$

where $\{\beta_i\}$ are independent standard Brownian motions. By straightforward computation, it is easy to see that

$$\mathbb{E}(z_2(t), e_n)_H^2 = \sum_{i=1}^{\infty} \mathbb{E} \int_0^t (C^*(x(r))S^*(t-r)e_n, u_i)^2 dr.$$

Thus,

$$\mathbb{E}|z_2(t)|_H^2 = \mathbb{E}\sum_{n=1}^{\infty} (z_2(t), e_n)_H^2 = \mathbb{E}\int_0^t \sum_{n=1}^{\infty} \| C^*(x(r))S^*(t-r)e_n \|_U^2 \, dr$$

$$= \mathbb{E}\int_0^t \| C^*(x(r))S^*(t-r) \|_{\mathcal{L}_2(H,U)}^2 \, dr. \tag{1.24}$$

Hence it follows from assumption (A) and (C) that there exists a constant $M_c > 0$ such that

$$\mathbb{E}|z_2(t)|_H^2 \leq M_c^2 \int_0^t (1 + \mathbb{E}|x(r)|_H^2)dr. \tag{1.25}$$

Thus we have proved that $Gx \in B_\infty^a(I, L_2(\Omega, H))$ for every $x \in B_\infty^a(I, L_2(\Omega, H))$ showing that G is a bounded operator in $B_\infty^a(I, L_2(\Omega, H))$.

By following similar steps and using the Lipschitz properties as stated in assumptions (B) and (C), we can prove that there exists a constant $\gamma > 0$ dependent on $\{M, b, c, T\}$ such that

$$\mathbb{E}|(Gx)(t) - (Gy)(t)|^2_H \leq \gamma \int_0^t \mathbb{E}|x(r) - y(r)|^2_H dr \qquad (1.26)$$

for every $x, y \in B^a_\infty(I, L_2(\Omega, H))$. Thus G is Lipschitz. Define the metric $\rho \equiv \rho_T$ derived from

$$\rho^2_t(x, y) \equiv \sup_{0 \leq s \leq t} \mathbb{E}\{|x(s) - y(s)|^2_H\}, t \in I.$$

With respect to this metric, $B^a_\infty(I, L_2(\Omega, H))$ is a complete metric space. By repeated substitution of (1.26) into itself, we obtain

$$\rho(G^n x, G^n y) \leq \left((\gamma T)^n / n!\right)^{1/2} \rho(x, y). \qquad (1.27)$$

So for n sufficiently large, the n-th iterate G^n of G is a contraction. Thus it follows from Banach fixed point theorem that G^n and hence G itself has one and the same fixed point which we may denote by $x \in B^a_\infty(I, L_2(\Omega, H))$. This proves the existence and uniqueness of a (mild) solution for the stochastic differential equation (1.18). Under certain additional assumptions, pathwise continuity follows from the factorization technique coupled with Yosida approximation $A_n \equiv nAR(n, A)$ of A and stochastic Fubini's theorem. For details see [Proposition 7.3, p184] Da Prato and Zabczyk [Da Prato (1992)]. This completes the proof. ●

In the above theorem we assumed that both B and C are uniformly Lipschitz. This assumption can be relaxed to allow local Lipschitz property. Under this relaxed assumption, the existence and uniqueness proof is slightly modified by use of stopping time.

In some applications, the drift vector B may be controlled or may depend on some parameters. Here we present a result on continuous dependence of solutions with respect to such parameters.

Corollary 1.4.3 *Let S be a compact Hausdorff space and suppose that the operator*

$$B : S \times H \longrightarrow H$$

is jointly continuous and satisfies the condition (B) of Theorem 1.4.2 uniformly with respect $s \in S$. Suppose the remaining assumptions of Theorem 1.4.2 hold. Then $s \longrightarrow x(\cdot, s)$ is sequentially continuous from S to $B_\infty^a(I, L_2(\Omega, H))$.

In some applications it is possible to choose the unbounded operator A from a given class. For example, let $\mathcal{G}_0(M, \omega)$ denote the class of infinitesimal generators of C_0 semigroups $\{S_A(t), t \geq 0\}$ of bounded linear operators in H. Suppose $G_0 \subset \mathcal{G}_0(M, \omega)$ with a fixed pair of stability parameters $\{M, \omega\}$ in the sense that

$$\| S_A(t) \|_{\mathcal{L}(H)} \leq M e^{\omega t}, t \geq 0,$$

for all $A \in G_0$

Corollary 1.4.4 *Suppose the assumptions (A), (B), (C) hold. Let $G_0 \subset \mathcal{G}_0(M, \omega)$ satisfy the following property: Every sequence $A_n \in G_0$ has a subsequence, relabeled as the original sequence, and an element $A_o \in G_0$ such that for each $\lambda \in \rho(A_o) \cap \rho(A_n)$ for all $n \in N$,*

$$R(\lambda, A_n) \xrightarrow{\tau_{so}} R(\lambda, A_o)$$

as $n \to \infty$ where τ_{so} denotes the strong operator topology on $\mathcal{L}(H)$. Let $x_n, x_o \in B_\infty^a(I, L_2(\Omega, H))$ denote the mild solutions of equation (1.18) corresponding to $A = A_n$ and $A = A_o$ respectively with the same initial state x_0. Then, as $n \to \infty$, $x_n \xrightarrow{s} x_o$ in $B_\infty^a(I, L_2(\Omega, H))$.

Proof We give a brief outline. The proof is based on a fundamental result from semigroup theory, in particular, Theorem 4.5.4 and Remark 4.5.5 [[Ahmed (1991)], p133] along with Lebesgue dominated convergence theorem and Gronwall lemma. Under the given assumptions, according to Theorem 4.5.4 of [[Ahmed (1991)], p133], the semigroup $S_n \equiv \{S_n(t), t \geq 0\}$, corresponding to A_n, converges in the strong operator topology to the C_0 semigroup $S_o = \{S_o(t), t \geq 0\}$ corresponding to A_o uniformly in $t \in I$. Using this result in equation (1.19) for S_n and S_o and taking the difference and then rearranging terms in a suitable way, one can readily verify that the assertion follows from Lebesgue dominated convergence theorem and Gronwall lemma. Here the assumptions (B) and (C) are used. •

1.4.1   *Measure Solutions*

Results presented in the preceding subsection with respect to the SDE (1.18), require that the nonlinear operators B and C are lipschitz having at most linear growth. These are results that guarantee solutions which are stochastic processes with values in H. If the operators B and C are merely continuous and possibly bounded only on bounded sets, there may not exist any solution in the sense of any of the standard notions like strong, mild, weak or martingale. However, if H is finite dimensional local solutions may exist with finite blowup time. In contrast, there are examples of deterministic systems in infinite dimensional spaces with continuous vector fields [Godunov (1974)], [Fattorini (1997)], [Ahmed (1999)] which do not have solutions. For detailed justification, why even a martingale solution may not exist see [Ahmed (1999)]. In recent years a new notion of solution has been introduced. This is known as measure solution or relaxed solution. In other words, instead of a path process $x(t), t \in I$, taking values from H, it is a measure valued stochastic process $\{\mu_t, t \in I\}$ taking values from $\mathcal{M}(H)$, the space of probability measures on the Borel subsets of the Hilbert space H.

Instead of going into details, we present the appropriate evolution equation corresponding to our system (1.18) on the space of measures. For details, see [Ahmed (2005)],[Ahmed (1999)], [Ahmed (2006)] and the references therein. The evolution equation is written in the weak form as follows

$$d\mu_t(\varphi) = \mu_t(\mathcal{A}\varphi)dt + \mu_t(\mathcal{B}\varphi)dt + <\mu_t(\mathcal{C}\varphi), dW(t)> \text{ for } t \geq 0, \quad (1.28)$$

where $\mu_0(\varphi) = \nu(\varphi)$ and the operators $\{\mathcal{A}, \mathcal{B}, \mathcal{C}\}$ are given by

$$(\mathcal{A}\varphi)(\xi) \equiv (1/2)Tr((D^2\varphi)CC^*)(\xi) \quad (1.29)$$

$$(\mathcal{B}\varphi)(\xi) \equiv (A^*D\varphi(\xi), \xi) + (B(\xi), D\varphi(\xi)) \quad (1.30)$$

$$(\mathcal{C}\varphi)(\xi) \equiv C^*(\xi)D\varphi(\xi), \quad (1.31)$$

with the test function φ belonging to the class

$$\Psi \equiv \{\varphi \in BC(H) : D\varphi, D^2\varphi \text{ exist, continuous, and bounded on } H\}$$

and the domains of $\mathcal{A}$ and $\mathcal{B}$ are given by

$$D(\mathcal{A}) \equiv \{\varphi \in \Psi : (1/2)Tr((D^2\varphi)CC^*) \in BC(H^+)\} \tag{1.32}$$

$$D(\mathcal{B}) \equiv \{\varphi \in \Psi : (A^*D\varphi(\xi), \xi) + (B(\xi), D\varphi(\xi)) \in BC(H^+)\}, \tag{1.33}$$

where $H^+ \equiv \beta H$ is the Stone-Čech compactification of H turning it into a compact Hausdorff space with H being homeomorphic with a dense subspace of H^+. Stone-Čech compactification is a technique for embedding a Tychonoff topological space in a compact Hausdorff space. For details on Stone-Čech compactification the reader is referred to [Willard (1970)]. The technical reason for introducing the space H^+ is to capture the support of the measure solution which may otherwise escape H. For details see [Ahmed (2001)] and [Ahmed (2005)]. Readers feeling uneasy with the notion of Stone-Čech compactification may simply disregard this and assume the original state space to be a compact Hausdorff space denoted by H^+.

Let $\mathcal{M}_{rba}(H^+)$ denote the space of regular bounded finitely additive measures on H^+ and $\Pi_{rba}(H^+) \subset \mathcal{M}_{rba}(H^+)$ the subset of probability measures. Let

$$M_{1,2}(I \times \Omega, BC(H^+)) \subset L_{1,2}(I \times \Omega, BC(H^+))$$

denote the space of $\mathcal{F}_t$ adapted random processes with values in $BC(H^+)$ furnished with the norm topology given by

$$\| \varphi \| \equiv \int_I \left(E(|\varphi(t,\omega,\cdot)|_{BC(H^+)})^2 \right)^{1/2} dt.$$

The dual of this space is given by $M_{\infty,2}^w(I \times \Omega, \mathcal{M}_{rba}(H^+))$ which consists of weak star measurable and $\mathcal{F}_t$-adapted processes taking values in $\mathcal{M}_{rba}(H^+)$. In other words, for each $\varphi \in M_{1,2}(I \times \Omega, BC(H^+))$

$$t \longrightarrow \int_{H^+} \varphi(t,x)\mu_t(dx)$$

is an $\mathcal{F}_t$ measurable scalar valued random process. Here we have suppressed the ω variable. Clearly, for each $\phi \in BC(H^+)$, the process $\{\mu_t(\phi), t \in I\}$ is an element of $B_\infty^a(I, L_2(\Omega))$. For more details, see [Ahmed (2005)]. With this introduction we are now prepared to give a formal definition of measure solution for the SDE (1.18).

Definition 1.4.5 *A random process $\{\mu_t, t \in I\}$ satisfying*

$$\mu \in M_{\infty,2}^w(I \times \Omega, ,\Pi_{rba}(H^+)) \subset M_{\infty,2}^w(I \times \Omega, \mathcal{M}_{rba}(H^+))$$

is said to be a measure solution of equation (1.18), equivalently (1.28), if for every $\varphi \in D(\mathcal{A}) \cap D(\mathcal{B})$ the following identity holds with probability one

$$\mu_t(\varphi) = \nu(\varphi) + \int_0^t \mu_s(\mathcal{A}\varphi)ds + \int_0^t \mu_s(\mathcal{B}\varphi)ds + \int_0^t <\mu_s(\mathcal{C}\varphi), dW(s) >,$$

$$(1.34)$$

for all $t \in I$. In case the system starts from state x_0, the initial measure ν is a Dirac measure and so $\nu(\varphi) = \varphi(x_0)$.

Remark 1.4.6 Note that if both B and C turn out to be sufficiently smooth, the system (1.18) has a solution in the usual sense (a path process) which is an H valued process $\{x(t), t \in I\}$. Then the measure valued process $\mu_t(d\xi) \equiv \delta_{x(t)}(d\xi)$ is a solution of equation (1.34) and hence $\mu_t(\varphi) = \varphi(x(t))$ and the identity (1.28) reduces to the Itô differential of the process $\varphi(x(t))$. For a detailed discussion on the deep distinction that exists between martingale and measure solutions see [Ahmed (1999)].

For details on existence and uniqueness of measure solutions, interested readers may see Theorems 3.3 and Theorem 4.1 of [Ahmed (2005)] and Theorem 3.2 of [Ahmed (1999)]. Here we present only the statement of the theorem without proof.

Theorem 1.4.7 *Suppose A is the infinitesimal generator of a C_0-semigroup in H and the maps $B : H \longrightarrow H$ and $C : H \longrightarrow \mathcal{L}(U, H)$ are continuous and bounded on bounded sets satisfying certain approximation properties and W is an U-valued cylindrical Brownian motion. Then, for every x_0 for which $P\{|x_0|_H < \infty\} = 1$, with law ν, the evolution equation (1.18), equivalently (1.28), has at least one measure valued solution*

$$\mu \in M_{\infty,2}^\omega(I \times \Omega, \Pi_{rba}(H^+))$$

in the sense of Definition 1.4.5 and further, $t \to \mu_t$ is weak star continuous P-as.

Here, we are only interested in the fact that for each $\varphi \in BC(H^+)$ with compact support, the process $t \longrightarrow \mu_t(\varphi)$ is an element of $L_\infty^a(I, L_2(\Omega))$.

This will allow us to construct canonical Wiener-Itô expansions. In fact, we go beyond this and admit much larger class of measure valued process containing the classical Wiener-Itô class as mentioned above. This is discussed in Chapters 6 and 7.

Remark 1.4.8 The notion of measure valued solution was first introduced for deterministic systems on Banach spaces [Fattorini (1997)], [Ahmed (1999)], [Ahmed (2006)], [Ahmed (2000)] admitting discontinuous vector fields (see also the references therein). For example, equation (1.18), with $C = 0$ and $x_0 \in H$ a fixed element, reduces to a deterministic system with B a nonlinear Borel measurable map from H to H bounded on bounded sets (not subject to classical Lipschitz and growth conditio). In this case equation (1.28) reduces to a deterministic differential equation on the space $\Pi_{rba}(H)$, the space of regular bounded finitely additive probability measures. This is written in the weak form:

$$(d/dt)\mu_t(\varphi) = \mu_t(\mathcal{B}\varphi), \mu_0(\varphi) = \delta_{x_0}(\varphi) = \varphi(x_0)$$

for $\varphi \in D(\mathcal{B})$. For existence of measure solutions for such deterministic systems see the references [Fattorini (1997)], [Ahmed (1999)], [Ahmed (2006)], [Ahmed (2000)] and also the citations therein.

1.5 Nonlinear Filtering

In this section we present the basic equations for nonlinear filtering both in finite and infinite dimensional spaces.

1.5.1 *Finite Dimensional Filtering*

Before we present the fundamental equations of nonlinear filtering, we give a brief outline of the original filtering problem. Consider the process $\{\xi(t), t \geq 0\}$ governed by the SDE in R^n,

$$d\xi = b(\xi)dt + \sigma(\xi)dW, \xi(0) = x \in R^n, t \in I, \tag{1.35}$$

where $b : R^n \longrightarrow R^n$ is the drift, $\sigma : R^n \longrightarrow \mathcal{L}(R^p, R^n)$ is a $n \times p$ matrix valued function determining the diffusion and W is a p-dimensional Wiener process with incremental covariance given by $Q \in M(p \times p)$. The process ξ is not directly accessible to the observer, but a noisy measurement of it is available and this is given by the SDE in R^m

$$dy = h(\xi)dt + \sigma_0(t)dV, y(0) = 0 \in R^m, t \in I, \tag{1.36}$$

where $h : R^n \longrightarrow R^m$ and $\sigma_0 : I \longrightarrow \mathcal{L}(R^m, R^m)$ and V is an m-dimensional Wiener process independent of the Wiener process W.

The problem is to find the best estimate of $\xi(t)$ or, more generally, a functional of $\xi(t)$ $\varphi(\xi(t))$, given the observed history $\{y(s), s \le t\}$ up to time t. In general, by a best estimate we mean an estimate that is unbiased and has minimum variance. Let $\mathcal{F}_t^y, t \ge 0$, denote the smallest family of (complete) sub sigma algebras of the sigma algebra $\mathcal{F}$ with respect to which the process $\{y(t), t \ge 0\}$ is measurable. It is well known that the best estimate is given by the conditional expectation of $\varphi(\xi(t))$ relative to the sigma algebra $\mathcal{F}_t^y$. Indeed, define

$$\hat{\varphi}(t) \equiv \mathbb{E}\{\varphi(\xi(t))|\mathcal{F}_t^y\}.$$

Since $\mathbb{E}\hat{\varphi}(t) = \mathbb{E}\{\varphi(\xi(t))\}$, it is clear that this estimate is unbiased. We verify that it has the minimum variance. Let $\eta(t)$ be any other $\mathcal{F}_t^y$ measurable estimate of $\varphi(\xi(t))$. Then the reader can easily verify that

$$\mathbb{E}\{(\eta(t) - \varphi(\xi(t)))^2|\mathcal{F}_t^y\} = (\eta(t) - \hat{\varphi}(t))^2 - \hat{\varphi}(t)^2 + \mathbb{E}\{\varphi(\xi(t))^2|\mathcal{F}_t^y\}.$$

Clearly, it follows from this expression that it attains its minimum if, and only if, $\eta(t) = \hat{\varphi}(t)$. Hence, we conclude that the conditional expectation is the best unbiased minimum variance estimate. Let $Q_t^y(\cdot)$ denote the conditional probability measure of $\xi(t)$ given the history of y till time t. That is, for any Borel set $K \subset R^n$,

$$Q_t^y(K) \equiv Prob.\{\xi(t) \in K|\mathcal{F}_t^y\}.$$

Hence

$$\hat{\varphi}(t) = \int_{R^n} \varphi(z)Q_t^y(dz),$$

and therefore, it is the conditional probability measure that determines the optimal filter. It is well-known that $Q_t^y, t \ge 0$, satisfies Kushner's equation which is a nonlinear stochastic partial differential equation [[Ahmed

(1998)], Theorem 3, p179]. On the other hand, the unnormalized measure μ_t, defined by $Q_t^y(\cdot) = \mu_t(\cdot)/\mu_t(R^n)$, satisfies a linear stochastic partial differential [[Ahmed (1998)], Theorem 2, p176]. It is the linearity of this equation that makes it so attractive. It is however, not a probability measure valued process, but a bounded and positive measure valued process. This equation, written in the weak form, is given by

$$d\mu_t(\varphi) = \mu_t(A\varphi)dt + (\mu_t(R_0^{-1}h\varphi), dy)_{R^m}, \mu_0(\varphi) = \nu(\varphi), t \geq 0, \quad (1.37)$$

where φ is any test function belonging to the domain of the operator A. The matrix $R_0 \equiv \sigma_0 R \sigma_0^*$ where R is the incremental covariance of the Wiener process V representing the measurement noise. It is assumed that R_0 is a nonsingular matrix valued function for all $t \in I$. The operator A is the infinitesimal generator of the Markov process $\{\xi(t), t \geq 0\}$. It is given by the partial differential operator

$$(A\phi)(x) \equiv (b(x), D\phi(x)) + (1/2)Tr((D^2\phi)(x)a(x)), \text{ with } a(x) \equiv \sigma(x)Q\sigma^*(x).$$

In the sequel, we use the linear stochastic partial differential equation (1.37), known as the Zakai equation [Ahmed (1998)].

Recall that the pair $\{\xi, y\}$, solving the equations (1.35) and (1.36) respectively, is continuous (or has continuous versions). Thus, it belongs to the Banach space $C \equiv C(I, R^{n+m})$ P-a.s. Let μ^1 denote the probability measure induced by the pair $\{\xi, y\}$ on the Borel field $\mathcal{B}(C)$ of C and μ^0 the measure induced when the vector h of equation (1.36) is set equal to zero. It is well known [[Ahmed (1998)], p179] that μ^1 is absolutely continuous with respect to measure μ^0 and the Radon-Nikodym derivative of μ^1 with respect to μ^0 is given by

$$d\mu^1 = q \, d\mu^0$$

where

$$q \equiv \exp\left\{\int_I (R_0^{-1}h, dy) - (1/2)\int_I (R_0^{-1}h, h)ds\right\}.$$

Under the measure μ^0, the process y is a Brownian motion with covariance R_0. Thus, we can rewrite the equation (1.37) in terms of an innovation process $\hat{V}$, with incremental covariance R again. This is given by

$$d\mu_t(\varphi) = \mu_t(A\varphi)dt + (\mu_t(R_0^{-1}h\varphi), d\hat{V})_{R^m}, \mu_0(\varphi) = \nu(\varphi), t \geq 0, \quad (1.38)$$

for every $\varphi \in D(A)$. For details on this topic, interested reader may refer to [Ahmed (1998)].

1.5.2   *Infinite Dimensional Filtering*

Here the system is given by

$$d\xi = A\xi dt + B(\xi)dt + R_1^{1/2}dW_1, \xi(0) = x_0, t \geq 0. \tag{1.39}$$

We introduce the following assumptions:

(H1): A is the infinitesimal generator of a C_0-semigroup $S(t), t \geq 0$, in H and there exist $M > 0, \omega > 0$ such that

$$\| S(t) \|_{\mathcal{L}(H)} \leq Me^{-\omega t}.$$

(H2): The operator $R_1 \in \mathcal{L}_1^+(H)$ (positive nuclear).
(H3): $B : H \longrightarrow H$ is Lipschitz continuous.

Under the above assumptions, for every $\mathcal{F}_0$ measurable H valued random variable x_0, equation (1.39) has a unique mild solution. If in addition $x_0 \in L_2(\Omega, H)$, then the solution $\xi \in B_\infty^a(I, L_2(\Omega, H))$. Later in the sequel, it is the strong solution that we are interested in.

The observation process (measurement process) is governed by a finite dimensional SDE given by

$$dy = h(\xi(t))dt + R_2^{1/2}dW_2, y(0) = 0, t \geq 0. \tag{1.40}$$

For the measurement dynamics we assume:

(H4): $h : H \longrightarrow R^d$ is a continuous bounded map,
(H5): $R_2 \in \mathcal{L}(R^d, R^d)$ is invertible.

Another technical assumption of significant importance for the existence of an invariant measure is

(H6): The operator $Q_\infty \in \mathcal{L}_1^+(H)$, where

$$Q_\infty \equiv \int_0^\infty S(r)R_1 S^*(r)dr.$$

As our basic reference measure, we choose the Gaussian measure $\mu^o \equiv \mathcal{N}(0, Q_\infty)$ which has mean zero and covariance Q_∞. This is the invariant measure induced by the solution of the linear evolution equation

$$dη = Aηdt + R_1^{1/2}dW_1, η(0) = x_0. \tag{1.41}$$

Clearly, the solution of this equation is given by

$$η(t) = S(t)x_0 + \int_0^t S(t-r)R_1^{1/2}dW_1(r), t \geq 0,$$

which is also known as the Ornstein-Uhlenbeck process. And it follows from the exponential stability of the semigroup S (see assumption (**H1**)) that $μ^o$ is its invariant measure. For any $φ \in UC_b(H, R)$, the Banach space of bounded uniformly continuous functions on H, for which the function $x \longrightarrow φ(Ax)$ has a continuous extension to all of H, is denoted by $φ_A(\cdot)$. Now define the set

$$D_0 \equiv \{φ \in UC_b^2(H, R) : φ_A \in UC_b^2(H, R), D^2φ \in UC_b(H, \mathcal{L}_1(H))\}$$

and the operators

$$\mathcal{A}φ \equiv \mathcal{A}_0φ + \mathcal{B}_0φ, φ \in D_0, \tag{1.42}$$
$$\mathcal{A}_0φ = (1/2)tr(D^2φ(x)R_1) + (x, A^*Dφ(x)), x \in H, φ \in D_0, \tag{1.43}$$
$$\mathcal{B}_0φ = (B(x), Dφ(x)), x \in H, φ \in D_0. \tag{1.44}$$

The Zakai equation for the infinite dimensional filtering problem, written in the weak form, is then given by

$$dμ_t(φ) = μ_t(\mathcal{A}φ)dt+ < μ_t(φR_2^{-1}h), dy >, μ_0(φ) = ν(φ), φ \in D_0. \tag{1.45}$$

Note that this equation has exactly the same form as that of the finite dimensional case.

For applications, we consider only the measure solutions. However, using the invariant measure $μ^o$ and some additional assumptions, one can prove the existence of Radon-Nikodym derivative of the measure solution $\{μ_t, t \geq 0\}$ with respect to the invariant measure $μ^o$ giving a density valued solution q in the Hilbert space $L_2(μ^o, H)$. The measure valued process $μ$ is then given by $μ_t(dx) = q(t, x)μ^o(dx)$. For more details on this topic see [Ahmed (1997)], in particular the Proposition 3.1.

1.6   Elements of Vector Measures

In this section we present only very basic elements of the theory of vector measures. Let X be any Banach space and S any set with $\mathcal{S}$ denoting an algebra of subsets of the set S and Σ denoting the sigma algebra of subsets of S (also known as Borel field). In the literature a set function $\mu : \mathcal{S} \longrightarrow X$ is called a vector measure and it is said to be bounded if its range $\mu(\mathcal{S})$ is a bounded subset of X. It is said to be additive if, for every disjoint pair $D, E \in \mathcal{S}$,

$$\mu(D \cup E) = \mu(D) + \mu(E).$$

Clearly, this property holds for any finite family of disjoint $\mathcal{S}$- measurable sets $\{D_i\}_{1 \leq i \leq n}$. The class of vector measures satisfying this property is known as the space of finitely additive vector measures.

An X valued vector measure defined on $\mathcal{S}$ is said to be countably additive if, for every pairwise disjoint $\mathcal{S}$ measurable sets $\{D_i\}_{i \geq 1}$ satisfying $\bigcup D_i \in \mathcal{S}$,

$$\mu\left(\bigcup_{i=1}^{\infty} D_i\right) = \sum_{i=1}^{\infty} \mu(D_i).$$

Definition 1.6.1 *A finitely additive X valued vector measure μ is said to have bounded variation on $D \in \mathcal{S}$ if*

$$|\mu|(D) \equiv \sup_{\pi} \sum_{\sigma \in \pi} |\mu(\sigma)|_X < \infty$$

where π is any finite disjoint $\mathcal{S}$ measurable partition of the set D and the supremum is taken over all such finite partitions π. The total variation of μ is then defined as $|\mu|(S) \equiv \sup\{|\mu|(D), D \in \mathcal{S}\}$.

Clearly, if the measure μ has finite variation it is a bounded vector measure. A more relaxed notion of variation is the semivariation as stated below.

Definition 1.6.2 *An X valued vector measure μ is said to have finite semivariation on $D \in \mathcal{S}$, if*

$$\| \mu \| (D) \equiv \sup\{|x^*\mu|(D) : x^* \in B_1(X^*)\} < \infty,$$

where $B_1(X^*)$ denotes the closed unit ball of the dual X^* and $|x^*\mu|$ denotes the variation of the scalar measure $x^*\mu$. The semivariation of μ is then given by

$$\| \mu \| (S) \equiv \sup\{\| \mu \| (D), D \in \mathcal{S}\}.$$

Remark 1.6.3 We give here an example of a vector measure which has finite semivariation but ∞ variation. Let $f : I \equiv [0, T] \longrightarrow X$ be a weakly measurable (but not strongly measurable) and Pettis integrable function on I. Then the measure μ given by

$$\mu(\sigma) \equiv \int_\sigma f(t)dt$$

is an X valued countably additive vector measure defined on the sigma algebra Σ of subsets of the interval I. Since for each $x^* \in X^*$, $t \longrightarrow |x^* f(t)|$ is measurable and integrable, it is clear that, for each $\sigma \in \Sigma$,

$$\| \mu \| (\sigma) \equiv \sup\{|x^*(\mu)|(\sigma) : x^* \in B_1(X^*)\} < \infty$$

and

$$\sup_\pi \left\{ \sum_{\sigma \in \pi} \| \mu \| (\sigma) \right\} \leq \sup\left\{ \int_I |x^* f| dt, x^* \in B_1(X^*) \right\} < \infty$$

where the supremum on the left hand side is taken over all finite disjoint measurable partitions of the interval I. Hence μ has finite semivariation but since f is not (strongly) measurable and therefore not Bochner integrable (see Definition 1.6.7), the measure μ has infinite variation.

Here is an elementary example. Let $I = [0, 1]$, $\Sigma \equiv \sigma(I)$, λ Lebesgue measure, $X = L_p(I), 1 < p < \infty$, and $\nu : \Sigma \longrightarrow X$ is a vector measure given by $\nu(E) \equiv \chi_E(\cdot)$, characteristic function of the set $E \in \Sigma$. Clearly $\nu(E) \in X$ for each $E \in \Sigma$. Let $F \in \Sigma$ with $\lambda(F) > 0$ and $F = \cup_{i=1}^n F_i$ with $\{F_i\}$ being mutually disjoint Σ measurable partition of F with $\lambda(F_i) = \lambda(F)/n, n \in N$. Clearly, for every $n \in N$,

$$|\nu|(F) \geq \sum_{i=1}^n |\chi_{F_i}|_X = \sum (\lambda(F_i))^{1/p} = n^{(p-1)/p}(\lambda(F))^{1/p}.$$

Since $p > 1$, this shows that ν does not have finite variation. But it has finite semivariation. Indeed, the semivariation of ν on F is given by

$$\| \nu \| (F) \equiv \sup\{|(x^*\nu)|(F), x^* \in B_1(X^*)\}$$

where $X^* = L_q$. Using the same partition, it is easy to see that

$$\sum_{i=1}^{n} |(x^*\nu)(F_i)| = \sum \left| \int_I x^*(s)\chi_{F_i}(s)\lambda(ds) \right| \leq \| x^* \|_{L_q(F)} (\lambda(F))^{1/p}.$$

Since this is valid for any such partition, taking the supremum of the sum on the left over all such partitions, we have $|x^*\nu|(F) \leq \| x^* \|_{L_q(F)} (\lambda(F))^{1/p}$. Clearly, it follows from this that

$$\| \nu \| (F) \leq (\lambda(F))^{1/p} < \infty$$

verifying that ν has finite semivariation.

Let $M_{fabv}(\Sigma, X)$ denote the vector space of finitely additive bounded X valued vector measures. It is well known that $M_{fabv}(\Sigma, X)$, furnished with the total variation norm, is a Banach space. This is easy to verify since X is a Banach space, see Diestel [Diestel (1977)].

Let $M_{cabv}(\Sigma, X)$ denote the class of countably additive bounded vector measures. It is a closed linear subspace of $M_{fabv}(\Sigma, X)$ and hence it is also a Banach space.

Lemma 1.6.4 *An X valued vector measure of bounded variation is countably additive if and only if its variation is countably additive.*

Proof See [[Diestel (1977)], Proposition 1.1.9, p3].

It is clear from this result that if $\mu \in M_{cabv}(\Sigma, X)$ then the positive measure $\nu(\cdot) \equiv |\mu|(\cdot)$ is an element of $M_{cabv}^+(\Sigma)$.

Like the uniform boundedness principle for linear operators, there is a similar result for vector measures originally proved by Nikodym.

Theorem 1.6.5 *(Nikodym Uniform Boundedness Principle) Let Λ be any index set and suppose $\{\mu_\alpha, \alpha \in \Lambda\}$ is a family of X valued bounded vector measures defined on Σ. If for each $\sigma \in \Sigma$, $\sup\{\| \mu_\alpha(\sigma) \|, \alpha \in \Lambda\} < \infty$, then the family $\{\mu_\alpha, \alpha \in \Lambda\}$ is uniformly bounded in semivariation that is,*

$$\sup\{\| \mu_\alpha \| (S), \alpha \in \Lambda\} < \infty.$$

Proof See [[Diestel (1977)], Theorem 1.3.1, p14].

Definition 1.6.6 *Let* (S, Σ, ν) *be a finite positive measure space. A function* $f : S \longrightarrow X$ *is said to be ν-measurable if there exists a sequence of ν-simple functions f_n of the form*

$$f_n \equiv \sum_i^n x_i \chi_{\sigma_i}$$

with $x_i \in X$ *and* $\sigma_i \in \Sigma$ *such that* $f_n \xrightarrow{s} f$ *(strongly) for ν almost all* $s \in S$, *that is,* $\| f_n(s) - f(s) \|_X \longrightarrow 0$ *for ν-almost all* $s \in S$.

Definition 1.6.7 *(Bochner Integral) A ν-measurable function f is said to be Bochner integrable if there exists a sequence of ν-simple functions $\{f_n\}$ such that*

$$\int_S \| f(s) - f_n(s) \|_X \, \nu(ds) \longrightarrow 0.$$

In this case the integral of f on any set $D \in \Sigma$ is given by

$$\int_D f(s)\nu(ds) = \lim_{n\to\infty} \int_D f_n(s)\nu(ds).$$

Note that the limit of the integral on the right hand side is the limit of a finite sum of elements of X weighted by the measure ν. The space of Bochner integrable functions is denoted $L_1(S, \Sigma, \nu; X) \equiv L_1(\nu, X)$.

Similarly, $L_p(\nu, X)$, $1 \leq p < \infty$, denotes the space of ν-measurable X valued functions $\{f\}$ such that

$$\| f \|_{L_p(\nu, X)} \equiv \left(\int_S \| f(s) \|_X^p \, \nu(ds) \right)^{1/p} < \infty.$$

These are normed vector spaces, and, with respect to the norm topology, they are Banach spaces. For $p = \infty$, $L_\infty(\nu, X)$ is defined in the same way as scalar valued functions.

In general, for infinite dimensional Banach spaces X, the topological (continuous) dual of $L_p(\nu, X), 1 \leq p < \infty$ is not $L_q(\nu, X^*)$ $\big((1/p) + (1/q) = 1\big)$, and the dual of $L_1(\nu, X)$ is not $L_\infty(\nu, X)$ unless the Banach space X satisfies some additional properties.

An important notion that arises in functional analysis and, in particular, in the study of vector measures, is the concept of Radon-Nikodym property.

Definition 1.6.8 *(Radon-Nikodym Property) Let (S, Σ, ν) be a finite measure space and X a Banach space. The Banach space X is said to have the Radon-Nikodym Property with respect to the measure ν, if for every ν-continuous vector measure $\mu : \Sigma \longrightarrow X$, there exists an $f \in L_1(\nu, X)$ such that $d\mu = fd\nu$ or equivalently,*

$$\mu(D) = \int_D f(s)\nu(ds),$$

for every $D \in \Sigma$. A Banach space X is said have the Radon-Nikodym Property (RNP) if it has this property for every finite measure space.

Theorem 1.6.9 *For $1 \leq p < \infty$, the dual of the Banach space $L_p(\nu, X)$ is $L_q(\nu, X^*)$ where q is the conjugate of p, $((1/p) + (1/q) = 1)$, if, and only if, the dual X^* of X has the RNP with respect to ν.*

Proof See Diestel [Diestel (1977)], [Theorem IV.1.1, p98].

Well known examples of Banach spaces which satisfy the Radon-Nikodym property are reflexive Banach spaces and separable dual spaces. Clearly all Hilbert spaces have the RNP.

In the absence of Radon-Nikodym property, the duality results do not hold. However by use of the theory of "lifting" [Tulcea (1969)] Tulcea (A. Ionescu and C. Ionescu) have proved a sort of weak duality result. For $1 < p < \infty$ and $\{p, q\}$ the conjugate pair, let $L_q^w(\nu, X^*)$ denote the class of weak star measurable X^* valued functions which are scalarly q-th power integrable.

Theorem 1.6.10 *Let X be a Banach space with dual X^* and $\{p, q\}$ the conjugate pair and $1 \leq p < \infty$. Then, for every continuous linear functional $L \in (L_p(\nu, X))^*$, there exists a unique $g \in L_q^w(\nu, X^*)$ such that*

$$L(f) = \int_S < g(s), f(s) >_{X^*, X} \nu(ds)$$

for all $f \in L_p(\nu, X)$.

Proof See Tulcea [Tulcea (1969)] Theorem 7, p94; and Theorem 9, p97.

In particular, the dual of $L_1(\nu, X)$ is precisely $L_\infty^w(\nu, X^*)$. This result is very important for vector measures as seen later. It has also found applications in the study of measure solutions [Ahmed (2007)] for evolution equations for which there is no solution in the usual sense such as mild, strong, classical. It has also found applications in control theory [Ahmed (2003)].

Characterization of weakly compact sets in Banach spaces is particularly useful in the study of optimization and optimal controls. The most often used Banach space is $L_1(\nu, X)$. The following result characterizes weakly compact sets in $L_1(\nu, X)$. This is the celebrated Dunford Theorem.

Theorem 1.6.11 (Dunford) *Let (S, Σ, ν) be a finite measure space and both X and its dual X^* satisfy the RNP. Then, a set $K \subset L_1(\nu, X)$ is relatively weakly compact if the following conditions are satisfied:*

(A): K is bounded

(B): the set K is uniformly (Bochner) integrable in the sense that

$$\lim_{\nu(\sigma) \to 0} \int_\sigma |f(s)|_X \nu(ds) = 0 \quad \text{uniformly in } f \in K.$$

(C): The set $\{\int_\sigma f(s)\nu(ds), f \in K\}$, for any $\sigma \in \Sigma$, is a relatively weakly compact subset of X.

Proof For proof see Diestel [Diestel (1977)] Theorem IV.2.1, p101. •

Further, in addition to the conditions (A)-(C), if K is also closed, then K is weakly compact. Note that, for reflexive Banach spaces, condition (C) holds automatically, and so in this case, the result is similar to the well known result in finite dimensional case.

This theorem has been generalized to cover the spaces of vector measures. Let S be an arbitrary set and Σ the sigma algebra subsets of the set S. Let $M_{cabv}(\Sigma, X)$ denote the space of countably additive X-valued vector measures of bounded variation. The following result characterizes weakly compact sets in this space.

Theorem 1.6.12 (Bartle-Dunford-Schwartz) *Consider the Banach space $M_{cabv}(\Sigma, X)$ with both X and its dual X^* satisfying the RNP. Then, a set $\Gamma \subset M_{cabv}(\Sigma, X)$ is weakly relatively compact if, and only if, the following conditions hold:*

($\tilde{A}$): The set Γ is bounded.

$(\tilde{B})$: *The set Γ is uniformly countably additive or equivalently, there exists a nonnegative countably additive measure β such that*

$$\lim_{\beta(\sigma)\to 0} |\mu|(\sigma) = 0 \text{ uniformly in } \mu \in \Gamma.$$

$(\tilde{C})$: *For each $\sigma \in \Sigma$, the set $\{\mu(\sigma), \mu \in \Gamma\}$ is a relatively weakly compact subset of X.*

Proof See Diestel [Diestel (1977)] Theorem IV.2.5, p105. See also Brooks [Brooks (1972)].

A finitely additive version of the above result was proved by Brooks and Dinculeanu.

Theorem 1.6.13 (Brooks-Dinculeanu) *Let $\mathcal{F}$ be an algebra of subsets of the set S and consider the Banach space $M_{fabv}(\mathcal{F}, X)$ with both X and its dual X^* satisfying the RNP. Then, a set $\Gamma \subset M_{fabv}(\mathcal{F}, X)$ is weakly relatively compact if, and only if, the following conditions hold:*

$(\hat{A})$: *The set Γ is bounded.*

$(\hat{B})$: *There exists a nonnegative finitely additive bounded measure β such that*

$$\lim_{\beta(\sigma)\to 0} |\mu|(\sigma) = 0 \text{ uniformly in } \mu \in \Gamma.$$

$(\hat{C})$: *For each $\sigma \in \mathcal{F}$, the set $\{\mu(\sigma), \mu \in \Gamma\}$ is a relatively weakly compact subset of X.*

Proof See Diestel [Diestel (1977)] Corollary IV.2.6, p105. See also Brooks and Dinculeanu [Brooks (1974)].

Remark 1.6.14 Theorem 1.6.12 is intimately related to Theorem 1.6.11. In fact, due to assumption $(\tilde{B})$ and the Radon-Nikodym property, to every $\mu \in \Gamma$ there corresponds a unique $g_\mu \in L_1(\beta, X)$ such that $d\mu = g_\mu d\beta$. This establishes an isometric isomorphism between $\Gamma \subset M_{cabv}(\Sigma, X)$ and a set $G \subset L_1(\beta, X)$. The conditional weak compactness of Γ is then equivalent to that of G.

The result of Theorem 1.6.12 has been extensively used in infinite dimensional control theory [Ahmed (2003)], [Ahmed (2008)], and the references therein. In the sequel, we will also have occasions to use them.

1.7 Some Problems for Exercise

P1: Consider the example (E1) of section 1.3. Verify that

$$\mathbb{E}J \equiv \mathbb{E} \int_0^T w(t)dw(t) \in [0, T].$$

If the integral is understood in the Itô sense it equals 0, if it is understood in the sense of Stratonovich it equals $(T/2)$. Give the physical reasons for any preference.

P2: Under the assumptions (1.10) and (1.11) of Theorem 1.3.2, prove that there exists a number $n_0 \in N$ such that the n-th iterate of the operator G given by the expression (1.12) is a contraction for all $n \geq n_0$.

P3: Consider the Black-Scholes model for stock price as given in example (E2). Suppose $\alpha \in L_1(I), \beta \in L_2(I)$ and $\{\xi_0, W\}$ are independent. Prove that if ξ_0 has p-th moment then $\xi(T)$ also has p-th moment for any $p > 0$. Find the mean of $\xi(T)$ and its p-th moment. Verify if the p-th moment is given by

$$\mathbb{E}|\xi(T)|^p = \mathbb{E}|\xi_0|^p \exp\left(p \int_0^T \alpha(t)dt + (1/2)p(p-1) \int_0^T \beta^2(t)dt \right).$$

P4: Give a detailed proof of Theorem 1.3.3 following the hints given in the brief outline of the proof.

P5: The problems indicated in Remarks 1.3.4 and 1.3.5 are of significant interest in control theory. Find the largest class of bounded Borel measurable state feedback controls (including the topology) for which the set of attainable measures for every $t \in I$ is weakly compact.

P6: Give a detailed proof of Corollary 1.4.3.

P7: Let $\{U, H\}$ be a pair of real separable Hilbert spaces. Consider the linear system on H given by

$$dx = Axdt + C(t)dW(t), x(0) = \xi$$

with W being a U valued cylindrical Brownian motion independent of the initial state $\xi \in L_2(\Omega, H)$ ($\mathcal{F}_0$ measurable) and $C \in B_\infty(I, \mathcal{L}_2(U, H))$. Let $\{x_n, x_o\}$ denote the (mild) solutions of the above equation corresponding to $\{A_n, A_o\}$ satisfying the resolvent convergence property as in Corollary 1.4.4. Prove that $x_n \xrightarrow{s} x_o$ uniformly on I in $L_2(\Omega, H)$ ($\mathcal{F}_t$ measurable).

P8: Consider the system of problem **P7**, suppose $C \in B_\infty(I, \mathcal{L}_{so}(U, H))$, not Hilbert-Schmidt, and ξ is a weakly second order H valued random variable. Show that the solution processes $\{x_n, x_o\}$ are also weakly second order H valued (weakly) $\mathcal{F}_t$ adapted random processes and that $x_n(t) \xrightarrow{w} x_o(t)$ in the meansquare sense, that is, for each $h \in H$, $E(x_n(t) - x_o(t), h)^2 \longrightarrow 0$ as $n \to \infty$.

P9: Consider the controlled semilinear evolution equation

$$dx = Axdt + F(x)dt + B(t)\nu(dt) + C(t)dW, x(0) = \xi,$$

on H where $\{A, C(t)\}$ and $\{\xi, W\}$ are as in problem **P7**, F is uniformly Lipschitz with at most linear growth, $B \in B_\infty(I, \mathcal{L}(E, H))$ uniformly norm bounded, and $\nu \in \Gamma \subset M_{cabv}(\Sigma_I, E)$ admissible set of E (another Hilbert space) valued vector measures as controls. (a): Prove existence of solutions and their regularity properties, (b): Give sufficient conditions for tightness of the attainable set of probability measures

$$\mathcal{R}(t) \equiv \{\mu \in \mathcal{M}_1(H) : \mu = \mu_t^\nu, \nu \in \Gamma\},$$

where $\mathcal{M}_1(H)$ is the space of countably additive probability measures on the Borel subsets of H.

Chapter 2

Regular Functionals of Brownian Motion

2.1 Introduction

It is known that Volterra series can be used to approximate any continuous functional on a Banach space. Similarly, it was Wiener who first proved that any L_2 functional of Brownian motion can be approximated in the limit in the mean by sums of multiple stochastic integrals. These are regular functionals of Brownian motion. Later in the eighties, Hida introduced an interesting class of generalized functionals of the Wiener process by use of a notion of Fourier transform on the dual of a Nuclear space which supports the white noise measure. This generalization is significant and applies to problems where the regular functionals are insufficient. In fact, white noise itself is a generalized functional of Brownian motion and does not belong to the class of regular functionals, L_2. We shall discuss this later in the sequel. In the early eighties [Ahmed (1983)], the author introduced a class of generalized functionals of white noise which covers those of Hida as special case. Further, the method of construction used in the author's paper [Ahmed (1983)] is direct and simple and does not use the Fourier transform as employed by Hida. In any case both the methods are instructive and will be discussed in the sequel.

In this chapter, we present several families of regular functionals based on scalar Brownian motion, vector Brownian motion, scalar Gaussian random fields, multidimensional Gaussian random fields and some generalizations thereof including Hilbert space valued Brownian motion. We develop Fourier analysis of these functionals on Wiener measure space.

Some authors prefer to start with the general (abstract) theory and then discuss the special cases following from it. We prefer the other way around because we feel that this way is much easier for new readers to follow.

2.2   Functionals of Scalar Brownian Motion

Let $I \equiv [0, T]$ be a closed bounded interval and $C(I)$ the space of continuous real valued functions defined on I. Let $\Omega = C_0(I)$ denote the canonical sample space where $C_0(I) \equiv \{x \in C(I) : x(0) = 0\}$, and let $\mathcal{B}$ denote the Borel σ-algebra of subsets of the set Ω and let $\mu \equiv \mu^w$ denote the Wiener measure. The class of regular homogeneous functionals of Brownian motion of degree n is defined by

$$g_n(w) \equiv \int_{I^n} K_n(\tau_1, \tau_2, \cdots, \tau_n) dw(\tau_1) dw(\tau_2) \cdots dw(\tau_n), \qquad (2.1)$$

for any $K_n \in L_2(I^n)$ where I^n denotes the Cartesian product of n-copies of I. These are the multiple Wiener integrals based on Brownian motion and are known as polynomial chaos of degree n. For each $K_n \in L_2(I^n)$, the functional (2.1) is well defined in the Wiener-Itô sense and belongs to $L_2(\Omega, \mathcal{B}, \mu)$. For each $n \in N_+$, let $\hat{L}_2(I^n)$ denote the class of symmetric L_2 kernels. In order to emphasize the dependence of g_n on the kernels K_n we shall often write $g_n(w)$ as $g_n(K_n, w)$. Given the space of Brownian motions, later we shall omit the display of w-dependence altogether and consider them as functions defined on Wiener measure space. We state some basic properties of these functionals:

(P1) For each $K_n \in \hat{L}_2(I^n)$, $w \longrightarrow g_n(K_n, w)$ is a μ measurable function on Ω and belongs to $L_2(\Omega, \mathcal{B}, \mu)$.

(P2) For each $K_n \in \hat{L}_2(I^n)$ and $L_m \in \hat{L}_2(I^m)$

$$< g_n(K_n, \cdot), g_m(L_m, \cdot) >_\mu \equiv \int_\Omega g_n \, g_m d\mu = 0, \quad \text{for} \ \ m \neq n.$$

(P3) For each pair $K_n, L_n \in \hat{L}_2(I^n)$,

$$< g_n(K_n, \cdot), g_n(L_n, \cdot) >_\mu \equiv \int_\Omega g_n(K_n, w) \, g_n(L_n, w) d\mu(w)$$

$$= n! \int_{I^n} K_n(\tau_1, \tau_2, \cdots, \tau_n) L_n(\tau_1, \tau_2, \cdots, \tau_n) d\tau_1 d\tau_2 \cdots d\tau_n.$$

(P4) The map $K_n \longrightarrow g_n(K_n, \cdot)$ from $\hat{L}_2(I^n)$ to $L_2(\Omega, \mathcal{B}, \mu)$ is linear and further it is continuous in the sense that whenever the sequence $K_n^r \xrightarrow{s} K_n^0$ in $\hat{L}_2(I^n)$ as $r \to \infty, g_n(K_n^r, \cdot) \xrightarrow{s} g_n(K_n^0, \cdot)$ in $L_2(\Omega, \mathcal{B}, \mu)$.

The most important properties are (P2) and (P3) and they state that the functionals $\{g_n\}$ are orthogonal in $L_2(\Omega, \mathcal{B}, \mu)$. Further, it was proved by Wiener [Wiener (1938)] [Wiener (1958)] that they are complete in the class $L_2(\Omega, \mathcal{B}, \mu)$ [see also, [Ahmed (1970)]]. Hence, they form a basis for the Hilbert space $L_2(\Omega, \mathcal{B}, \mu)$. Thus, these functionals play a similar role as the sine cosine functions do in the Fourier analysis of L_2 functions on finite dimensional spaces. Several years ago the author proved some basic results on the Fourier analysis on Wiener measure space [Ahmed (1968)] such as Bessel's inequality, Parseval's equality, and Riesz-Fischer theorem. It is interesting to observe the strong similarity of these results with their counterparts on finite dimensional spaces. We present these results here.

From now on we shall suppress the w variable and write $g_n(K_n)$ in place of $g_n(K_n, w)$.

Lemma 2.2.1 *For each $f \in L_2(\Omega, \mathcal{B}, \mu)$ and each $s \in N_+$, there exists a unique kernel $L_s \in \hat{L}_2(I^s)$, with respect to the fundamental basis $\{g_n\}$, such that*

$$< f, g_s(K_s) >_\mu = s! \int_{I^s} L_s(\tau_1, \tau_2, \cdots, \tau_s) K_s(\tau_1, \tau_2, \cdots, \tau_s) d\tau_1 d\tau_2 \cdots d\tau_s,$$

$$(2.2)$$

for all $K_s \in L_2(I^s)$.

Proof Clearly, for each $s \in N$, $K \longrightarrow g_s(K)$ is linear from $\hat{L}_2(I^s)$ to $L_2(\Omega, \mathcal{B}, \mu)$ and it is also continuous. Indeed, if $K^n \xrightarrow{s} K^o$ (strongly) in $\hat{L}_2(I^s)$, we have $g_s(K^n) \xrightarrow{s} g_s(K^o)$ in $L_2(\Omega, \mathcal{B}, \mu)$. Thus, for each $f \in L_2(\Omega, \mathcal{B}, \mu)$,

$$K \longrightarrow \ell(K) \equiv < f, g_s(K) >_\mu$$

is a continuous linear functional on $\hat{L}_2(I^s)$ and hence bounded. Indeed, it follows from Schwartz inequality and the property (P3) that

$$|\ell(K)| \leq \| f \|_\mu (\sqrt{s!}) \| K \|_{L_2(I^s)} .$$

Therefore, $K \longrightarrow \ell(K)$ is a bounded continuous linear functional on the space $\hat{L}_2(I^s)$. Hence, it follows from Riesz representation theorem for L_p spaces that there exists an $L_s \in \hat{L}_2(I^s)$ such that

$$\ell(K) = s! \, (L_s, K_s)_{L_2(I_s)}$$
$$= s! \int_{I^s} L_s(\tau_1, \tau_2, \cdots, \tau_s) K_s(\tau_1, \tau_2, \cdots, \tau_s) d\tau_1 d\tau_2 \cdots d\tau_s$$

for every $K_s \in \hat{L}_2(I^s)$. The kernel L_s is uniquely determined by f alone. If not, there exists another kernel $\tilde{L}_s \in \hat{L}_2(I^s)$ that satisfies the same identity. This implies that

$$0 = \int_{I^s} [L_s(\tau_1, \tau_2, \cdots, \tau_s) - \tilde{L}_s(\tau_1, \tau_2, \cdots, \tau_s)] K_s(\tau_1, \tau_2, \cdots, \tau_s) d\tau_1 d\tau_2 \cdots d\tau_s$$

for all $K_s \in \hat{L}_2(I^s)$. This is possible if, and only if, $\tilde{L}_s$ equals L_s almost everywhere on I^s. This completes the proof. •

This result suggests that, given the Wiener measure space $(\Omega, \mathcal{B}, \mu)$ and the fundamental set $\{g_n\}$, the Fourier-Wiener kernels for any $f \in L_2(\Omega, \mathcal{B}, \mu)$ are uniquely determined by f alone.

Next, we present Bessel's inequality. For convenience of notation, we let $L_2(I^0)$ denote the scalars.

Theorem 2.2.2. *Let $\{g_n(K_n), K_n \in \hat{L}_2(I^n)\}$ be the fundamental set as defined above. Then for each $f \in L_2(\Omega, \mathcal{B}, \mu)$, and for all $n \in N$,*

$$< f, f >_\mu \equiv \| f \|^2 \geq \sum_{s=0}^{s=n} s! \int_{I^s} |L_s(\tau_1, \tau_2, \cdots, \tau_s)|^2 d\tau_1 d\tau_2 \cdots d\tau_s, \quad (2.3)$$

where $L_s \in \hat{L}_2(I^s), s \in N$, are the Fourier-Wiener kernels uniquely determined by f through the relation (2.2).

Proof For detailed proof see [1]. Here we present a brief outline. Define

$$J_n \equiv \int_\Omega |f - \sum_{s=0}^{n} g_s(K_s)|^2 d\mu.$$

Using the orthogonality property of the fundamental set $\{g_s\}$ we have

$$J_n = \| f \|_\mu^2 - 2 \sum_{s=0}^{n} < f, g_s(K_s) >_\mu + \sum_{s=0}^{n} s! \| K_s \|_{L_2(I^s)}^2 .$$

Now recalling that $\{L_s\}$ are the Fourier-Wiener kernels of f, it follows from Lemma 2.2.1, in particular the identity (2.2), that J_n can be written as

$$J_n = \| f \|_\mu^2 + \sum_{s=0}^n \| L_s - K_s \|_{L_2(I^s)}^2 - \sum_{s=0}^n s! \| L_s \|_{L_2(I^s)}^2.$$

Since J_n is nonnegative for all n and the sequence of kernels $\{K_s\}$ in the expressions for $\{g_s(K_s)\}$ is arbitrary elements of the Hilbert spaces $\hat{L}_2(I^s)$, we can choose them as $\{L_s\}$. This then leads to the following expression

$$J_n = \| f \|_\mu^2 - \sum_{s=0}^n s! \| L_s \|_{L_2(I^s)}^2 \geq 0$$

which holds for all $n \in N$. Hence, we have the inequality (2.3). This completes the outline of our proof. •

The inequality (2.3) is true for every $n \in N$. As a consequence, one can assert that, for a given function $f \in L_2(\Omega, \mathcal{B}, \mu)$ to be approximated (in the limit in the mean) by a suitable choice of L_2 kernels $\{K_s \in \hat{L}_2(I^s)\}$ for the orthogonal system $\{g_s\}$, the Parseval's equality

$$\| f \|^2 = \sum_{s=0}^\infty s! \int_{I^s} |K_s(\tau_1, \tau_2, \cdots, \tau_s)|^2 d\tau_1 d\tau_2 \cdots d\tau_s \qquad (2.4)$$

must hold.

Now we present a result for the space $L_2(\Omega, \mathcal{B}, \mu)$ which is similar in spirit to the celebrated Riesz-Fischer theorem for ordinary $L_2(\Sigma, \lambda)$ space with Σ contained in a finite dimensional space like R^n and λ the Lebesgue measure.

Theorem 2.2.3 *Given the orthogonal set $\{g_s(K_s), s \in N_+\}$ with $\{K_s \in \hat{L}_2(I^s)\}$, a necessary and sufficient condition that there exists an $L_2(\Omega, \mathcal{B}, \mu)$ function with $\{K_s\}$ as it's Fourier-Wiener kernels is that the series*

$$\sum_{s=0}^\infty s! \int_{I^s} |K_s(\tau_1, \tau_2, \cdots, \tau_s)|^2 d\tau_1 d\tau_2 \cdots d\tau_s \qquad (2.5)$$

converges.

Proof For detailed proof see [[Ahmed (1968)], Theorem 2]. We present a brief outline. Let $f \in L_2(\Omega, \mathcal{B}, \mu)$ with $\{K_s\}$ as its Fourier-Wiener kernels.

Then by Lemma 2.2.1,

$$< f, g_s(K_s) >_\mu = s! \int_{I^s} |K_s(\tau_1, \tau_2, \cdots, \tau_s)|^2 d\tau_1 d\tau_2 \cdots d\tau_s.$$

Thus, it follows from Bessel's inequality (Lemma 2.2.2) that

$$\| f \|_\mu^2 \geq \sum_{s=0}^{n} s! \| K_s \|_{L_2(I^s)}^2$$

for every $n \in N$. This proves that the series converges and hence the necessity. For the proof of the sufficient condition, suppose the series given by (2.5) is convergent. Define the sequence $\{f_n\}$ by

$$f_n \equiv \sum_{s=0}^{n} g_s(K_s).$$

Clearly, this sequence belongs to $L_2(\Omega, \mathcal{B}, \mu)$. For any integer $p \geq 1$, it follows from straight forward computation that

$$\| f_{n+p} - f_n \|_\mu^2 = \sum_{s=n+1}^{n+p} (s!) \| K_s \|_{L_2(I^s)}^2 .$$

Since the series (2.5) converges, it follows from the above expression that $\{f_n\}$ is a Cauchy sequence in $L_2(\Omega, \mathcal{B}, \mu)$. Thus it converges to a unique limit, say $f \in L_2(\Omega, \mathcal{B}, \mu)$. The question is whether or not this limit has the same set of Fourier-Wiener kernels as those given. Suppose, instead, that $\{L_s\}, L_s \in \hat{L}_2(I^s)$, are the Fourier-Wiener kernels of f. Then by straight forward computation, one can verify that

$$\| f - f_n \|_\mu^2 = \left\{ \| f \|_\mu^2 - \sum_{s=0}^{n} s! \| L_s \|_{L_2(I^s)}^2 \right\} + \sum_{s=0}^{n} s! \| L_s - K_s \|_{L_2(I^s)}^2 .$$

By virtue of Bessel's inequality (2.3), the expression within the curly bracket is nonnegative. Thus, it follows from the above identity that

$$\| f - f_n \|_\mu^2 \geq \sum_{s=0}^{n} s! \| L_s - K_s \|_{L_2(I^s)}^2$$

for all $n \in N$. Since $f_n \overset{s}{\longrightarrow} f$ in $L_2(\Omega, \mathcal{B}, \mu)$, it follows from the above inequality that L_s must coincide with the given set of Fourier-wiener kernels K_s as elements of the Hilbert space $\hat{L}_2(I^s)$ for all $s \in N$. This proves the sufficient condition and hence the outline of our proof. •

Remark 2.2.4 The content of this theorem is in the spirit of the celebrated Riesz-Fischer theorem arising in Fourier analysis of functions defined on finite dimensional spaces. This is discussed further below.

Consider the product space $\mathcal{K}_\alpha \equiv \prod_{n=0}^{\infty} \hat{L}_2(I^n)$, where $\hat{L}_2(I^0)$ stands for scalars. This is a linear vector space. We furnish this with a family of seminorms $Q^\alpha \equiv \{q_r\}$ given by $q_r(K) \equiv \| K_r \|_{L_2(I^r)}, r \in N_0 = \{0\} \cup N$, for $K \in \mathcal{K}_\alpha$. Clearly, this is a countable family of seminorms and so $\mathcal{K}_\alpha \equiv (\mathcal{K}_\alpha, Q^\alpha)$ is a seminrmed space. Using this family of seminorms, the space $\mathcal{K}_\alpha$ is metrizable and the metric is given by

$$d(K, L) \equiv \sum_{r=0}^{\infty} (1/2^r) \frac{q_r(K - L)}{(1 + q_r(K - L))}. \tag{2.6}$$

Since each of the component spaces are Hilbert spaces and so complete, it is clear that $\mathcal{K}_\alpha$, furnished with this metric topology, is a complete (linear) metric space and hence Fréchet space. Define $G_\alpha \subset \mathcal{K}_\alpha$ by

$$G_\alpha \equiv \{K \in \mathcal{K}_\alpha : \sum_{n=0}^{\infty} n! \, |K_n|_{L_2(I^n)}^2 < \infty\}.$$

Clearly, this is a linear subspace of the Frèchet space $\mathcal{K}_\alpha$. The space G_α, completed with respect to the scalar product

$$< K, L > \equiv \sum_{s=0}^{\infty} s! \int_{I^s} K_s(\tau_1, \tau_2, \cdots, \tau_s) L_s^*(\tau_1, \tau_2, \cdots, \tau_s) d\tau_1 d\tau_2 \cdots d\tau_s, \tag{2.7}$$

is a Hilbert space. We denote its completion by the same symbol G_α. Clearly, the norm of an element $K \in G_\alpha$ is given by

$$\| K \|_{G_\alpha} \equiv \Big(\sum_{s=0}^{\infty} s! \int_{I^s} |K_s(\tau_1, \tau_2, ..., \tau_s)|^2 d\tau_1 d\tau_2 \cdots d\tau_s \Big)^{1/2}. \tag{2.8}$$

In view of the Riesz-Fischer Theorem 2.2.3, we see that $L_2(\Omega, \mathcal{B}, \mu)$ is isometrically isomorphic to G_α. More specifically, define the map $\Phi_\alpha : G_\alpha \longrightarrow L_2(\Omega, \mathcal{B}, \mu)$ by

$$\Phi_\alpha(K) \equiv \sum_{s=0}^{\infty} g_s(K_s). \tag{2.9}$$

Clearly,

(1): $\Phi_\alpha(aK) = a\Phi_\alpha(K)$ for all scalars a and $K \in G_\alpha$

(2): $\Phi_\alpha(K + L) = \Phi_\alpha(K) + \Phi_\alpha(L)$ for all $K, L \in G_\alpha$.

(3): $\| \Phi_\alpha(K) \|_{L_2(\Omega, \mathcal{B}, \mu)} = \| K \|_{G_\alpha}$.

Thus, Φ_α is an isometric isomorphism of G_α onto $L_2(\Omega, \mathcal{B}, \mu)$ and we express this by writing $G_\alpha \cong L_2(\Omega, \mathcal{B}, \mu)$. For convenience of notations we use $G_\alpha \leftrightarrow L_2(\Omega, \mathcal{B}, \mu)$ instead.

Set $\mathcal{G}_\alpha \equiv L_2(\Omega, \mathcal{B}, \mu)$ and define the linear subspace $\mathcal{G}_{\alpha,n}$, spanned by the set $\{g_n(K), K \in \hat{L}_2(I^n)\}$ as follows:

$$\mathcal{G}_{\alpha,n} \equiv \{g_n(K_n) : \sqrt{n!} \, K_n \in \hat{L}_2(I^n)\}. \tag{2.10}$$

This is the space of Wiener's homogeneous chaos (of degree n). The celebrated Wiener-Itô decomposition follows as a corollary from the above discussions.

Corollary 2.2.5 *The space G_α is isometrically isomorphic to $\mathcal{G}_\alpha$ and this is denoted by $L_2(\Omega, \mathcal{B}, \mu) \equiv \mathcal{G}_\alpha \leftrightarrow G_\alpha$ and further, $\mathcal{G}_\alpha = \sum \bigoplus_{n=0}^\infty \mathcal{G}_{\alpha,n}$.*

2.3 Functionals of Vector Brownian Motion

The results of the previous section were stated for one dimensional Brownian motion. In fact they are also valid for multidimensional case. This requires use of symmetric multi linear forms for the multiple Wiener integrals. This is presented here briefly. Further, the results also apply to multi parameter Brownian motion.

Let $(\Omega, \mathcal{B}, P)$ be a probability space and $I \equiv (0, T)$ an open bounded interval and $\{w(t), t \in I\}$, a d-dimensional standard Brownian motion on $(\Omega, \mathcal{B}, P)$ starting from the origin with probability one. We set $\Omega \equiv C_0(I, R^d)$, the canonical sample space, where $C_0(I, R^d) \equiv \{x \in C(I, R^d) : x(0) = 0\}$, and let $\mathcal{B}$ denote the Borel σ-algebra generated by open or closed subsets of the set Ω and let μ^w denote the canonical Wiener measure. For each nonnegative integer n, let $M_n \equiv \mathcal{L}(R^{d \otimes n}, R^\ell)$ denote the space of symmetric n-linear forms (multilinear forms) from R^d to R^ℓ. For $B_1, B_2 \in M_n$

define

$$< B_1, B_2 > = Tr(B_1^* B_2) = Tr(B_2^* B_1), \quad \text{and} \quad \| B \|_{M_n}^2 \equiv Tr(B^* B).$$

Now we define a regular homogeneous functional of Brownian motion of degree n by

$$f_n(K_n)(w) \equiv \int_{I^n} K_n(\tau_1, \tau_2, \cdots, \tau_n) dw(\tau_1) \otimes dw(\tau_2) \cdots \otimes dw(\tau_n),$$

$$(2.11)$$

for any $K_n \in L_2(I^n, M_n)$. These are known as the Polynomial Chaos of degree n taking values from R^ℓ. For each $K_n \in L_2(I, M_n)$, the functional f_n is well defined in the Wiener-Itô sense and belongs to $L_2(\Omega, \mathcal{B}, P, \mu^w; R^\ell) \equiv L_2(\mu^w, R^\ell)$. Again, let $\hat{L}_2(I^n, M_n)$ denote the class of symmetric L_2 kernels on I^n taking values from M_n. Like in the scalar case, these homogeneous functionals $\{f_n\}$ satisfy the following properties:

(P1) For each $K_n \in \hat{L}_2(I^n, M_n)$, $f_n(K_n)$ is a μ^w-measurable function on Ω and belongs to $L_2(\mu^w, R^\ell)$.

(P2) For each $K_n \in \hat{L}_2(I^n, M_n)$ and $L_m \in \hat{L}_2(I^m, M_m)$

$$< f_n(K_n), f_m(L_m) >_{\mu^w} \equiv \int_\Omega < f_n, f_m >_{R^\ell} d\mu^w = 0 \text{ for } n \neq m.$$

(P3) For each $K_n \in \hat{L}_2(I^n, M_n)$ and $L_n \in \hat{L}_2(I^n, M_n)$

$$< f_n(K_n), f_n(L_n) >_{\mu^w} \equiv \int_\Omega < f_n, f_n >_{R^\ell} d\mu^w$$

$$= n! \int_{I^n} Tr(K_n^* L_n) d\tau_1 d\tau_2 \cdots d\tau_n.$$

(P4) The map $K_n \longrightarrow f_n(K_n)$ from $\hat{L}_2(I^n, M_n)$ to $L_2(\mu^w; R^\ell)$ is linear and, further, it is continuous in the sense that whenever $K_n^r \xrightarrow{s} K_n^o$ in $\hat{L}_2(I^n, M_n)$, as $r \to \infty$,

$$f_n(K_n^r) \xrightarrow{s} f_n(K_n^o) \text{ in } L_2(\Omega, \mathcal{B}, \mu^w; R^\ell).$$

The most fundamental property of the sequence $\{f_n(K_n), K_n \in \hat{L}_2(I^n, M_n), n \in N\}$, as in the scalar case, is that they are orthogonal and complete in the class $L_2(\mu^w, R^\ell)$. For convenience of notation we shall use μ for μ^w.

As in the scalar case, we have the following result.

Lemma 2.3.1 *For each $f \in L_2(\Omega, \mathcal{B}, \mu; R^\ell) \equiv L_2(\mu, R^\ell)$ and each $s \in N$, there exists a unique kernel $L_s \in \hat{L}_2(I^s, M_s)$, with respect to the fundamental basis $\{f_n\}$, such that*

$$< f, f_s(K_s) >_\mu = s! \int_{I^s} Tr((L_s^* K_s)(\tau_1, \tau_2, \cdots, \tau_s)) d\tau_1 d\tau_2 \cdots d\tau_s,$$

$$= (s)! \int_{I^s} < L_s(\tau_1, \tau_2, \cdots, \tau_s), K_s(\tau_1, \tau_2, \cdots, \tau_s) >_{M_s} d\tau_1 d\tau_2 \cdots d\tau_s,$$

$$(2.12)$$

for all $K_s \in \hat{L}_2(I^s, M_s)$.

Proof. Proof is similar to that of the scalar case Lemma 2.2.1. •

Next, we have the Bessel's inequality as stated in the following Lemma.

Lemma 2.3.2 *Let $\{f_n(K_n)\}$ with $K_n \in \hat{L}_2(I^n, M_n)$ denote the fundamental set as defined above. Then, each $f \in L_2(\Omega, \mathcal{B}, \mu; R^\ell) \equiv L_2(\mu, R^\ell)$, with Fourier-Wiener Kernels $\{L_n\}$ uniquely determined by the relation (2.12), satisfies the following inequality:*

$$< f, f >_\mu$$

$$\equiv \| f \|_{L_2(\mu, R^\ell)}^2 \geq \sum_{n=0}^{m} n! \int_{I^n} |L_n(\tau_1, \cdots, \tau_n)|_{M_n}^2 d\tau_1 \cdots d\tau_n, \quad (2.13)$$

for all $m \in N$.

Proof The proof is identical to that of Lemma 2.2.2.

Remark 2.3.3 As a consequence of this result we observe that, for a function $f \in L_2(\mu, R^\ell)$ to be approximated (in the mean) by a suitable choice of kernels $\{K_s\} \subset \hat{L}_2(I^s, M_s)$ for the fundamental system $\{f_s\}$, the Parseval's equality must hold.

$$\| f \|_\mu^2 = \sum_{s=0}^{\infty} s! \int_{I^s} |K_s(\tau_1, \tau_2, \cdots, \tau_s)|_{M_s}^2 d\tau_1 d\tau_2 \cdots d\tau_s. \quad (2.14)$$

Again, as in the scalar case, we have the celebrated Riesz-Fischer theorem for the space $L_2(\Omega, \mathcal{B}, \mu; R^\ell) \equiv L_2(\mu, R^\ell)$.

Theorem 2.3.4 *Given the orthogonal system* $\{f_s(K_s), K_s \in \hat{L}_2(I^s, M_s), s \in N\}$, *the necessary and sufficient condition that there exists an element* $f \in L_2(\mu, R^\ell)$ *with* $\{K_s\}$ *as its Fourier-Wiener kernels is that the series*

$$\sum_{s=0}^{\infty} s! \int_{I^s} |K_s(\tau_1, \tau_2, \cdots, \tau_s)|_{M_s}^2 d\tau_1 d\tau_2 \cdots d\tau_s \qquad (2.15)$$

converges.

Proof The proof is identical to that of Theorem 2.2.3. •

Now, we introduce $\mathcal{K}_\beta$ to denote the product space

$$\mathcal{K}_\beta \equiv \prod_{n=0}^{\infty} \hat{L}_2(I^n, M_n).$$

Note that this is the vector version of $\mathcal{K}_\alpha$. Exactly as in the scalar case, one can introduce a countable family of seminorms on it turning it into a Fréchet space $(\mathcal{K}_\beta, Q^\beta)$. Let G_β be a linear subspace of $\mathcal{K}_\beta$ completed with respect to the scalar product

$$< K, L >_{G_\beta} \equiv \sum_{s=0}^{\infty} s! \int_{I^s} < K_s(\cdots), L_s(\cdots) >_{M_s} d\tau_1 d\tau_2 \cdots d\tau_s$$

$$\equiv \sum_{s=0}^{\infty} s! \int_{I^s} Tr(L_s^* K_s)(\tau_1, \tau_2, \cdots, \tau_s) d\tau_1 d\tau_2 \cdots d\tau_s. \qquad (2.16)$$

Naturally, G_β is a Hilbert space with respect to the norm topology,

$$\| K \|_{G_\beta} \equiv \left(\sum_{s=0}^{\infty} s! \int_{I^s} |K_s(\tau_1, \tau_2, \cdots, \tau_s)|_{M_s}^2 d\tau_1 d\tau_2 \cdots d\tau_s \right)^{1/2}, (2.17)$$

which is inherited from the scalar product given by the expression (2.16). In view of Theorem 2.3.4, we see that the space $L_2(\Omega, \mathcal{B}, \mu; R^\ell) \equiv L_2(\mu, R^\ell)$ is isometrically isomorphic to G_β. Indeed, define the map $\Phi_\beta : G_\beta \longrightarrow L_2(\mu, R^\ell)$ by

$$\Phi_\beta(K) \equiv \sum_{s=0}^{\infty} f_s(K_s). \qquad (2.18)$$

Clearly,

(1) $\Phi_\beta(aK) = a\Phi_\beta(K)$ for all scalars a and $K \in G_\beta$

(2) $\Phi_\beta(K + L) = \Phi_\beta(K) + \Phi_\beta(L)$ for all $K, L \in G_\beta$.

(3) $\| \Phi_\beta(K) \|_{L_2(\mu, R^\ell)} = \| K \|_{G_\beta}$.

Thus Φ_β is an isometric isomorphism of G_β onto $L_2(\Omega, \mathcal{B}, \mu; R^\ell) \equiv L_2(\mu, R^\ell)$. Set $\mathcal{G}_\beta \equiv L_2(\Omega, \mathcal{B}, \mu; R^\ell)$ and define, for each integer n, the linear subspace

$$\mathcal{G}_{\beta,n} \equiv \{f_n(K) : K \in \hat{L}_2(I^n, M_n)\} \subset \mathcal{G}_\beta \qquad (2.19)$$

representing the space of Wiener's homogeneous chaos of degree n. Using these subspaces we obtain the Wiener-Itô decomposition as well as the isometric isomorphism as stated in the following corollary.

Corollary 2.3.5 $\mathcal{G}_\beta = \sum_{n=0}^{\infty} \oplus \mathcal{G}_{\beta,n}$ *and further,* $\mathcal{G}_\beta$ *is isometrically isomorphic to* G_β *and this is symbolically represented by* $G_\beta \leftrightarrow \mathcal{G}_\beta$.

Orthogonal Functionals of Nonstandard Brownian Motion: So far we have considered standard Brownian motion which has the property:

$$\mathbb{E}\{(w(t), \xi)(w(s), \eta)\} = (t \wedge s)(\xi, \eta)$$

for any $\xi, \eta \in R^d$. This means that the corresponding white noise has normalized power density 1. In fact, the power density of each of the components is one and so for vector case it equals the dimension d. We consider it normalized by dividing the true power density by the dimension of the vector Brownian motion. For arbitrary power density, we may consider Brownian motions satisfying

$$\mathbb{E}\{(w(t), \xi)(w(s), \eta)\} = (1/2\lambda)(t \wedge s)(\xi, \eta)$$

for any $\lambda > 0$. The Wiener measure μ is now given by the indexed family of Wiener measures $\mu_\lambda, \lambda > 0$. In this case the orthogonal functionals $\{f_n(K_n), K_n \in \hat{L}_2(I^n, M_n)\}$ have exactly the same form as given by (2.11) with w being now the nonstandard Brownian motion. All the properties (P1), (P2), (P4) remain valid and the property (P3) changes to

$$\overline{(\text{P3})}: \quad < f_n(K_n), f_n(L_n) >_{\mu_\lambda} = (n!/(2\lambda)^n) \int_{I^n} Tr(L_n^* K_n) d\tau_1 d\tau_2 \cdots d\tau_n.$$

In this case the Hilbert space G_β is replaced by the indexed family of Hilbert spaces $\{F_\lambda, \lambda > 0\}$, given by

$$F_\lambda \equiv \{K \in \mathcal{K}_\beta : \| K \|_\lambda \equiv \Big(\sum_{n=0}^{\infty} (n!/(2\lambda)^n) \| K_n \|_{L_2(I^n, M_n)}^2 \Big)^{1/2} < \infty\}.$$

Note that for $\lambda = 1/2$ we get back G_β. Again by use of the isomorphism Φ_β we obtain a new class of functionals of Brownian motion given by

$$\mathcal{F}_\lambda \equiv \Phi_\beta(F_\lambda).$$

All the results presented in this section remain valid for the spaces F_λ and the space of functionals of Brownian motion given by $\mathcal{F}_\lambda \equiv L_2(\Omega, \mathcal{B}, \mu_\lambda; R^\ell)$. Following result has interesting significance in applications.

Theorem 2.3.6 *The system of Hilbert spaces $\{F_\lambda, \lambda > 0\}$ forms a totally ordered lattice with $F_\lambda \subset F_\rho$ for $0 < \lambda < \rho$. By virtue of the isomorphism Φ_β, preserving the lattice property, $\mathcal{F}_\lambda \subset \mathcal{F}_\rho$ for $\lambda < \rho$ with continuous injection.*

For details the reader is referred to [Ahmed, [Ahmed (1968)]], [Ahmed (1995)], [Ahmed (1973)], [Ahmed (1969)], [Ahmed (1994)].

Physical significance of the above result is the following. The smaller the λ is, the larger is the power density of the Brownian motion. Hence the class of kernels $\{K\} \subset \mathcal{K}_\beta$, for which $f \equiv \Phi_\beta(K) \equiv \sum_{n=0}^\infty f_n(K_n)$ is well defined and has finite energy, is smaller. The smaller the noise intensity is, the faster is the rate of convergence of the representing Wiener series. And in applications this presents an advantage: a small number of kernels are required to be constructed.

Another interpretation is: for a given nonstandard Brownian motion, the maximal set of admissible nonlinear transformations for which the output has finite energy is limited and is determined by the family $\{\mathcal{F}_\lambda, \lambda > 0\}$.

2.4 Functionals of Gaussian Random Field (GRF)

Let D be an open bounded connected subset of R^d (with smooth boundary) and let $\mathcal{B}(D)$ denote the class of Borel subsets of D, and ℓ the Lebesgue measure on $\mathcal{B}(D)$. Let $W(dz)$ denote the Gaussian Random measure (or Field) defined on $\mathcal{B}(D)$ satisfying the following basic properties:

(1) For any $\Gamma \in \mathcal{B}$, $W(\Gamma)$ is a Gaussian random variable with mean $\mathbb{E}\{W(\Gamma)\} = 0$.

(2) For any two sets, $D_1, D_2 \in \mathcal{B}$, $\mathbb{E}\{W(D_1)W(D_2)\} = \ell(D_1 \cap D_2)$, and this implies that $\mathbb{E}(W(\Gamma))^2 = \ell(\Gamma)$ for any $\Gamma \in \mathcal{B}$.

(3) For any two disjoint sets $\Gamma, \Delta \in \mathcal{B}$, $W(\Gamma \cup \Delta) = W(\Gamma) + W(\Delta)$, and $W(\Gamma \cap \Delta) \equiv W(\emptyset) = 0$ with probability one.

Let $\mathcal{M}_s \equiv \mathcal{M}_s(D)$ denote the space of signed measures on Borel subsets of D and let $X(\omega, d\xi)$ denote a random Gaussian measure on $\mathcal{B}(D)$ satisfying the properties (1)-(3). In other words X is a random variable with values in $\mathcal{M}_s$, that is

$$X : (\Omega, \mathcal{F}, P) \longrightarrow \mathcal{M}_s.$$

Under the canonical map $X(\omega, d\xi) = \omega(d\xi)$, X induces a Wiener measure μ on $\mathcal{M}_s$. We call this space the Wiener measure space and denote it by

$$(\mathcal{M}_s, \mathcal{B}_s, \mu) \equiv (\mathcal{M}_s, \mu).$$

We are interested in the Hilbert space $L_2(\mathcal{M}_s, \mu)$. This is the space of equivalence classes of μ measurable functions defined on the space of signed measures $\mathcal{M}_s$, and square integrable with respect to the measure μ.

As in the preceding sections, for consistency of notations, we introduce the product space

$$\mathcal{K}_\gamma \equiv \prod_{n=0}^{\infty} \hat{L}_2(D^n)$$

where D^n denotes the Cartesian product of n-copies of D. Then we define a similar family of seminorms Q^γ so that we have the Fréchet space $\mathcal{K}_\gamma \equiv (\mathcal{K}_\gamma, Q^\gamma)$. We use systematically the following subspaces:

$$G_\gamma \equiv \{K \in \mathcal{K}_\gamma : \Big(\sum_{n=0}^{\infty} n! \parallel K_n \parallel^2_{L_2(D^n)}\Big)^{1/2} < \infty\},$$

$$H_\gamma \equiv \{K \in \mathcal{K}_\gamma : \Big(\sum_{n=0}^{\infty} \parallel K_n \parallel^2_{L_2(D^n)}\Big)^{1/2} < \infty\},$$

whose completions, with respect to the norm topologies indicated above, are Hilbert spaces.

For each $K_n \in \hat{L}_2(D^n)$, we introduce the fundamental sequence of homogeneous functionals $\{e_n\}$ of Gaussian random fields given by:

$$e_n(K_n) \equiv \int_{D^n} K_n(\xi_1, \xi_2, ..., \xi_n) W(d\xi_1) W(d\xi_2) \cdots W(d\xi_n). \quad (2.20)$$

For each integer $n \geq 0$, define

$$\mathcal{G}_{\gamma,n} \equiv \{e_n(K), K \in \hat{L}_2(D^n)\}. \quad (2.21)$$

The space $\mathcal{G}_{\gamma,n}$ is a linear vector space of n-th order polynomial (Wiener) Chaos. We can introduce a scalar product in $\mathcal{G}_{\gamma,n}$ as before. For $\phi, \psi \in \mathcal{G}_{\gamma,n}$, we have

$$< \phi, \psi >_\mu = \int_{\mathcal{M}_s} \phi \, \psi \, d\mu \equiv \int_{\mathcal{M}_s} \phi \, \psi \, d\mu = n! \, (K_n, L_n)_{L_2(D^n)}$$

$$\equiv n! \int_{D^n} K_n(\xi_1, \cdots, \xi_n) \, L_n(\xi_1, \xi_2, \cdots, \xi_n) \, \ell(d\xi_1) \cdots \ell(d\xi_n). \quad (2.22)$$

Just as in section 2, we have the isometric isomorphism,

$$\mathcal{G}_{\gamma,n} \leftrightarrow \sqrt{n!} \, \hat{L}_2(D^n). \quad (2.23)$$

Similar to the case of functionals of Brownian motion, as seen in section 2.2, we have the Wiener-Itô decomposition:

$$L_2(\mathcal{M}_s, \mu) = \sum_{n=0}^{n=\infty} \oplus \mathcal{G}_{\gamma,n} \equiv \mathcal{G}_\gamma. \quad (2.24)$$

Further, Riesz-Fischer Theorem holds as stated below.

Theorem 2.4.1 *Given the orthogonal system $\{e_s(K_s), K_s \in \hat{L}_2(D^s), s \in N\}$, the necessary and sufficient condition that there exists an element $e \in L_2(\mathcal{M}_s, \mu)$ with $\{K_s\}$ as its Fourier-Wiener kernels is that the series*

$$\sum_{s=0}^{\infty} \int_{D^s} |K_s(\xi_1, \xi_2, \cdots, \xi_s)|^2 \ell(d\xi_1) \ell(d\xi_2) \cdots \ell(d\xi_s) \quad (2.25)$$

converges.

Proof The proof is identical to that of Theorem 2.2.3. •

Introduce the Hilbert space

$$G_\gamma \equiv \sum_{n=0}^{\infty} \oplus \sqrt{n!} \hat{L}_2(D^n)$$

and let Π_n denote the projection (operator) from $L_2(\mathcal{M}_s, \mu)$ to $\mathcal{G}_{\gamma,n}$. In view of Riesz-Fischer Theorem 2.4.1, each $e \in L_2(\mathcal{M}_s, \mu)$ has the representation

$$e = \sum_{n=0}^{\infty} \Pi_n(e) = \sum_{n=0}^{\infty} e_n(L_n), \qquad (2.26)$$

where

$$e_n(L_n) \equiv \int_{D^n} L_n(\xi_1, \xi_2, \cdots, \xi_n) W(d\xi_1) W(d\xi_2) \cdots W(d\xi_n), \quad (2.27)$$

with

$$L \equiv \{L_n \in \hat{L}_2(D^n), n \geq 0\} \in G_\gamma.$$

Again, as seen in section 2.2, the Fourier-Wiener kernel L is uniquely determined by e alone and we have

$$\| e \|_{\mathcal{G}_\gamma} \equiv \| e \|_{L_2(\mathcal{M}_s, \mu)} = \left(\sum_{n=0}^{\infty} n! \| L_n \|_{\hat{L}_2(D^n)}^2 \right)^{1/2} \equiv \| L \|_{G_\gamma}. \quad (2.28)$$

It follows from the above discussion that the map $\Phi_\gamma : G_\gamma \longrightarrow \mathcal{G}_\gamma = L_2(\mathcal{M}_s, \mu)$ given by

$$\Phi_\gamma(K) \equiv \sum_{n=0}^{\infty} e_n(K_n)$$

is an isometric isomorphism between G_γ and $\mathcal{G}_\gamma$ satisfying similar properties as those of Φ_α and Φ_β.

From these observations, we obtain the following corollary.

Corollary 2.4.2 $\mathcal{G}_\gamma$ *is the space of regular functionals of Gaussian random fields on D having the Wiener-Itô decomposition $\mathcal{G}_\gamma = \sum_{n=0}^{\infty} \oplus \mathcal{G}_{\gamma,n}$. With respect to the norm topologies as given above, the spaces G_γ and $\mathcal{G}_\gamma$ are Hilbert spaces and they are isometrically isomorphic, $G_\gamma \leftrightarrow \mathcal{G}_\gamma$.*

2.5   Functionals of Multidimensional Gaussian Random Fields

The results presented above have been extended to multidimensional Gaussian Random Fields. Interested readers may check with [Ahmed (1995)].

We present briefly the modifications necessary to extend the results of section 2.4 to the vector case. The space of scalar signed measures $\mathcal{M}_s(D)$ is replaced by the space of vector measures $\mathcal{M}(D, R^k)$. Let $X(\omega, d\xi)$ denote a k-dimensional Gaussian random measure on $\mathcal{B} \equiv \mathcal{B}(D)$ satisfying similar basic properties as in section 2.4 with slight modifications.

(1) For each $\Gamma \in \mathcal{B}$, $W(\Gamma)$ is an R^k-valued Gaussian random variable and for every $\eta \in R^k$, $\mathbb{E}\{(W(\Gamma), \eta)\} = 0$.

(2) For any two sets $D_1, D_2 \in \mathcal{B}$, and $\eta, \zeta \in R^k$

$$\mathbb{E}\{(W(D_1), \eta)(W(D_2), \zeta)\} = (\eta, \zeta)\ell(D_1 \cap D_2),$$

and this implies that $\mathbb{E}(W(\Gamma), \zeta)^2 = |\zeta|^2_{R^k} \ell(\Gamma)$ for any $\Gamma \in \mathcal{B}$.

(3) For any two disjoint sets $\Gamma, \Delta \in \mathcal{B}$, $W(\Gamma \cup \Delta) = W(\Gamma) + W(\Delta)$, and $W(\Gamma \cap \Delta) \equiv W(\emptyset) = 0$ with probability one.

Under the canonical map $X(\omega, d\xi) \longrightarrow \omega(d\xi)$, X induces a Wiener measure $\mu^w \equiv PX^{-1}$ on $\mathcal{M} \equiv \mathcal{M}(D, R^k)$ which we call the canonical Wiener measure. We call the space $(\mathcal{M}, \mathcal{B}(\mathcal{M}), \mu^w)$ the Wiener measure space to be briefly denoted by $(\mathcal{M}, \mu)$.

Like in section 2.3, for any pair of fixed nonnegative integers k and s, and for each nonnegative integer n, let $M_n \equiv \mathcal{L}(R^{k \otimes n}, R^s)$ denote the space of symmetric n-linear forms (multilinear forms) from R^k to R^s. For $C, B \in M_n$ define

$$< C, B >_{M_n} = Tr(C^*B) = Tr(B^*C).$$

Now using standard tensor notation as in section 2.3, we define a regular homogeneous functional of Brownian fields of degree n by $h_n(K_n) \equiv h_n(K_n)(W)$ where

$$h_n(K_n)(W) \equiv \int_{D^n} K_n(\xi_1, \xi_2, \cdots, \xi_n) W(d\xi_1) \otimes W(d\xi_2) \cdots \otimes W(d\xi_n),$$

$$(2.29)$$

for any $K_n \in \hat{L}_2(D^n, M_n)$. These are known as the polynomial chaos of degree n taking values from R^s. The space spanned by these elementary functionals is denoted by

$$\mathcal{G}_{\delta,n} \equiv \{h_n(K), K \in \hat{L}_2(D^n, M_n)\}.$$

Clearly, the space $\hat{L}_2(D^n, M_n)$ has the natural norm topology given by

$$\| K_n \|_{L_2(D^n, M_n)} \equiv \left(\int_{D^n} |K_n(\xi_1, \xi_2, \cdots, \xi_n)|^2_{M_n} \ell(d\xi_1) \ell(d\xi_2) \cdots \ell(d\xi_n) \right)^{1/2}$$

for $K_n \in \hat{L}_2(D^n, M_n)$.

The functionals $\{h_n(K_n)\}$ introduced above have exactly similar properties as those of the scalar case treated in section 2.4. Without going into details we simply state the results. Define the Hilbert space

$$G_\delta \equiv \sum_{n=0}^{\infty} \oplus \sqrt{n!} \, \hat{L}_2(D^n, M_n) \subset \mathcal{K}_\delta \equiv \prod_{n=0}^{\infty} \hat{L}_2(D^n, M_n) \qquad (2.30)$$

where $\mathcal{K}_\delta$ is the Fréchet space. The space G_δ is furnished with the norm topology

$$\| L \|_{G_\delta} \equiv \left(\sum_{n=0}^{\infty} n! \, \| L_n \|^2_{L_2(D^n, M_n)} \right)^{1/2}.$$

Again, we can introduce the map $\Phi_\delta : G_\delta \longrightarrow \mathcal{G}_\delta \equiv L_2(\mathcal{M}, \mu)$ through the expression

$$\Phi_\delta(L) \equiv \sum_{n=0}^{\infty} h_n(L_n).$$

This leads us to the following result.

Theorem 2.5.1 *Given the orthogonal system*

$$\{h_s(K_s), K_s \in \hat{L}_2(D^s, M_s), s \in N\},$$

the necessary and sufficient condition that there exists an element $h \in L_2(\mathcal{M}, \mu)$ with $\{K_s\}$ as its Fourier-Wiener kernels is that the series

$$\sum_{s=0}^{\infty} s! \int_{D^s} |K_s(\xi_1, \xi_2, \cdots, \xi_s)|^2_{M_s} \ell(d\xi_1) \ell(d\xi_2) \cdots \ell(d\xi_s) \qquad (2.31)$$

converges.

Using this result, we obtain the following corollary.

Corollary 2.5.2 *The space $\mathcal{G}_\delta$ is the space of regular functionals of Gaussian random fields on D having the Wiener-Itô decomposition $\mathcal{G}_\delta = \sum_{n=0}^{\infty} \oplus \mathcal{G}_{\delta,n}$. With respect to the norm topologies as given above, the* *spaces G_δ and $\mathcal{G}_\delta$ are Hilbert spaces and they are isometrically isomorphic: $G_\delta \leftrightarrow \mathcal{G}_\delta$.*

Summary 2.5.3 So far in this chapter, we have introduced the regular functionals of scalar Brownian motions, vector Brownian motions, scalar Gaussian random fields and vector valued Gaussian random fields. These are elements of Hilbert spaces $\mathcal{G}_\alpha, \mathcal{G}_\beta, \mathcal{G}_\gamma, \mathcal{G}_\delta$ which are isomorphic (isometric) images of the Hilbert spaces $G_\alpha, G_\beta, G_\gamma, G_\delta$ respectively.

Remark 2.5.4 (Extension of $\mathcal{G}_\gamma$) In view of the above results, it is evident that there is no particular problem considering space-time orthogonal Gaussian random fields (measures) and Wiener-Itô multiple integrals with respect to such measures. For example, considering the scalar case, the homogeneous chaos of degree n with respect to space time orthogonal Gaussian measure is written as

$$\gamma_n(K_n) \equiv \gamma_n(K_n, W)$$
$$\equiv \int_{D_T^n} K_n(\tau_1, \xi_1; \tau_2, \xi_2; \cdots; \tau_n, \xi_n) W(d\tau_1 \times d\xi_1) \cdots W(\tau_n \times d\xi_n)$$

$$(2.32)$$

where $D_T \equiv [0, T] \times D$ with $T < \infty$. Assuming that, for each $n \in N$, the kernel $K_n \in \hat{L}_2(D_T^n)$ and the Gaussian random measure is standard with zero mean, the set

$$\{\gamma_n(K_n), K_n \in \hat{L}_2(D_T), n \geq 0\}$$

forms an orthogonal system of functionals of space-time Gaussian random field. For each $n \in N$, the norm square of the random variable γ_n is given by the usual expression

$$\mathbb{E}|\gamma_n|^2 = n! \parallel K_n \parallel_{L_2(D_T^n)}^2.$$

Clearly, this is an extension of the class $\mathcal{G}_\gamma$ and it has application to stochastic partial differential equations driven by space time white noise. This is discussed in section 6.6 of Chapter 6.

2.6   Functionals of ∞-Dimensional Brownian Motion

Let H and U be a pair of separable Hilbert spaces each furnished with a complete orthonormal basis $\{e_i\}$ and $\{u_i\}$ respectively. Let $W = \{W(t), t \geq 0\}$, with $P\{W(0) = 0\} = 1$, be an U valued Brownian motion with (incremental) covariance $Q \in \mathcal{L}_1^+(U)$, the space of nonnegative symmetric nuclear operators in the Hilbert space U. That is, for each $u, v \in U$,

$$\mathbb{E}\{(W(t), u)(W(t), v)\} = t(Qu, v), t \geq 0.$$

Without any loss of generality we may assume that the Wiener process W has the representation

$$W(t) \equiv \sum_{i=1}^{\infty} \sqrt{q_i} u_i \beta_i(t)$$

where $\{\beta_i\}$ is a system of pairwise independent scalar Brownian motions and $\{q_i \geq 0\}$ are the eigen values of the operator Q with the corresponding eigen functions $\{u_i\}$. Let $U^{\otimes n}$ denote the n-fold tensor product of U and $\mathcal{L}_2(U^{\otimes n}, H)$ the space of symmetric n-linear forms from U to H. This is endowed with the scalar product given by

$$< L, K >_{\mathcal{L}_2(U^{\otimes n}, H)} \equiv Tr(K^*L)$$

$$\equiv \sum_{i_1, \cdots, i_n \geq 1}^{\infty} (L u_{i_1} \otimes u_{i_2} \cdots \otimes u_{i_n}, K u_{i_1} \otimes u_{i_2} \cdots \otimes u_{i_n})_H$$

for $L, K \in \mathcal{L}_2(U^{\otimes n}, H)$. Naturally, the associated norm is given by $\| K \|_{\mathcal{L}_2(U^{\otimes n}, H)} \equiv (Tr(K^*K))^{1/2}$ with respect to which $\mathcal{L}_2(U^{\otimes n}, H)$ is a Hilbert space, called the space of Hilbert-Schmidt n-linear forms. Let $(t_1, t_2, \cdots, t_n) \longrightarrow K_n(t_1, t_2, \cdots, t_n)$ be a Borel measurable function defined on $I^n \equiv [0, T]^n$ and taking values in the Hilbert space $\mathcal{L}_2(U^{\otimes n}, H)$. Define the multiple Wiener-Itô integral

$$J_n(K_n) \equiv J_n(K_n, W)$$

$$\equiv \int_{I^n} K_n(t_1, t_2, \cdots, t_n) dW(t_1) \otimes dW(t_2) \cdots \otimes dW(t_n). \quad (2.33)$$

Let $\Omega \equiv C_0(I, U)$ denote the space of continuous functions with values in U and vanishing at zero; and μ the canonical Wiener measure defined on $\mathcal{B}$, the Borel subsets of the set Ω, giving the canonical Wiener measure space

$(\Omega, \mathcal{B}, \mu)$. Let $L_2(\Omega, \mu; H)$ denote the Hilbert space of H-valued random variables whose norms are square integrable with respect to the Wiener measure μ. For each $K_n \in \hat{L}_2(I^n, \mathcal{L}_2(U^{\otimes n}, H))$, it is easy to verify that $J_n(K_n) \in L_2(\Omega, \mu; H)$. Indeed, by direct computation, we have

$$\mathbb{E}|J_n(K_n)|_H^2 = \sum_{\ell=1}^{\infty} \mathbb{E}\{(J_n(K_n), e_\ell)_H^2\} = n! \int_{I^n} \sum_{i_1, i_2, \cdots, i_n \geq 1} q_{i_1} q_{i_2} \cdots q_{i_n}$$

$$\bullet \sum_{\ell}^{\infty} (K_n(t_1, t_2, \cdots, t_n) u_{i_1} \otimes u_{i_2} \cdots \otimes u_{i_n}, e_\ell)_H^2 \, dt_1 \, dt_2 \cdots t_n. \quad (2.34)$$

In case W is a cylindrical Brownian motion, $Q = I$ and the above expression simplifies to

$$\mathbb{E}|J_n(K_n)|_H^2$$

$$= n! \int_{I^n} \sum_{i_1, i_2, \cdots, i_n} \sum_{\ell}^{\infty} (K_n(t_1, t_2, \cdots, t_n) u_{i_1} \otimes \cdots \otimes u_{i_n}, e_\ell)_H^2 dt_1 dt_2 \cdots dt_n$$

$$= n! \int_{I^n} \| K_n(t_1, t_2, \cdots, t_n) \|_{\mathcal{L}_2(U^{\otimes n}, H)}^2 \, dt_1 \, dt_2 \cdots t_n. \quad (2.35)$$

In the general case, we write this as

$$\mathbb{E}|J_n(K_n)|_H^2 = n! \int_{I^n} \| K_n(t_1, t_2, \cdots, t_n) \|_{\mathcal{L}_2^Q(U^{\otimes n}, H)}^2 \, dt_1 \, dt_2 \cdots t_n, \quad (2.36)$$

where $\mathcal{L}_2^Q(U^{\otimes n}, H)$ denotes the space of weighted Hilbert-Schmidt n-linear forms. The family of functionals $\{J_n(K_n), K_n \in \hat{L}_2(I^n, \mathcal{L}_2^Q(U^{\otimes n}, H))\}$ satisfies similar properties like **(P1)-(P4)** as in the scalar case given in section 2.2. In particular, they are orthogonal in the Hilbert space $L_2(\Omega, \mu; H)$. As in the preceding sections, we introduce the product space

$$\mathcal{K}_\varpi \equiv \prod_{n=0}^{\infty} \hat{L}_2(I^n, \mathcal{L}_2^Q(U^{\otimes n}, H))$$

where for $n = 0$, $\hat{L}_2(I^0, \mathcal{L}_2^Q(U^{\otimes 0}, H)) = H$. Then we introduce a similar family of seminorms $\mathcal{Q}^\varpi \equiv \{p_n\}$ where, for each $K \in \mathcal{K}_\varpi$,

$$p_n(K) \equiv \| K_n \|_{L_2(I^n, \mathcal{L}_2^Q(U^{\otimes n}, H))}.$$

Endowed with this family of seminorms, $\mathcal{K}_\varpi \equiv (\mathcal{K}_\varpi, \mathcal{Q}^\varpi)$ is a Fréchet space. Let $\mathcal{L}_1^+(U)$ denote the class of bounded positive selfadjoint nuclear operators on the Hilbert space U. For any given $Q \in \mathcal{L}_1^+(U)$, we use systematically the following subspaces:

$$G_\varpi \equiv \left\{ K \in \mathcal{K}_\varpi : \left(\sum_{n=0}^\infty n! \parallel K_n \parallel^2_{L_2(I^n, \mathcal{L}_2^Q(U^{\otimes n}, H))} \right)^{1/2} < \infty \right\},$$

$$H_\varpi \equiv \left\{ K \in \mathcal{K}_\varpi : \left(\sum_{n=0}^\infty \parallel K_n \parallel^2_{L_2(I^n, \mathcal{L}_2^Q(U^{\otimes n}, H))} \right)^{1/2} < \infty \right\},$$

whose completions, with respect to the norm topologies indicated above, are Hilbert spaces.

Like in the finite dimensional cases, the Riesz-Fischer theorem remains valid. For completeness, this is stated below.

Theorem 2.6.1 *Given the orthogonal system*

$$\{J_s(K_s), K_s \in \hat{L}_2(I^s, \mathcal{L}_2^Q(U^{\otimes s}, H)), s \in N\},$$

the necessary and sufficient condition that there exists an element $J \in L_2(\Omega, \mu; H)$ *with* $\{K_s\}$ *as its Fourier-Wiener kernels is that the series*

$$\sum_{s=0}^\infty s! \int_{I^s} |K_s(t_1, t_2, \cdots, t_s)|^2_{\mathcal{L}_2^Q(U^{\otimes s}, H)} dt_1 \, dt_2 \cdots dt_s \qquad (2.37)$$

converges.

Based on this result, we have the following corollary.

Corollary 2.6.2 *The space* $\mathcal{G}_\varpi = L_2(\Omega, \mu; H)$ *represents the class of regular functionals of* U-*valued Wiener process having the Wiener-Itô decomposition* $\mathcal{G}_\varpi = \sum_{n=0}^\infty \oplus \mathcal{G}_{\varpi,n}$. *With respect to the norm topologies as given above, the spaces* G_ϖ *and* $\mathcal{G}_\varpi$ *are Hilbert spaces and they are isometrically isomorphic:* $G_\varpi \leftrightarrow \mathcal{G}_\varpi$.

It is interesting to note that the result stated in Theorem 2.3.6 for the scalar case can be partially extended to the infinite dimensional case. First, note that for $Q \in \mathcal{L}_1^+(U)$ and $a > 0$, $aQ \geq 0$; for $Q_1, Q_2 \in \mathcal{L}_1^+(U)$, $Q_1 + Q_2 \geq 0$. For $0 \leq Q_1 \leq Q_2$ we have $Q_2 - Q_1 \geq 0$. Thus $\mathcal{L}_1^+(U)$ forms a partially ordered lattice in the real Banach space of all nuclear operators in the Hilbert space U.

Note that the Hilbert spaces G_ϖ and $\mathcal{G}_\varpi$ are very much dependent on the covariance operator Q and hence we may denote them by $G_\varpi^Q, \mathcal{G}_\varpi^Q$.

Corollary 2.6.3 *The system of Hilbert spaces* $\{G_{\varpi}^Q, Q \in \mathcal{L}_1^+(U)\}$ *forms a partially ordered lattice with* $G_{\varpi}^{Q_2} \subset G_{\varpi}^{Q_1}$ *for* $0 < Q_1 \leq Q_2$. *The isomorphism* Φ_{ϖ} *preserves the lattice property and hence* $\mathcal{G}_{\varpi}^{Q_2} \subset \mathcal{G}_{\varpi}^{Q_1}$ *for* $0 < Q_1 \leq Q_2$ *with continuous injection turning the family of Hilbert spaces* $\{\mathcal{G}_{\varpi}^Q, Q \in \mathcal{L}_1^+(U)\}$ *into a partially ordered lattice.*

This result provides a large family of Hilbert spaces of functionals of infinite dimensional Brownian motion. For application to stochastic differential equations on infinite dimensional Hilbert spaces, one can choose the one that is compatible with the given problem.

Remark 2.6.4 *Suppose* $\{Q_n, Q_o\} \subset \mathcal{L}_1^+(U)$ *such that* $Q_n \xrightarrow{s} Q_o$ *in the Banach space* $\mathcal{L}_1(U)$. *The problem is to verify that the sequence of Hilbert spaces* $\mathcal{G}_{\varpi}^{Q_n}$ *converges to the Hilbert space* $\mathcal{G}_{\varpi}^{Q_o}$ *in some sense. It would be interesting to determine the appropriate topology for this convergence.*

2.7 Fr-Br. Motion and Regular Functionals Thereof

Let $(\Omega, \mathcal{F}, \mathcal{F}_t \uparrow, P)$ denote the standard probability space with a filtration. Fractional Brownian motion is a particular convolution integral of the standard Brownian motion W with respect to a suitable kernel. It is given by

$$B_H(t) \equiv \int_0^t K_H(t,s) dW(s), t \geq 0 \tag{2.38}$$

where, for simplicity, we may choose

$$K_H(t,s) = \frac{1}{\Gamma(H + (1/2))} (t-s)^{H-(1/2))}, t \geq s \geq 0,$$

for the kernel K_H. The parameter $H \in (0,1)$ is called the Hurst parameter. It satisfies the following basic properties:

(P1): $P\{B_H(0) = 0\} = 1$,

(P2): For each $t \in R_+ = [0, \infty), B_H(t)$ is an $\mathcal{F}_t$ measurable random variable having Gaussian distribution with $\mathbb{E}B_H(t) = 0$.

(P3): For each $t, s \in R_+$,

$$\mathbb{E}\{B_H(t)B_H(s)\} = (1/2)\{t^{2H} + s^{2H} - |t-s|^{2H}\}. \tag{2.39}$$

(P4): Form the above expression it is easy to verify that $B_H(t), t \geq 0$, is a self similar process in the sense that the law of the process $B_H(\alpha t), t \geq$

0, is the same as that of the process $\alpha^H B_H(t), t \geq 0$. For the standard Brownian motion the law of $W(\alpha t)$ coincides with that of $\sqrt{\alpha} W(t)$.

Clearly, $\mathbb{E}\{|B_H(t)|^2\} = |t|^{2H}$ and so $B_H(t)$ is a second order Gaussian process and it converges in law to the standard Brownian motion as $H \to (1/2)$. Comparing the regularity of Brownian motion with its fractional counterpart, it is interesting to note that, with probability one, $W(t)$ is Hölder continuous exponent $\gamma < (1/2)$ while $B_H(t)$ is Hölder continuous exponent $\gamma < H$. Thus, for $H \in (1/2, 1)$, fractional Brownian motion is smoother. It is clear that B_H is not a martingale while Brownian motion is. These long range dependence and self-similarity properties make the fractional Brownian motion very useful in the study of Network traffic and hydrology [Mandelbrot (1968)].

In recent years fractional Brownian motion has been used in the study of filtering problems [Ahmed, Charalambous, [Ahmed (2002)]] and network traffic dynamics [Ahmed, Chen [Ahmed (2004)]]. Duncan and his colleagues [Duncan (2000)] have used fractional Brownian motion to develop stochastic calculus similar to that of Itô. They have also studied multiple Wiener-Itô integrals based on the fractional Brownian motion. Here we present very briefly some important properties of these functionals and the vector spaces they belong to. It is easy to verify that

$$\mathbb{E}\{B_H(t)B_H(s)\} = \int_0^t \int_0^s \varphi_H(\tau - \theta) d\theta ds. \tag{2.40}$$

Thus a function that plays an important role in the construction of functionals of fractional Brownian motion is given by

$$\varphi \equiv \varphi_H(t) = H(2H - 1)|t|^{2H-2}, t \in R_+. \tag{2.41}$$

We are interested in any finite time interval $I = [0, T]$ and from now one we may assume that $\varphi(t) = 0$ for all $t \leq 0$.

Now we can introduce the class of (deterministic) functions which are integrable with respect to the fractional Brownian motion. This class depends very much on the function φ given by the expression (2.41). Let $L_2^\varphi(I)$ denote the class of Borel measurable real valued functions $\{K\}$ defined on I such that

$$\int_{I^2} K(t)K(s)\varphi(t - s) ds dt \equiv \parallel K \parallel_{L_2^\varphi(I)}^2 < \infty. \tag{2.42}$$

With respect to the scalar product,

$$\int_{I^2} K(t)L(s)\varphi(t-s)dsdt \equiv (K,L)_{L_2^\varphi(I)}, \tag{2.43}$$

$L_2^\varphi(I)$ is a Hilbert space. By straightforward computation it is easy to see that for $K \in L_2(I)$, with $I \equiv [0,T]$ any finite interval,

$$\| K \|_{L_2^\varphi}^2 \leq (HT^{2H-1})^2 \| K \|_{L_2(I)}^2 .$$

Thus for $H \geq (1/2)$, the inclusion $L_2(I) \hookrightarrow L_2^\varphi(I)$ holds and continuous (see also [Ahmed (2002)]) and therefore the class of integrands for fractional Brownian motion is larger than that for standard Brwonian motion as expected (at least for finite intervals).

We start with the linear functional. Let $\beta \in L_2^\varphi(I)$ and consider the functional

$$I_1 \equiv \int_I \beta(t)dB_H(t).$$

Clearly, I_1 is a Gaussian random variable with mean and variance given by

$$\mathbb{E}I_1 \equiv M = 0$$

$$\mathbb{E}(I_1)^2 \equiv \sigma^2 = \int_{I^2} \varphi(t-s)\beta(t)\beta(s)ds.dt.$$

Thus the characteristic functional of the random variable I_1 is given by

$$C_H(\xi) \equiv \mathbb{E}\{e^{i\xi I_1}\} = \exp\{-(1/2)\xi^2 \int_{I\times I} \beta(t)\beta(s)\varphi(t-s)dtds\}.$$

Note that as the Hurst parameter $H \to (1/2)$, the function $\varphi(t) \to \delta(t)$, the Dirac measure, and the characteristic functional converges to

$$\lim_{H\to(1/2)} C_H = \exp\{-(1/2)\xi^2 \int_I \beta^2(t)dt\} = C_{1/2}(\xi),$$

which is the correct characteristic functional of the random variable I_1 with B_H replaced by the standard Brownian motion W. Now we can introduce the Wiener-Itô homogeneous chaos generated by fractional Brownian motion. Let $L_2^\varphi(I^n)$ denote the class of real valued Borel measurable symmetric functions defined on I^n such that

$$\| K_n \|_{L_2^\varphi(I^n)}^2 \equiv \int_{I^{2n}} \varphi(t_1-s_1)\cdots\varphi(t_n-s_n)K_n(t_1,\cdots,t_n)K_n(s_1,\cdots,s_n)$$

$$dt_1\cdots dt_n;ds_1\cdots ds_n < \infty. \tag{2.44}$$

Completed with respect to the natural scalar product,

$$< K_n, L_n >_{L_2^\varphi(I^n)}$$
$$\equiv \int_{I^{2n}} \varphi(t_1 - s_1) \cdots \varphi(t_n - s_n) K_n(t_1, \cdots, t_n) L_n(s_1, \cdots, s_n)$$
$$dt_1 \cdots dt_n; ds_1 \cdots ds_n,$$

$L_2^\varphi(I^n)$ is a Hilbert space. By letting the Hurst parameter $H \downarrow (1/2)$, it follows from equation (2.44) that

$$\lim_{H \downarrow (1/2)} \| K_n \|_{L_2^\varphi(I^n)}^2 = n! \, \| K_n \|_{L_2(I^n)}^2 .$$

In view of the above results one can introduce a class of homogeneous chaos of degree n based on fractional Brownian motion just as we have done for standard Brownian motion. This is given by

$$g_n(K_n) \equiv \int_{I^n} K_n(t_1, t_2, \cdots, t_n) dB_H(t_1) dB_H(t_2) \cdots dB_H(t_n), \quad (2.45)$$

for $K_n \in L_2^\varphi(I^n)$ and $n \in N$. Again these functionals satisfy the orthogonality property

$$< g_n(K_n), g_m(L_m) >_{L_2^\varphi(\Omega)} = \begin{cases} 0 & \text{for } n \neq m \\ < K_n, L_n >_{L_2^\varphi(I^n)}. \end{cases}$$

Thus we may conclude that, other than the norm topology, there is really no significant difference between the functionals of standard Brownian motion and its fractional counterpart. For more details see Duncan and his school [Duncan (2000)] where they consider also stochastic Itô and Stratonovich integrals of random processes which are not necessarily adapted to the filtration $\mathcal{F}_t$ induced by the fractional Brownian motion.

In view of the above results, following the same procedure as in the preceding sections, we can now introduce the Hilbert spaces G_φ and $\mathcal{G}_\varphi$ as follows. Let $\mathcal{K}_\varphi \equiv \prod_{n=0}^\infty L_2^\varphi(I^n)$ denote the Fréchet space endowed with a family of seminorms $\{q_n\}$ given by $q_n(K) \equiv \| K_n \|_{L_2^\varphi(I^n)}$. Then we introduce the vector space G_φ characterized by

$$G_\varphi \equiv \{ K \in \mathcal{K}_\varphi : \sum_{n=0}^\infty \| K_n \|_{L_2^\varphi(I^n)}^2 < \infty \} \subset \mathcal{K}_\varphi. \quad (2.46)$$

Furnished with the norm topology $\| K \|_{G_\varphi} \equiv (\sum_{n=0}^\infty \| K_n \|^2_{L_2^\varphi(I^n)})^{1/2}$ (and completed with respect to this topology) G_φ is a Hilbert space. The vector space of functionals of fractional Brownian motion is characterized by

$$\mathcal{G}_\varphi \equiv \{ f \in L_2^\varphi(\Omega) : f = \sum_{n=0}^\infty g_n(K_n), K_n \in L_2^\varphi(I^n) \}. \qquad (2.47)$$

Furnished (and completed) with the norm topology,

$$\| f \|_{\mathcal{G}_\varphi} \equiv \left(\sum_{n=0}^\infty \| g_n(K_n) \|^2_{L_2^\varphi(\Omega)} \right)^{1/2},$$

$\mathcal{G}_\varphi$ is a Hilbert space. Let μ_H (Hurst parameter H) denote the Gaussian measure induced by the fractional Brownian motion on the measurable space $(\Omega, \mathcal{B})$ where $\Omega = C_0(I) \equiv \{ x \in C(I) : x(0) = 0 \}$ and $\mathcal{B}$ is the sigma algebra of subsets of the set Ω generated by all cylinder sets. Let $L_2^\varphi(\Omega, \mathcal{B}, \mu_H) \equiv L_2^\varphi(\Omega)$ denote the class of measurable functions on Ω which are square integrable with respect to the measure μ_H. Now we can state two main results similar to those of Theorem 2.2.3 and Corollary 2.2.5.

Theorem 2.7.1 *Given the orthogonal set $\{ g_s(K_s), s \in N_+ \}$ with $\{ K_s \in L_2^\varphi(I^s) \}$, a necessary and sufficient condition that there exists an $f \in L_2^\varphi(\Omega) = L_2^\varphi(\Omega, \mathcal{B}, \mu_H)$ with $\{ K_s \}$ as it's Fourier-Wiener kernels is that the series*

$$\sum_{s=0}^\infty \| K_s \|^2_{L_2^\varphi(I^s)} \qquad (2.48)$$

converges.

Corollary 2.7.2 *The space G_φ is isometrically isomorphic to $\mathcal{G}_\varphi$.*

The notion of absolute continuity of one measure with respect to another and the associated Radon-Nikodym derivative plays an important role in the study of stochastic differential equations. In the case of Wiener measure space, the RND of the shifted measure μ^h, for $h \in L_2(I)$, is given by the well known Girsanov formula,

$$\frac{d\mu^h}{d\mu} = \exp\{ \int_I h dw - (1/2) \int |h|^2 dt \}. \qquad (2.49)$$

The associated SDE is given by

$$dx(t) = h(t)dt + dw(t), x(0) = 0. \qquad (2.50)$$

A similar formula defines the density of a shifted measure with respect to the measure μ_H induced by the fractional Brownian motion on the canonical path space. However, these shifts cannot be arbitrary; they must belong to certain class as discussed below. Let $h \in L_2^\varphi(I)$ and define the convolution

$$(\varphi \star h)(t) \equiv \int_I \varphi(t-s)h(s)ds.$$

The set of admissible shifts is given by

$$\mathcal{A}_s \equiv \left\{ L_\varphi(h) : (L_\varphi(h))(t) = \int_0^t (\varphi \star h)(s)ds, t \in I; h \in L_2^\varphi(I) \right\}.$$

In this case the RND is given by

$$\frac{d\mu_H^h}{d\mu_H} = \exp\{ \int_I h \, dB_H - (1/2) \int |h|_{L_2^\varphi}^2 dt \}, \tag{2.51}$$

and the corresponding SDE is given by

$$dx_H = (\varphi \star h)(t)dt + dB_H(t), x_H(0) = 0. \tag{2.52}$$

Remark 2.7.3 Letting the Hurst parameter $H \downarrow 1/2$, the reader can verify that the RND given by the expression (2.51) reduces to the RND given by (2.49). Also note that the SDE driven by fractional Brownian motion given by equation (2.52) reduces to the SDE given by (2.50) which is driven by standard Brownian motion.

2.8   Lévy Process and Regular Functionals Thereof

Let $\Omega \equiv (\Omega, \mathcal{F}, \mathcal{F}_{t\geq0}, P)$ denote a complete filtered probability space supporting all random processes discussed below. Here we consider the Lévy process, in particular, the centered Poisson random process. This is a counting (jump) process that evolves purely by jumps of varied sizes. The size distribution is characterized by a nonnegative regular countably additive bounded measure on the state space say $R^n \setminus 0 \equiv Z$. This measure is known as the Lévy measure which we denote by Π. For simplicity, and without any loss of generality, we consider only the scalar case ($n = 1$). The Poisson counting measure, denoted by $p(dt \times dz)$, is a Poisson random variable with mean and variance $dt \times \Pi(dz)$ where $dt \equiv \ell(dt)$ is the

Lebesgue measure and Π is the Lévy measure. For example, for any Borel set like $J \times D \subset I \times Z$,

$$Prob.\{p(J \times D) = n\} = \frac{(\ell(J)\Pi(D))^n \exp\{-\ell(J)\Pi(D)\}}{n!}. \quad (2.53)$$

This is just the probability that over the time period J, there are exactly n jumps of sizes or intensities lying in the set D. The centered random Poisson measure denoted by $q(dt \times dz)$ is given by

$$q(dt \times dz) = p(dt \times dz) - dt \times \Pi(dz). \quad (2.54)$$

Often it is this random measure that is known as Lévy measure. We will use this terminology interchangeably.

Remark 2.8.1 Note that if $Z = \{z^*\}$ is a singleton, the process reduces to the classical Poisson random variable with rate $\lambda = \Pi(z^*)$.

We are interested in functionals of Poisson random process p under the assumption that it has independent increments. Let $\mathcal{F}_t$ and $\mathcal{F}^t$ denote the sigma algebra of events $\sigma([0, t] \times Z)$ and $\sigma([t, T] \times Z)$ respectively. For any $A \in \mathcal{F}_t$, $p(A)$ is an $\mathcal{F}_t$ measurable nonnegative random variable and, for any $B \in \mathcal{F}^t$, $p(A)$ is independent of $p(B)$. Define the mean measure ν on $(I \times Z)$ by setting $\nu(A) \equiv \int_{A \subset I \times Z} dt\Pi(dz)$, for $A \in \mathcal{B}(I \times Z)$. This turns the centered Poisson measure q, as defined by the expression (2.54), into an orthogonal measure in the sense that (1): $\mathbb{E}q(A) = 0$ for all $A \in \mathcal{B}$ and (2): $\mathbb{E}q^2(A) = \nu(A)$, and (3) $\mathbb{E}\{q(A)q(B)\} = \nu(A \cap B)$.

Now let $L_2^\Pi(I \times Z) \equiv L_2(I \times Z, dt \times \Pi(dz))$ denote the space of Borel measurable functions which are square integrable on $I \times Z$ with respect to the product of Lebesgue and Lévy measure $dt \times \Pi(dz)$. Let $K \in L_2^\Pi(I \times Z)$ and consider the random variable

$$X_p = \int_{I \times Z} K(t, z)p(dt \times dz).$$

The characteristic functional of this random variable is given by

$$C_p(\xi) = \exp\left\{\int_{I \times Z} (e^{i\xi K} - 1)dt\Pi(dz)\right\}, \quad (2.55)$$

while the characteristic functional of the random variable $X_q = \int_{I \times Z} K(t,z) q(dt \times dz)$ is given by

$$C_q(\xi) = \exp\left\{ \int_{I \times Z} \left(e^{i\xi K} - 1 - \xi K\right) dt \Pi(dz) \right\}. \tag{2.56}$$

The reader can easily verify these facts by using a sequence of simple functions K_n that converges to K in $L_2^\Pi(I \times Z)$. Using the characteristic functional $C_p(\xi)$ one can easily verify that the second moment of X_p is given by

$$\mathbb{E}X_p^2 = \int_{I \times Z} |K(t,z)|^2 \, dt \Pi(dz) + \left(\int_{I \times Z} K(t,z) dt \Pi(dz) \right)^2. \tag{2.57}$$

In contrast, using the characteristic functional $C_q(\xi)$ one can easily verify that the second moment of the random variable X_q is given by

$$\mathbb{E}X_q^2 = \int_{I \times Z} |K(t,z)|^2 \, dt \Pi(dz). \tag{2.58}$$

Clearly, $\mathbb{E}X_p^2 \geq \mathbb{E}X_q^2$. Now we consider the multiple integrals with respect to the centered Poisson process $q(dt \times dz)$. As usual in this book, we denote the Cartesian product of n copies of $I \times Z$ by $(I \times Z)^n$ with the product measure. Let $L_2^\Pi((I \times Z)^n)$ denote the class of symmetric kernels which are square integrable with respect to the Lebesgue-Lévy measure, that is,

$$\| K_n \|^2_{L_2^\Pi((I \times Z)^n)}$$
$$\equiv \int_{(I \times Z)^n} |K_n(t_1, z_1; t_2, z_2; \cdots ; t_n, z_n)|^2 dt_1 \Pi(dz_1) \cdots dt_n \Pi(dz_n). \tag{2.59}$$

Completed with respect to the natural scalar product, this is a Hilbert space. Then the homogeneous Poisson chaos of degree n, denoted by $g_n(K_n, q)$, is defined by the multiple integral with respect to the Lévy process $\{q\}$,

$$g_n(K_n, q))$$
$$\equiv \int_{(I \times Z)^n} K_n(t_1, z_1; t_2, z_2; \cdots ; t_n, z_n) q(dt_1 \times dz_1) \cdots q(dt_n \times dz_n). \tag{2.60}$$

From now on we assume that q is fixed and so we can suppress the q variable and use simply the standard notation $g_n(K_n)$ for the n-th order homogeneous chaos corresponding to $K_n \in L_2^\Pi((I \times Z)^n)$. Let $L_2^\Pi(\Omega)$ denote the Hilbert space of square integrable random variables defined on the probability space Ω. Again these functionals satisfy the orthogonality property

$$< g_n(K_n), g_m(L_m) >_{L_2^\Pi(\Omega)} = \begin{cases} 0, & \text{if } n \neq m \\ n! < K_n, L_n >_{L_2^\Pi((I \times Z)^n)}, & n = m. \end{cases}$$

Following the same procedure as in the preceding sections, we can now introduce the Hilbert spaces G_Π and $\mathcal{G}_\Pi$ as follows. Let

$$\mathcal{K}_\Pi \equiv \prod_{n=0}^{\infty} L_2^\Pi((I \times Z)^n)$$

denote the Fréchet space endowed with a family of seminorms $\{q_n\}$ given by $q_n(K) \equiv \| K_n \|_{L_2^\Pi((I \times Z)^n)}$. Then we introduce the vector space G_Π characterized by

$$G_\Pi \equiv \Big\{ K \in \mathcal{K}_\Pi : \sum_{n=0}^{\infty} n! \| K_n \|_{L_2^\Pi((I \times Z)^n)}^2 < \infty \Big\} \subset \mathcal{K}_\Pi. \tag{2.61}$$

Furnished with the norm topology

$$\| K \|_{G_\Pi} \equiv \Big(\sum_{n=0}^{\infty} n! \| K_n \|_{L_2^\Pi((I \times Z)^n)}^2 \Big)^{1/2}$$

(and completed with respect to this topology) G_Π is a Hilbert space. The vector space of functionals of the centered Poisson process is characterized by

$$\mathcal{G}_\Pi \equiv \Big\{ f \in L_2^\Pi(\Omega) : f = \mathbb{E}f + \sum_{n=1}^{\infty} g_n(K_n), K_n \in L_2^\Pi((I \times Z)^n) \Big\}.$$

Furnished (and completed) with the norm topology,

$$\| f \|_{\mathcal{G}_\Pi} \equiv \Big(\sum_{n=0}^{\infty} \| g_n(K_n) \|_{L_2^\Pi(\Omega)}^2 \Big)^{1/2}, \tag{2.62}$$

$\mathcal{G}_\Pi$ is a Hilbert space. Now we can state two results similar to those of Theorem 2.2.3 and Corollary 2.2.5.

Theorem 2.8.2 *Given the orthogonal set* $\{g_s(K_s), s \in N_+\}$ *with* $\{K_s \in L_2^{\Pi}((I \times Z)^s)\}$, *a necessary and sufficient condition that there exists an* $f \in L_2^{\Pi}(\Omega))$ *with* $\{K_s\}$ *as it's Fourier-Wiener-Lévy kernels is that the series*

$$\sum_{s=0}^{\infty} s! \parallel K_s \parallel^2_{L_2^{\Pi}((I \times Z)^s)} \tag{2.63}$$

converges.

Corollary 2.8.3 *The space* G_{Π} *is isometrically isomorphic to* $\mathcal{G}_{\Pi}$ *with* Φ_{Π} *denoting the isometric map* $G_{\Pi} \longrightarrow \mathcal{G}_{\Pi}$.

Remark 2.8.4 If the state space Z consists of a finite set of points, that is, $Z \equiv Z_f \equiv \{\zeta_1, \zeta_2, \cdots, \zeta_k\}$, the multiple integrals (2.59)-(2.60) reduce to the multiple (n-fold) sums over Z_f of multiple integrals over I.

It is clear from the results presented in this chapter that nonlinear functionals of any of the processes like Brownian motion, Gaussian random fields, fractional Brownian motion, Poisson random processes all have similar representation provided the random variables they represent are square integrable.

2.9   Some Problems for Exercise

P1: Verify the basic properties **(P1)-(P4)** of the orthogonal functionals $\{g_n\}$ stated in section 2.2. Hint: Use orthonormal expansion of symmetric kernels $K_n \in \hat{L}_2(I^n)$ by any orthonormal set $\{\psi_i\} \subset L_2(I)$.

P2: Read the statements of Lemma 2.2.1, Theorem 2.2.2 and Theorem 2.2.3 and compare them with similar results known in regards to the famous Riesz-Fischer theorem for the classical pair of (separable) Hilbert spaces L_2 and ℓ_2.

P3: Prove that the statement of Theorem 2.2.3 is equivalent to the following statement

$$L_2(\Omega, \mathcal{B}, \mu) \cong \ell_2(\Pi_{n=1}^{\infty} \sqrt{n!} \hat{L}_2(I^n))$$

where $\cong$ denotes the metric and topological isomorphism.

P4: Consider the nonstandard Brownian motion of section 2.3 parameterized by $\lambda > 0$. Verify the Lattice property stated in Theorem 2.3.6.

P5: Let $\beta \in L_2(I)$ and W a scalar Brownian motion on some probability space $(\Omega, \mathcal{F}, P)$. Prove that the random variable $\Theta \equiv \exp\{\int_I \beta(t)dW(t)\} \in L_p(\Omega, \mathcal{F}, P)$ for all $p \in [0, \infty)$. Hint: Define $X \equiv \int_I \beta(t)dW(t)$. Can be done in two ways.

Direct: First note that $\mathbb{E}(X) = 0$ and $\mathbb{E}|X|^2 \equiv \sigma^2 = \int_I |\beta(t)|^2 dt$. Compute

$$\mathbb{E}\{|Z|^p\} = \int_{-\infty}^{\infty} e^{px} \frac{e^{-(1/2)(x/\sigma)^2}}{\sqrt{2\pi\sigma^2}} dx$$

and show that $\mathbb{E}\{|Z|^p\} = e^{(1/2)p^2\sigma^2}$.

Indirect: Expand the exponential function

$$|Z|^p = e^{p\int \beta(t)dW} = \sum_{n=0}^{\infty} (1/n!) \left(\int_I p\beta(t)dW \right)^n$$

and use the properties of multiple Wiener-Itô integrals to show that

$$\mathbb{E} \left(\int_I p\beta(t)dW \right)^{2n} = \frac{p^{2n}(2n)!}{2^n n!} \left(\int_I |\beta(t)|^2 dt \right)^n.$$

P6: Use the results from the problem (P5) and show that the process $\xi(t), t \geq 0$, given by $d\xi = \alpha(t)\xi dt + \beta(t)\xi dW, \xi(0) = \xi_0$, has p-th moment whenever ξ_0 has the p-th moment. Verify that for $p \geq 0$, this is given by

$$\mathbb{E}\{|\xi(t)|^p\} = \mathbb{E}\{|\xi_0|^p\} \exp\left\{ p \int_I \alpha(t)dt + (1/2)p(p-1) \int_I |\beta(t)|^2 dt \right\}.$$

P7: Verify that all the abstract spaces $\mathcal{G} \equiv \{\mathcal{G}_\alpha, \mathcal{G}_\beta, \mathcal{G}_\gamma, \mathcal{G}_\delta, \mathcal{G}_\varpi, \mathcal{G}_\varphi, G_\Pi\}$ introduced in this chapter satisfy the following isometric isomorphism

$$\mathcal{G} \cong \ell_2(\mathcal{K})$$

for a suitable $\mathcal{K}$ representing an infinite product of a countable family of suitable Hilbert spaces.

P8: Given $P, Q \in \mathcal{L}_1^+(U)$ with $0 \leq Q \leq P$, prove that the embedding $G_\varpi^P \hookrightarrow G_\varpi^Q$ is continuous and dense. Hint: Follow the discussion around the expressions (2.34)-(2.36).

P9: Consider the Remark 2.6.4 and propose a suitable topology and a technique for the convergence. Hint: The embedding result from problem (P8) may have significance in this respect.

P10: Consider the statements of Remark 2.7.3 and verify their correctness. Hint: Show that φ converges to the Dirac measure and then verify all the assertions.

P11: Verify that the Poisson random process p and its centered counter part q have characteristic functionals given by the expressions (2.55) and (2.56) respectively. Hint: Use the basic fact that for any Borel set $J \times \Gamma \subset I \times Z$ the random variable $p(J \times \Gamma)$ is a classical Poisson random variable with mean and variance $\ell(J) \times \Pi(\Gamma)$.

Chapter 3

Generalized Functionals of the First Kind I

3.1 Introduction

In this chapter, using the Hilbert spaces $\{G_\alpha, G_\beta, G_\gamma, G_\delta, G_\varpi, G_\varphi, G_\Pi\}$ and their corresponding isomorphic images $\{\mathcal{G}_\alpha \equiv \Phi_\alpha(G_\alpha), \mathcal{G}_\beta \equiv \Phi_\beta(G_\beta), \mathcal{G}_\gamma \equiv \Phi_\gamma(G_\gamma), \mathcal{G}_\delta \equiv \Phi_\delta(G_\delta), \mathcal{G}_\varpi \equiv \Phi_\varpi(G_\varpi), \mathcal{G}_\varphi \equiv \Phi_\varphi(G_\varphi), \mathcal{G}_\Pi \equiv \Phi_\Pi(G_\Pi)\}$, we construct a large class of generalized functionals of Brownian motion, fractional Brownian motion, Gaussian random fields and Lévy (Poisson counting) process. These generalized functionals are much more general and they contain the regular or the Wiener-Itô class as a special case. These are constructed by use of duality arguments.

3.2 Mild Generalized Functionals I

Here, we construct a class of mild generalized functionals which we call generalized functionals of the first kind. These are strongly related to the regular functionals discussed in Chapter 2. Recall the Fréchet space $\mathcal{K}_\alpha$ and the Hilbert space G_α, a subspace of $\mathcal{K}_\alpha$, as introduced in section 2.2. Let H_α denote the vector space

$$H_\alpha \equiv \{K \in \mathcal{K}_\alpha : \| K \|_{H_\alpha} \equiv \big(\sum_{n=0}^{\infty} \| K_n \|_{L_2(I^n)}^2\big)^{1/2} < \infty\}. \qquad (3.1)$$

We assume that H_α has been completed with respect to the scalar product

$$(K, L) \equiv \sum_{n=0}^{\infty} (K_n, L_n)_{L_2(I^n)},$$

so that it becomes a real Hilbert space. It is easy to verify that for each $L \in G_\alpha$, we have

$$\| L \|_{H_\alpha} \leq \| L \|_{G_\alpha}.$$

In other words $G_\alpha \subset H_\alpha$ but the reverse inclusion is not true and the injection $G_\alpha \hookrightarrow H_\alpha$ is continuous and dense. Let L be any element of H_α and define the linear functional on G_α by setting $\ell_L(K) = (L, K)_{H_\alpha}$. Clearly, for any $K \in G_\alpha$,

$$|\ell_L(K)| \leq \| L \|_{H_\alpha} \| K \|_{H_\alpha} \leq \| L \|_{H_\alpha} \| K \|_{G_\alpha}.$$

Thus, every $L \in H_\alpha$ determines a continuous linear functional on G_α and so $H_\alpha \subset G_\alpha^*$, where G_α^* is the topological dual of G_α. In fact, the space of continuous linear functionals on G_α is much larger. For each $L \in \mathcal{K}_\alpha$, a Fréchet space, and $K \in G_\alpha$, it is clear that the finite series,

$$< L, K >_n \; \equiv \sum_{s=0}^{s=n} (L_s, K_s)_{L_2(I^s)},$$

is well defined and finite. Indeed,

$$
\begin{aligned}
| < L, K >_n | &= \left| \sum_{s=0}^{n} (L_s, K_s)_{L_2(I^s)} \right| \\
&= \left| \sum_{s=0}^{n} ((1/\sqrt{s!}) L_s, (\sqrt{s!}) K_s)_{L_2(I^s)} \right| \\
&\leq \left(\sum_{s=0}^{n} (1/s!) \| L_s \|^2_{L_2(I^s)} \right)^{1/2} \left(\sum_{s=0}^{n} (s!) \| K_s \|^2_{L_2(I^s)} \right)^{1/2}.
\end{aligned}
$$

Since $K \in G_\alpha$, it follows from this that

$$| < L, K >_n \leq \left(\sum_{s=0}^{n} (1/s!) \| L_s \|^2_{L_2(I^s)} \right)^{1/2} \| K \|_{G_\alpha}. \tag{3.2}$$

This holds for every $n < \infty$. Thus, for each $K \in G_\alpha$, the limit,

$$| < L, K > | \equiv \lim_{n \to \infty} | < L, K >_n |$$

exists and is finite if, and only if, $L(\in \mathcal{K}_\alpha)$ is such that

$$\sum_{s=0}^{\infty} (1/s!) \| L_s \|^2_{L_2(I^s)} < \infty. \tag{3.3}$$

Hence, for every such L, we have

$$| < L, K > | \leq \left(\sum_{s=0}^{\infty} (1/s!) \parallel L_s \parallel_{L_2(I^s)}^2 \right)^{1/2} \parallel K \parallel_{G_\alpha} < \infty. \qquad (3.4)$$

Identify H_α with its dual H_α^*. Let G_α^* denote the dual of G_α. The inequality (3.4) characterizes the dual G_α^*. That is, continuous linear functionals on G_α are given by only those elements of $\mathcal{K}_\alpha$ which satisfy (3.3). Define $\parallel L \parallel_{G_\alpha^*} \equiv \left(\sum_{s=0}^{\infty} (1/s!) \parallel L_s \parallel_{L_2(I^s)}^2 \right)^{1/2}$. Thus, the bilinear form or the pairing on the left of (3.4) has a natural extension and can be interpreted as G_α^*-G_α pairing giving

$$| < L, K >_{G_\alpha^*, G_\alpha} | \leq \parallel L \parallel_{G_\alpha^*} \parallel K \parallel_{G_\alpha} . \qquad (3.5)$$

Then, G_α^* is also a Hilbert space with respect to the norm topology given by

$$\parallel L \parallel_{G_\alpha^*} \equiv \left(\sum_{n=0}^{\infty} (1/n!) \parallel L_n \parallel_{L_2(I^n)}^2 \right)^{1/2} . \qquad (3.6)$$

It is easy to show that G_α is dense in H_α and H_α is dense in G_α^*. We call the system $\{G_\alpha, H_\alpha, G_\alpha^*\}$ the Gelfand triple. Thus we have proved the following result.

Lemma 3.2.1 *Consider the triple* $G_\alpha, H_\alpha, G_\alpha^*$ *as defined above. They are all Hilbert spaces and the embeddings* $G_\alpha \hookrightarrow H_\alpha \hookrightarrow G_\alpha^*$ *are continuous and dense.*

Note that $G_\alpha^* - G_\alpha$ pairing is given by

$$< L^*, K >_{G_\alpha^*, G_\alpha} = \sum_{n=0}^{\infty} (L_n^*, K_n)_{L_2(I^n)}.$$

However, if $L^* \in H_\alpha$, this reduces to the standard scalar product in H_α as given by the pairing following the expression (3.1).

As a consequence of Hahn-Banach theorem, we expect that for every $K \in G_\alpha$ there exists an $L^* \in B_1(G_\alpha^*)$, the closed unit ball of the dual G_α^*, such that $< L^*, K >_{G_\alpha^*, G_\alpha} = \parallel K \parallel_{G_\alpha}$ and that $\parallel L^* \parallel_{G_\alpha^*} = 1$. Indeed, such an element is given by

$$L^* \equiv \{L_n^*, n \geq 0\}, \text{ with } L_n^* = \frac{n! K_n}{\parallel K \parallel_{G_\alpha}}, n \in N_0.$$

By direct computation, the reader can easily verify that, for this choice of L^*, we have

$$< L^*, K >_{G_\alpha^*, G_\alpha} \; = \; \| K \|_{G_\alpha}$$

and that $\| L^* \|_{G_\alpha^*} \; = \; 1$ as expected. It is also easy to verify that both the G_α norm and H_α norm of L^* are infinite. That is, $\| L^* \|_{G_\alpha} = +\infty$ and $\| L^* \|_{H_\alpha} = +\infty$. This also shows that G_α^* is a very large space.

Using the above results and the isomorphism Φ_α, as defined in Chapter 2, we can construct a class of generalized functionals as follows. Since the embeddings $G_\alpha \hookrightarrow H_\alpha \hookrightarrow G_\alpha^*$ are continuous and dense, by virtue of the principle of extension by continuity, we can extend Φ_α all the way up to G_α^* in such a way that it's restriction to G_α is Φ_α itself. Instead of introducing new notations, we shall denote this extension also by Φ_α itself. We define

$$\mathcal{G}_\alpha \equiv \Phi_\alpha(G_\alpha), \;\; \mathcal{H}_\alpha \equiv \Phi_\alpha(H_\alpha), \;\; \mathcal{G}_\alpha^* \equiv \Phi_\alpha(G_\alpha^*). \tag{3.7}$$

Hence we have the following result.

Theorem 3.2.2 *The class $\mathcal{G}_\alpha$ is the space of regular L_2 functionals of Brownian motion while the two classes $\mathcal{H}_\alpha$ and $\mathcal{G}_\alpha^*$ are generalized functionals of Brownian motion with $\mathcal{G}_\alpha$ being their test functionals and they satisfy the following diagram:*

$$G_\alpha \hookrightarrow H_\alpha \hookrightarrow G_\alpha^*$$

$$\updownarrow \qquad \updownarrow \qquad \updownarrow \tag{3.8}$$

$$\mathcal{G}_\alpha \hookrightarrow \mathcal{H}_\alpha \hookrightarrow \mathcal{G}_\alpha^*$$

where $\hookrightarrow$ denotes continuous and dense embedding and $\updownarrow$ stands for isometric isomorphism.

The action of an element of $\mathcal{G}_\alpha^*$ on those of the regular Brownian functionals $\mathcal{G}_\alpha = L_2(\Omega, \mathcal{B}, \mu)$, considered as test functionals, is given by the duality pairing

$$< f^*, f >_{\mathcal{G}_\alpha^*, \mathcal{G}_\alpha} = \sum_{n=0}^{\infty} (1/n!)(g_n(L_n), g_n(K_n))_{L_2(\Omega, \mathcal{B}, \mu)} = < L, K >_{G_\alpha^*, G_\alpha}$$

and clearly,

$$| < f^*, f >_{\mathcal{G}_\alpha^*, \mathcal{G}_\alpha} | \; \leq \; \| f^* \|_{\mathcal{G}_\alpha^*} \| f \|_{\mathcal{G}_\alpha} \; \leq \; \| L \|_{G_\alpha^*} \| K \|_{G_\alpha}. \tag{3.9}$$

Note that, again by Hahn-Banach theorem, for every $f \in \mathcal{G}_\alpha$ there exists an $f^* \in \mathcal{G}_\alpha^*$ such that

$$< f^*, f > = \parallel f \parallel_{\mathcal{G}_\alpha} . \tag{3.10}$$

For example, if $f \leftrightarrow K$, then f^* given by $f^* \equiv \Phi_\alpha(L^*)$ for $L^* \equiv (L_0, L_1, L_3, \cdots L_n, \cdots)$ with $L_n = (n!/ \parallel K \parallel_{\mathcal{G}_\alpha})K_n$ has the required property and, further, $\parallel f^* \parallel_{\mathcal{G}_\alpha^*} = 1$.

Clearly, as expected, it follows from this that $\mathcal{G}_\alpha^*$ separates points of $\mathcal{G}_\alpha$.

Corollary 3.2.3 *The dual $\mathcal{G}_\alpha^*$ of $\mathcal{G}_\alpha$ has a norm topology which is compatible with the isometric isomorphism between G_α^* and $\mathcal{G}_\alpha^*$. This is given by*

$$\parallel f \parallel_{\mathcal{G}_\alpha^*} = \sqrt{\sum_{n=0}^{\infty}(1/n!) \parallel K_n \parallel_{L_2(I^n)}^2} \equiv \parallel K \parallel_{G_\alpha^*} .$$

Proof The space $\mathcal{G}_\alpha^*$ can be furnished with a natural scalar product (not the duality pairing) that turns it into a (pre)-Hilbert space. Indeed, let Π_n denote the projection map $\mathcal{G}_\alpha^* \longrightarrow \Pi_n(\mathcal{G}_\alpha^*) = \mathcal{G}_{\alpha,n}^* \equiv \{g_n(R) : (1/\sqrt{n!})R \in \hat{L}_2(I^n)\}$. Let $f, g \in \mathcal{G}_\alpha^*$ be given by $f \equiv \Phi_\alpha(K), g \equiv \Phi_\alpha(L)$ for some $K, L \in G_\alpha^*$. Then introduce the scalar product in $\mathcal{G}_\alpha^*$ as follows:

$$< f, g >_{\mathcal{G}_\alpha^*} \equiv \sum_{n=0}^{\infty} < \Pi_n f, \Pi_n g >_{\mathcal{G}_{\alpha,n}^*}$$

$$= \sum_{n=0}^{\infty} < g_n(K_n), g_n(L_n) >_{\mathcal{G}_{\alpha,n}^*}$$

$$= \sum_{n=0}^{\infty}((1/n!)g_n(K_n), (1/n!)g_n(L_n))_{\mathcal{G}_{\alpha,n}}$$

$$= \sum_{n=0}^{\infty}(1/n!)(K_n, L_n)_{L_2(I^n)} = (K, L)_{G_\alpha^*} .$$

Hence the $\mathcal{G}_\alpha^*$ norm of f is given by

$$\parallel f \parallel_{\mathcal{G}_\alpha^*} = \sqrt{\sum_{n=0}^{\infty}(1/n!) \parallel K_n \parallel_{L_2(I^n)}^2} \equiv \parallel K \parallel_{G_\alpha^*} .$$

Clearly, this scalar product is compatible with the isometry. ●

A practical significance of this result is interesting. In case a given functional f of Brownian motion fails to have finite energy, that is, $\| f \|_{\mathcal{G}_\alpha} = +\infty$, it means it does not admit standard Wiener-Itô decomposition. However, it may be an element of the dual $\mathcal{G}_\alpha^*$ with an elevated energy given by its Hilbertian norm as given in the above corollary.

3.3   Mild Generalized Functionals II

The results presented in the preceding section also hold for multi dimensional Brownian motion as discussed in section 2.3. Without going into details, we restate the results for the multidimensional case. Here the (pivot) space is the Hilbert H_β given by

$$H_\beta \equiv \{K \in \mathcal{K}_\beta : \| K \|_{H_\beta} \equiv \big(\sum_{n=0}^{\infty} \| K_n \|_{L_2(I^n, M_n)}^2 \big)^{1/2} < \infty\}, \quad (3.11)$$

where $\mathcal{K}_\beta$ is the Fréchet space introduced earlier. Recall the Hilbert space G_β, a subspace of the Hilbert space H_β, with the norm topology given by (2.17). By virtue of the isometric map Φ_β as introduced in section 2.3, we have $\mathcal{G}_\beta = \Phi_\beta(G_\beta) = L_2(\Omega, \mathcal{B}, \mu; R^\ell) \equiv L_2(\mu, R^\ell)$ representing the regular functionals of vector Brownian motions. As seen before, continuous extension of this map, also denoted by Φ_β, gives the generalized functionals $\mathcal{H}_\beta = \Phi_\beta(H_\beta)$ and $\mathcal{G}_\beta^* = \Phi_\beta(G_\beta^*)$. By virtue of Theorem 2.3.4 and the Corollary 2.3.5 we have the following results parallel to those of Lemma 3.2.1 and Theorem 3.2.2. In particular, we have the Gelfand triple $\{G_\beta, H_\beta, G_\beta^*\}$.

Lemma 3.3.1 *Consider the triple $G_\beta, H_\beta, G_\beta^*$ as defined above. They are all Hilbert spaces and the embeddings $G_\beta \hookrightarrow H_\beta \hookrightarrow G_\beta^*$ are continuous and dense.*

Using the isometric isomorphism Φ_β, we obtain a broad class of generalized functionals as stated in the following theorem.

Theorem 3.3.2 *The class $\mathcal{G}_\beta$ is the space of regular $L_2 \equiv L_2(\mu, R^\ell)$ functionals of vector Brownian motion while the two classes $\mathcal{H}_\beta$ and $\mathcal{G}_\beta^*$ are generalized functionals of vector Brownian motion with $\mathcal{G}_\beta$ being their test* *functionals and they satisfy the following diagram:*

$$G_\beta \hookrightarrow H_\beta \hookrightarrow G_\beta^*$$
$$\updownarrow \qquad \updownarrow \qquad \updownarrow \qquad (3.12)$$
$$\mathcal{G}_\beta \hookrightarrow \mathcal{H}_\beta \hookrightarrow \mathcal{G}_\beta^*.$$

Corollary 3.3.3 *The dual $\mathcal{G}_\beta^*$ of $\mathcal{G}_\beta$ has a norm topology which is compatible with the isometry and isomorphism between G_β^* and $\mathcal{G}_\beta^*$.*

Proof The proof is similar to that of Corollary 3.2.3. •

Practical significance of this result is similar to that following Corollary 3.2.3. In case a given functional f of vector Brownian motion fails to have finite energy, that is, $\| f \|_{\mathcal{G}_\beta} = +\infty$, it means it does not have standard Wiener-Itô expansion. However, it may be an element of the dual $\mathcal{G}_\beta^*$ a larger class.

3.4 Generalized Functionals of GRF I

Here, we consider regular and generalized functionals of scalar Gaussian random field (GRF). Consider the vector space

$$\mathcal{K}_\gamma \equiv \prod_{s=0}^\infty \hat{L}_2(D^s)$$

where we interpret $L_2(D^0) \equiv R/C$. Again, as in Chapter 2, we can introduce a countable family of seminorms on $\mathcal{K}_\gamma$ with respect to which it is a Frèchet space. Define the set

$$G_\gamma \equiv \left\{ L \in \mathcal{K}_\gamma : \| L \|_{G_\gamma} \equiv \left(\sum_{n=0}^\infty n! \, \| L_n \|_{\hat{L}_2(D^n)}^2 \right)^{1/2} < \infty \right\}.$$

Clearly, this is a linear subspace of the Frèchet space $\mathcal{K}_\gamma$. Again, we use G_γ to denote its completion with respect to the norm topology as indicated above which is induced by the scalar product

$$< K, L >_{G_\gamma} \equiv \sum_{s=0}^\infty s! \int_{D^s} (K_s L_s^*)(\eta_1, \eta_2, ..., \eta_s) \ell(d\eta_1)...\ell(d\eta_s). \quad (3.13)$$

For convenience of systematic notations, let $\mathcal{G}_\gamma$ denote $L_2(\mathcal{M}_s, \mu)$, (see section 2.4) and Φ_γ the map that assigns to each element of G_γ an element of $\mathcal{G}_\gamma$ through the relation

$$\Phi_\gamma(K) = \sum_{n=0}^{\infty} e_n(K_n, W) \equiv \sum_{n=0}^{\infty} e_n(K_n)$$

where,

$$e_n(L_n) \equiv \int_{D^n} L_n(\xi_1, \xi_2, \cdots, \xi_n) W(d\xi_1) W(d\xi_2) \cdots W(d\xi_n), \quad (3.14)$$

as given by the expression (2.27). For convenience, we have suppressed W in the last expression. According to Riesz-Fischer theorem giving the identity (2.28), Φ_γ is an isometric isomorphism between G_γ and $\mathcal{G}_\gamma$. Define

$$H_\gamma \equiv \{K \in \mathcal{K}_\gamma : \| K \| \equiv (\sum_{n=0}^{\infty} \| K_n \|^2_{L_2(D^n)})^{1/2} < \infty\}. \quad (3.15)$$

Then, following similar approach and arguments as in section 3.2, we have the following results.

Lemma 3.4.1 *The Gelfand triple* $G_\gamma, H_\gamma, G_\gamma^*$, *as defined above, are all Hilbert spaces and the embeddings* $G_\gamma \hookrightarrow H_\gamma \hookrightarrow G_\gamma^*$ *are continuous and dense.*

Again, we use Φ_γ to denote it's continuous extension from G_γ to H_γ and G_γ^*. Writing $\mathcal{H}_\gamma \equiv \Phi_\gamma(H_\gamma)$ and $\mathcal{G}_\gamma^* \equiv \Phi_\gamma(G_\gamma^*)$ we obtain the following result.

Theorem 3.4.2 *The class* $\mathcal{G}_\gamma$ *is the space of regular* L_2 *functionals of the Gaussian random field* $\{W(A), A \in \mathcal{B}(D)\}$, *while* $\mathcal{H}_\gamma$ *and* $\mathcal{G}_\gamma^*$ *are generalized functionals with* $\mathcal{G}_\gamma$ *being the space of test functionals and they satisfy the following diagram:*

$$G_\gamma \hookrightarrow H_\gamma \hookrightarrow G_\gamma^*$$
$$\updownarrow \quad \updownarrow \quad \updownarrow \quad\quad (3.16)$$
$$\mathcal{G}_\gamma \hookrightarrow \mathcal{H}_\gamma \hookrightarrow \mathcal{G}_\gamma^*$$

with continuous and dense embeddings. Naturally, for $\phi^* \in \mathcal{G}_\gamma^*$, it's norm is defined by

$$\| \phi^* \|_{\mathcal{G}_\gamma^*} \equiv Sup\{< \phi^*, \phi >_{\mathcal{G}_\gamma^*, \mathcal{G}_\gamma}, \| \phi \|_{\mathcal{G}_\gamma} = 1\}.$$

Corollary 3.4.3 *The dual $\mathcal{G}_\gamma^*$ of $\mathcal{G}_\gamma$ has a norm topology which is compatible with the isometric isomorphism between G_γ^* and $\mathcal{G}_\gamma^*$.*

Proof The proof is similar to that of Corollary 3.2.3. ●

Practical significance of this result is similar to that following Corollary 3.2.3.

3.5 Generalized Functionals of GRF II

Here, we consider the multidimensional case. Since the basic procedure is the same, we simply state the results. Consider the product space

$$\mathcal{K}_\delta \equiv \prod_{s=0}^{\infty} \hat{L}_2(D^s, M_s).$$

One can introduce, as in Chapter 2, [see Ahmed [Ahmed (1973)]] a countable family of seminorms on $\mathcal{K}_\delta$ with respect to which it is a Frèchet space. Define the set

$$G_\delta \equiv \left\{ L \in \mathcal{K}_\delta : \| L \|_{G_\delta} \equiv \left(\sum_{n=0}^{\infty} n! \| L_n \|_{\hat{L}_2(D^n, M_n)}^2 \right)^{1/2} < \infty \right\}.$$

Clearly, this is a linear subspace of the Frèchet space $\mathcal{K}_\delta$. Again, we use G_δ to denote its completion with respect to the norm topology as indicated above. The associated scalar product is given by

$$< K, L >_{G_\delta} \equiv \sum_{s=0}^{\infty} s! \int_{D^s} Tr((K_s^* L_s)(\eta_1, \eta_2, \cdots, \eta_s)) \ell(d\eta_1) \cdots \ell(d\eta_s).$$

$$(3.17)$$

Let $\mathcal{G}_\delta$ denote $L_2(\mathcal{M}, \mu^w)$, and Φ_δ the map that assigns to each element of G_δ an element of $\mathcal{G}_\delta$ through the relation

$$\Phi_\delta(K) = \sum_{n=0}^{\infty} h_n(K_n)$$

where,

$$h_n(K_n) \equiv \int_{D^n} K_n(\xi_1, \cdots, \xi_n) W(d\xi_1) \otimes \cdots \otimes W(d\xi_n), \qquad (3.18)$$

which is similar to the expression (2.27) for the scalar case with $K_n \in \hat{L}_2(D^n, M_n)$ for $n \in N_0$. According to Riesz-Fischer theorem, giving the identity similar to (2.28), Φ_δ is an isometric isomorphism of G_δ onto $\mathcal{G}_\delta$. Define

$$H_\delta \equiv \{ K \in \mathcal{K}_\delta : \parallel K \parallel \ \equiv (\sum_{n=0}^{\infty} \parallel K_n \parallel_{\hat{L}_2(D^n, M_n)}^2)^{1/2} < \infty \}. \qquad (3.19)$$

Then, following similar approach and arguments as given in section 3.2, we have the following results.

Lemma 3.5.1 *The Gelfand triple $G_\delta, H_\delta, G_\delta^*$, as defined above, are all Hilbert spaces and the embeddings $G_\delta \hookrightarrow H_\delta \hookrightarrow G_\delta^*$ are continuous and dense.*

Recall that the updown arrows are used to represent isometric isomorphisms usually denoted by the symbol $\cong$.

Writing $\mathcal{H}_\delta \equiv \Phi_\delta(H_\delta)$ and $\mathcal{G}_\delta^* \equiv \Phi_\delta(G_\delta^*)$ we obtain the following result giving a very large class of generalized functionals of multidimensional Gaussian random fields.

Theorem 3.5.2 *The class $\mathcal{G}_\delta$ is the space of regular L_2 functionals of the vector Brownian field $\{W(A), A \in \mathcal{B}(D)\}$, while $\mathcal{H}_\delta$ and $\mathcal{G}_\delta^*$ are generalized functionals with $\mathcal{G}_\delta$ being the space of test functionals and they satisfy the following diagram:*

$$G_\delta \hookrightarrow H_\delta \hookrightarrow G_\delta^*$$
$$\updownarrow \quad \updownarrow \quad \updownarrow \qquad (3.20)$$
$$\mathcal{G}_\delta \hookrightarrow \mathcal{H}_\delta \hookrightarrow \mathcal{G}_\delta^*,$$

with all the embeddings being continuous and dense.

Corollary 3.5.3 *The dual $\mathcal{G}_\delta^*$ of $\mathcal{G}_\delta$ has a norm topology which is compatible with the isometric isomorphism between G_δ^* and the space of generalized functionals represented by $\mathcal{G}_\delta^*$.*

Proof The proof is similar to that of Corollary 3.2.3. •

In case a given functional f of vector Brownian field fails to have finite energy, that is, $\| f \|_{\mathcal{G}_\delta} = +\infty$, it means it does not have the standard Wiener-Itô decomposition. However, it may very well be an element of the space $\mathcal{G}_\delta^*$ of generalized functionals with an elevated energy measured in terms of its Hilbertian norm.

3.6 Generalized Functionals of ∞-Dim. Brownian Motion

In section 2.6, we introduced the Hilbert spaces G_ϖ and H_ϖ. The reader can easily verify that the injection $G_\varpi \hookrightarrow H_\varpi$ is continuous and dense. As seen in all the cases considered above, the dual G_ϖ^* is well defined and it is a very large space containing H_ϖ. As usual we identify the dual of H_ϖ with itself. Using similar arguments, as employed in the preceding sections, one can prove the following results.

Lemma 3.6.1 *The vector spaces in the Gelfand triple $\{G_\varpi, H_\varpi, G_\varpi^*\}$ are all Hilbert spaces and the embeddings $G_\varpi \hookrightarrow H_\varpi \hookrightarrow G_\varpi^*$ are continuous and dense.*

Let Φ_ϖ denote the isometric isomorphism between G_ϖ and $\mathcal{G}_\varpi$. Using the same notation for its extension, and writing $\mathcal{H}_\varpi \equiv \Phi_\varpi(H_\varpi)$ and $\mathcal{G}_\varpi^* \equiv \Phi_\varpi(G_\varpi^*)$, we obtain the following results giving a very large class of generalized functionals of infinite dimensional Brownian motion.

Theorem 3.6.2 *The class $\mathcal{G}_\varpi$ is the space of regular L_2 functionals of U-valued Brownian motion, while $\mathcal{H}_\varpi$ and $\mathcal{G}_\varpi^*$ are generalized functionals with $\mathcal{G}_\varpi$ being the space of test functionals and they satisfy the following diagram:*

$$G_\varpi \hookrightarrow H_\varpi \hookrightarrow G_\varpi^*$$

$$\updownarrow \quad \updownarrow \quad \updownarrow \qquad\qquad (3.21)$$

$$\mathcal{G}_\varpi \hookrightarrow \mathcal{H}_\varpi \hookrightarrow \mathcal{G}_\varpi^*,$$

with all the embeddings being continuous and dense.

Corollary 3.6.3 *The dual $\mathcal{G}_\varpi^*$ of $\mathcal{G}_\varpi$ has a norm topology which is compatible with the isometric isomorphism between G_ϖ^* and $\mathcal{G}_\varpi^*$.*

In case $\| f \|_{\mathcal{G}_\varpi} = +\infty$, it means f does not have standard Wiener-Itô decomposition. However, it may be an element of the dual $\mathcal{G}_\varpi^*$ having an elevated energy measured in terms of it's Hilbertian norm.

Remark 3.6.4 In view of the Corollary 2.6.3 and Theorem 3.6.2, the reader can easily verify that, for $Q_1, Q_2 \in \mathcal{L}_1^+(U)$ satisfying $Q_1 \leq Q_2$, the inclusion $\mathcal{G}_\varpi^{Q_2} \subset \mathcal{G}_\varpi^{Q_1}$ is reversed for the corresponding duals, that is, $(\mathcal{G}_\varpi^{Q_2})^* \supset (\mathcal{G}_\varpi^{Q_1})^*$.

We have seen that $\mathcal{L}_1^+(U)$ is a partially ordered lattice. It is easy to see that if $P, Q \in \mathcal{L}_1^+(U)$ and $Q \leq P$ and $P \leq Q$ then $P = Q$. Thus it follows from Zorn's Lemma that any chain $\{Q_n, \uparrow\} \equiv \{0 \leq Q_1 \leq Q_2 \leq \cdots \leq Q_n \leq \cdots\}$ has a unique least upper bound.

Let $\downarrow *$ denote the mapping that takes any Banach space X to its topological dual X^*. In view of the above observations, we can readily obtain the following result.

Corollary 3.6.5 *Corresponding to this chain* $\{Q_n, \uparrow\} \subset \mathcal{L}_1^+(U)$, *we have the following chains from the partially ordered lattice* $\{\mathcal{G}_\varpi^Q, Q \in \mathcal{L}_1^+(U)\}$ *and its dual* $\{(\mathcal{G}_\varpi^Q)^*, Q \in \mathcal{L}_1^+(U)\}$

$$
\begin{array}{ccccc}
\mathcal{G}_\varpi^{Q_n} & \hookrightarrow & \mathcal{G}_\varpi^{Q_{n-1}} \cdots \hookrightarrow & \mathcal{G}_\varpi^{Q_1} \\
\downarrow * & & \downarrow * & \downarrow * \\
(\mathcal{G}_\varpi^{Q_n})^* & \hookleftarrow & (\mathcal{G}_\varpi^{Q_{n-1}})^* \cdots \hookleftarrow & (\mathcal{G}_\varpi^{Q_1})^*
\end{array} \tag{3.22}
$$

and that they have unique least upper bound and greatest lower bound.

Proof Follows from Remark 3.6.4 and Zorn's Lemma. •

3.7 Generalized Functionals of Fr.Brownian Motion and Lévy Process

Following the same basic procedure we can also construct generalized functionals of fractional Brownian motion as well as Lévy process. These are presented below. The reader may like to carry out the details.

Considering the functionals of fractional Brownian motion, as seen in section 2.7, the vector space corresponding to regular functionals is given by $\mathcal{G}_\varphi$ which is isometrically isomorphic to the Hilbert space G_φ. Following the standard practice of this monograph, we denote this isometric isomorphism by $G_\varphi \leftrightarrow \mathcal{G}_\varphi$ and the corresponding isomorphic map by Φ_φ. Now using the Fréchet space $\mathcal{K}_\varphi \equiv \prod L_2^\varphi(I^n)$, we introduce the Hilbert space H_φ characterized by

$$H_\varphi \equiv \{K \in \mathcal{K}_\varphi : \| K \|_\varphi^2 = \sum_{n=0}^\infty (1/n!) \| K_n \|_{L_2^\varphi(I^n)}^2 < \infty\}. \tag{3.23}$$

The dual of G_φ denoted by G_φ^* is characterized as follows. For $K \in G_\varphi$ an element $L \in \mathcal{K}_\varphi$ belongs to the dual G_φ^* if the following duality pairing is well defined

$$< L, K >_{G_\varphi^*, G_\varphi} \equiv \sum_{n=0}^\infty (1/n!) < L_n, K_n >_{L_2^\varphi(I^n)}. \tag{3.24}$$

Clearly this is well defined if and only if

$$\sum_{n=0}^\infty (1/n!)^2 \| L_n \|_{L_2^\varphi(I^n)}^2 < \infty. \tag{3.25}$$

Again this shows that the dual is a very large space. The reader can easily verify that if $L \in H_\varphi$, then the pairing (3.24) reduces to the scalar product in the Hilbert space H_φ. Now using the isomorphism Φ_φ, (more precisely its continuous extension), we obtain the following result.

Theorem 3.7.1 *The class $\mathcal{G}_\varphi$ is the space of regular $L_2^\varphi(\Omega)$ functionals of fractional Brownian motion while the two classes $\mathcal{H}_\varphi$ and $\mathcal{G}_\varphi^*$ are generalized functionals of Fr.Brownian motion with $\mathcal{G}_\varphi$ being their test functionals and they satisfy the following diagram:*

$$G_\varphi \hookrightarrow H_\varphi \hookrightarrow G_\varphi^*$$
$$\updownarrow \qquad \updownarrow \qquad \updownarrow \tag{3.26}$$
$$\mathcal{G}_\varphi \hookrightarrow \mathcal{H}_\varphi \hookrightarrow \mathcal{G}_\varphi^*$$

where $\hookrightarrow$ denotes continuous and dense embedding and $\updownarrow$ stands for isometric isomorphism.

Remark 3.7.2 The results of Theorem 3.7.1 reduces to those of Theorem 3.2.2 as the Hurst parameter $H \downarrow (1/2)$. Recall that in the limit the correlation kernel $\varphi = \varphi_H$ converges to the Dirac measure.

Remark 3.7.3 It is interesting to note the difference in the characterization of the Hilbert space H_φ (3.23) and the Hilbert spaces $\{H_\alpha, H_\beta, H_\gamma, H_\delta, H_\varpi, H_\Pi\}$. All of the later spaces have identical characterization.

Similar results hold for functionals of Lévy process (particularly centered Poisson process). Following the same procedure the reader can prove the following result.

Theorem 3.7.4 *The class $\mathcal{G}_\Pi$ is the space of regular $L_2^\Pi(\Omega)$ functionals of the centered Poisson process q with Lévy measure Π while the two classes $\mathcal{H}_\Pi$ and $\mathcal{G}_\Pi^*$ are the generalized functionals thereof with $\mathcal{G}_\Pi$ being their test functionals and they satisfy the following diagram:*

$$G_\Pi \hookrightarrow H_\Pi \hookrightarrow G_\Pi^*$$

$$\updownarrow \qquad \updownarrow \qquad \updownarrow \qquad\qquad (3.27)$$

$$\mathcal{G}_\Pi \hookrightarrow \mathcal{H}_\Pi \hookrightarrow \mathcal{G}_\Pi^*$$

where $\hookrightarrow$ denotes continuous and dense embedding and $\updownarrow$ stands for isometric isomorphism.

3.8   Some Problems for Exercise

P1: Prove that the embeddings $G_\alpha \hookrightarrow H_\alpha \hookrightarrow G_\alpha^*$ are continuous. Find the embedding constants.

P2: Consider the triple of Hilbert spaces $G_\alpha \hookrightarrow H_\alpha \hookrightarrow G_\alpha^*$ introduced in section 3.2. Let $B_1(G_\alpha^*)$ denote the closed unit ball of the dual space G_α^*. Prove that for every $K \in G_\alpha$ there exists $L^* \in \partial B_1(G_\alpha^*)$ such that

$$< L^*, K >_{G_\alpha^*, G_\alpha} = \| K \|_{G_\alpha}$$

and that

$$\| L^* \|_{G_\alpha^*} = 1, \| L^* \|_{G_\alpha} = \infty, \| L^* \|_{H_\alpha} = \infty.$$

P3: Use the isomorphism Φ_α to reach similar conclusions for the problem (P2) with $G_\alpha \hookrightarrow H_\alpha \hookrightarrow G_\alpha^*$ replaced by the Gelfand triple $\mathcal{G}_\alpha \hookrightarrow \mathcal{H}_\alpha \hookrightarrow \mathcal{G}_\alpha^*$ which represent the Hilbert spaces of functionals of Brownian motion and their duals.

P4: Verify that similar conclusions, as in problems **(P2)-(P3)**, hold for all the Hilbert spaces $G_\alpha, G_\beta, G_\gamma, G_\delta, G_\varpi, G_\varphi, G_\Pi$ and their isomorphic images $\mathcal{G}_\alpha, \mathcal{G}_\beta, \mathcal{G}_\gamma, \mathcal{G}_\delta, \mathcal{G}_\varpi, \mathcal{G}_\varphi, \mathcal{G}_\Pi$.

P5: Consider the pair of infinite dimensional systems on a separable Hilbert space H

$$dx = Axdt + F(x)dt + C(t)dW^Q(t), x(0) = \xi, t \in I \equiv [0, T]$$
$$dx = Axdt + F(x)dt + C(t)dW^P(t), x(0) = \xi, t \in I \equiv [0, T],$$

where $\{W^Q, W^P\}$ are U valued Brownian motions with incremental covariance operators $Q, P \in \mathcal{L}_1^+(U)$. Suppose A generates a C_0-semigroup $S(t), t \geq 0$, on H, F is uniformly Lipschitz on H and $C \in \mathcal{L}(U, H)$. Let x^Q, x^P denote the mild solutions of the above systems corresponding to Brownian motions $\{W^Q, W^P\}$ respectively. Suppose ξ is an H valued second order random variable independent of the sigma algebra $\mathcal{F}_t, t > 0$, to which the Brownian motions $\{W^Q, W^P\}$ are adapted.

(a): Let $\mathcal{L}_2(H, U)$ denote the Hilbert space of Hilbert-Schmidt operators from H to U. Prove the following inequality

$$\sup_{t \in I} \mathbb{E}|x^Q(t) - x^P(t)|_H^2 \leq C \int_0^T \| (\sqrt{Q} - \sqrt{P})C^* S^*(t) \|_{\mathcal{L}_2(H, U)}^2 \, dt,$$

and verify that $C \equiv e^{TM^2 K^2}$ with K being the Lipschitz constant for F and M the norm bound of the semigroup over I.

(b): Prove the continuity of the map $Q \longrightarrow x^Q$ on $\mathcal{L}_1^+(U)$ with respect to the norm topologies on $\mathcal{L}_1(U)$ and $B_\infty^a(I, L_2(\Omega, H))$ respectively. Hint: Recall that $B_\infty^a(I, L_2(\Omega, H))$ is the Banach space of $\mathcal{F}_t$ adapted H valued random processes having second moments uniformly bounded on I.

P6: Consider the space G_φ^* and the related space $\mathcal{G}_\varphi^*$ of generalized functionals of fractional Brownian motion. (a): Show that the norm topology of the former is given by the expression (3.25). Give the norm topology of $\mathcal{G}_\varphi^*$ which is compatible with that of G_φ^*. (b): For any $K \in G_\varphi$, find an $L \in \partial B_1(\mathcal{G}_\varphi^*)$ (the unit sphere) such that

$$< L, K >_{G_\varphi^*, G_\varphi} = \| K \|_{G_\varphi} .$$

Chapter 4

Functional Analysis on $\{G, \mathcal{G}\}$ and Their Duals

4.1 Introduction

In this chapter we study the questions of compactness and weak compactness in the Fock spaces $\{G, \mathcal{G}\}$ and their duals $\{G^*, \mathcal{G}^*\}$ respectively. In other words we present results characterizing compact and weakly compact sets in these spaces. Here the spaces G and $\mathcal{G}$ stand for any member of the family of Fock spaces $\{G_\alpha, G_\beta, G_\gamma, G_\delta, G_\varpi, G_\varphi, G_\Pi\}$ and their isomorphic images $\{\mathcal{G}_\alpha, \mathcal{G}_\beta, \mathcal{G}_\gamma, \mathcal{G}_\delta, \mathcal{G}_\varpi, \mathcal{G}_\varphi, \mathcal{G}_\Pi\}$ respectively. Using the compactness results we consider some problems involving optimization. This is illustrated by several examples. In particular, they are applied to stochastic differential equations (SDE), input-output representation thereof, inverse problems and nonlinear filtering problems etc.

4.2 Compact and Weakly Compact Sets

Here we consider some basic questions of analysis on the vector space $\mathcal{G} = L_2(\Omega, \mathcal{B}, \mu)$ or equivalently G. Since G and $\mathcal{G}$ are Hilbert spaces, bounded subsets $A \subset G$ and $\hat{A} \subset \mathcal{G}$ are conditionally weakly compact. If closed, then they are also weakly compact. The interesting question is what are the necessary and sufficient conditions for (strong) compactness. These questions were addressed in a paper of the author [Ahmed (1973)] for the spaces G_α and $\mathcal{G}_\alpha$. We present here some of these results.

For convenience of presentation we shall replace I by R and extend the elements $\{K_n\}$ of $\hat{L}_2(I^n)$ or those of $\hat{L}_2(I^n, M_n)$ by setting $K_n \equiv 0$ outside the interval I. The result we are going to present here applies to the Hilbert spaces G_α G_β and G_ϖ and therefore to their isomorphic images $\mathcal{G}_\alpha$ $\mathcal{G}_\beta$ and $\mathcal{G}_\varpi$ respectively. For convenience, we use one pair of symbols, G and $\mathcal{G}$, to denote the corresponding spaces. Let Π_n denote the projection of G to

$$G^n \equiv \{K \in G : K = (K_0, K_1 \cdots K_n, 0, 0, 0, \cdots)\}.$$

Let $\Lambda_h, h \in R$, denote the translation operator:

$$\Lambda_h K = \big(K_0, K_1(\cdot + h), K_2(\cdot + h, \cdot + h) \cdots K_n(\cdot + h, \cdot + h, \cdots, \cdot + h), \cdots\big).$$

Clearly $\Lambda_h : G \longrightarrow G$.

Theorem 4.2.1 *A set $\Gamma \subset G$ is conditionally compact if, and only if,*

(1) Γ *is bounded,*

(2) $\lim_{h \to 0} \| \Lambda_h K - K \|_G = 0$ *uniformly with respect to $K \in \Gamma$,*

(3) $\lim_{n \to \infty} \| (1 - \Pi_n)K \|_G = 0$ *uniformly with respect to $K \in \Gamma$.*

Further, Γ is compact if it is also closed.

Proof For detailed proof see [[Ahmed (1973)], Theorem 4].

Next we present a parallel result for the function space $\mathcal{G}$ representing either of the spaces $L_2(\Omega, \mathcal{B}, \mu)$, $L_2(\Omega, \mathcal{B}, \mu; R^\ell)$ or $L_2(\Omega, \mathcal{B}, \mu; H)$ where we have used μ to denote the Wiener measure defined on the topological Borel field $\mathcal{B}$ of subsets of either of the path spaces $C_0(I)$, $C_0(I, R^d)$ or $C_0(I, U)$. Let $\{T_t, t \in R\}$ denote the ergodic group of measure preserving transformations of $\mathcal{B}$ into itself; and P_n denote the projection of $\mathcal{G}$ to $\mathcal{G}_n$ as defined below:

$$\mathcal{G}_n = P_n(\mathcal{G}) \equiv \bigg\{ f \in \mathcal{G} : f = \sum_{s=0}^{n} g_s(K_s), K_s \in L_2(I^s) \text{ or}$$

$$f = \sum_{s=0}^{n} f_s(K_s), K_s \in L_2(I^s, M_s), \text{ or}$$

$$f = \sum_{s=0}^{n} J_s(K_s), K_s \in L_2(I^s, \mathcal{L}_2^Q(U^{\otimes s}, H)), 0 \le s \le n \bigg\}.$$

Theorem 4.2.2 *A set* $F \subset \mathcal{G}$ *is conditionally compact if, and only if,*

(1) F *is bounded,*

(2) $\lim_{t \to 0} \int_{\Omega} |f(T_t w) - f(w)|^2 d\mu(w) = 0$ *uniformly with respect to* $f \in F$,

(3) $\lim_{n \to \infty} \int_{\Omega} |(1 - P_n)f(w)|^2 d\mu(w) = 0$ *uniformly with respect to* $f \in F$.

Proof The proof follows from Theorem 4.2.1 and the isomorphism $\Phi \equiv \{\Phi_\alpha \text{ or } \Phi_\beta \text{ or } \Phi_\varpi\}$ [see also [Ahmed (1973)], Corollary 2].

Theorem 4.2.3 *Let* $\Gamma \subset G$ *and consider its image* $\Xi \equiv \{\Phi(K), K \in \Gamma\}$. *Then* Ξ *is a weakly compact subset of* $\mathcal{G}$ *if, and only if,* Γ *is a weakly compact subset of* G.

Proof Since weak compactness is preserved under isomorphism and Φ is an isomorphism of G into $\mathcal{G}$, the set $\Xi \subset \mathcal{G}$ is weakly compact whenever $\Gamma \subset G$ is weakly compact and conversely. •

Similar results are valid for the dual spaces G^* and $\mathcal{G}^*$.

Theorem 4.2.4 *Let* $\tilde{\Gamma}$ *be a bounded, w^*-closed (weak star), convex subset of* G^*, *and consider its image* $\tilde{\Xi} \equiv \{\Phi(K), K \in \tilde{\Gamma}\}$. *Then,* $\tilde{\Xi}$ *is a weak star compact subset of* $\mathcal{G}^*$. *Conversely, if* $\tilde{\Xi} \subset \mathcal{G}^*$ *is a bounded, w^*-closed, convex set, then* $\tilde{\Gamma} \equiv \Phi^{-1}(\tilde{\Xi})$ *is a w^* compact subset of* G^*.

Proof Since $\tilde{\Gamma}$ is a bounded w^*-closed convex subset of a dual space (conjugate space), it follows from Alaoglu's theorem that $\tilde{\Gamma}$ is weak star compact. Weak and w^*-compactness are preserved under isomorphism, and so, we conclude that $\tilde{\Xi} = \Phi(\tilde{\Gamma})$ is a w^*-compact subset of $\mathcal{G}^*$. The converse follows since Φ is an isomorphism. •

Remark 4.2.5 Since for reflexive Banach spaces, the weak star topology on the dual is equivalent to the weak topology, a set $C \subset \mathcal{G}^*$ is weak star compact, if and only if, it is weakly compact.

4.3 Some Optimization Problems

Here we present some typical optimization results that may be applied to nonlinear system representation and identification (inverse) problems. First we prove the following result.

Theorem 4.3.1 *Let J be a weakly lower semicontinuous functional on G bounded away from $-\infty$ and suppose Ξ is a weakly compact subset of G. Then J attains it's infimum on Ξ.*

Proof Proof is classical. •

Theorem 4.3.2 *Let J be a weakly lower semicontinuous functional on G bounded away from $-\infty$ and suppose $\lim_{\|K\|_G \to \infty} J(K) = \infty$. Then J attains it's minimum on G. Further, if J is strictly convex it has a unique minimum.*

Proof Define $m \equiv \inf\{J(K), K \in G\}$. Clearly $m > -\infty$. Hence there exists a minimizing sequence $\{K^n\} \subset G$ such that $\lim_{n \to \infty} J(K^n) = m$. Since $\lim_{\|K\|_G \to \infty} J(K) = \infty$, it is clear that $\{K^n\}$ is a bounded sequence in G. In any Hilbert space, a bounded sequence has a weakly convergent subsequence. Relabeling the subsequence as the original sequence, we conclude that there exists a $K^o \in G$ such that $K^n \overset{w}{\longrightarrow} K^o$ in G. Hence, by weak lower semi continuity of J, we have

$$J(K^o) \leq \liminf_{n \to \infty} J(K^n) \leq \lim_{n \to \infty} J(K^n) = m.$$

Since m is the infimum of J on G, $m \leq J(K^o)$. In view of the opposite inequalities, we conclude that $J(K^o) = m$. This proves the existence. The uniqueness follows from strict convexity. Indeed, if there are two elements $L^o, K^o \in G$ at which minimum is attained, it follows from strict convexity of J that for $\alpha \in (0,1)$ we have

$$J(\alpha L^o + (1-\alpha)K^o) < \alpha J(L^o) + (1-\alpha)J(K^o) = m.$$

This contradicts the fact that m is the minimum value of J on G. This ends the proof. •

Theorem 4.3.3 *Theorems 4.3.1 and 4.3.2 remain valid if in their statement, G is replaced by its dual G^*, weak lower semi continuity is replaced* *by w^*-lower semi continuity and Ξ in Theorem 4.3.1 is a w^* (weak star) compact subset of G^*.*

For applications we may choose specific functionals for J.

Theorem 4.3.4 *Let $\Xi \subset G$ be a weakly compact set, $\tilde{K}$ a fixed element of G and $\varphi : [0, \infty] \longrightarrow [0, \infty]$ be a nondecreasing, continuous, and extended real valued function. Then the functional J given by*

$$J(K) \equiv \varphi(\| K - \tilde{K} \|_G)$$

attains its minimum on Ξ. Further, if Ξ is a convex set and φ is a strictly convex function, J has a unique minimum on Ξ.

Proof First note that the norm in any Banach space is a weakly lower semicontinuous functional. In particular, $K \longrightarrow \eta(K) \equiv \| K \|_G$ is weakly lower semicontinuous. Suppose $K^n \xrightarrow{w} K^o$ in G. Let $B_1(G^*)$ denote the closed unit ball in G^*, the dual of G. Corresponding to the element K^o, there exists an element $L^o \in B_1(G^*)$, (as seen in Chapter 2), such that $\| K^o \|_G = < L^o, K^o >_{G^*,G}$. Then,

$$\| K^o \|_G = < L^o, K^o >_{G^*,G} = < L^o, K^o - K^n >_{G^*,G} + < L^o, K^n >_{G^*,G} .$$

Since K^n converges weakly to K^o, it is clear from the above identity that for any $\varepsilon > 0$ there exists $n_\varepsilon \in N$ such that

$$\| K^o \|_G \leq \varepsilon + < L^o, K^n >_{G^*,G} \leq \varepsilon + \| K^n \|_G$$

for all $n \geq n_\varepsilon$. Hence, $\| K^o \|_G \leq \liminf_{n \to \infty} \| K^n \|_G$ and therefore the norm functional is weakly lower semicontinuous. Since φ is continuous and nondecreasing, it is clear that

$$J(K^o) = \varphi(\| K^o - \tilde{K} \|) \leq \liminf_{n \to \infty} \varphi(\| K^n - \tilde{K} \|) = \liminf_{n \to \infty} J(K^n).$$

Hence, J is weakly lower semicontinuous on G and, since Ξ is weakly compact by assumption, J attains its minimum on Ξ. We prove uniqueness by contradiction. Suppose the assumptions on φ and Ξ hold, but there are multiple elements at which J attains its minimum $m \equiv \inf\{J(K), K \in \Xi\}$. Let $K^1, K^2 \in \Xi$ be two such elements. Since φ is nondecreasing, it is clear that

$$J((1/2)(K^1 + K^2)) \leq \varphi((1/2) \| K^1 - \tilde{K} \| + (1/2) \| K^2 - \tilde{K} \|).$$

By strict convexity of φ, this leads to the strict inequality

$$J((1/2)(K^1 + K^2)) \; < \; (1/2)\varphi(\| \, K^1 - \tilde{K} \, \|) + (1/2)\varphi(\| \, K^2 - \tilde{K} \, \|).$$

In other words, we have

$$J((1/2)(K^1 + K^2)) \; < \; (1/2)J(K^1) + (1/2)J(K^2) = m.$$

By assumption Ξ is a convex subset of G and so $K^* \equiv (1/2)(K^1 + K^2) \in \Xi$ and hence $J(K^*) < m$. This contradicts the hypothesis that K^1, K^2 are minimizers. This completes the proof. •

An exact analog of this result in the context of the space of regular functionals of Brownian motion $\mathcal{G}$ is given in the following corollary.

Corollary 4.3.5 *Let $\Gamma \subset \mathcal{G}$ be a weakly compact set, $\tilde{f}$ a given element of $\mathcal{G}$ and $\varphi : [0, \infty] \longrightarrow [0, \infty]$ a nondecreasing, continuous, extended real valued function. Then, the functional J given by*

$$J(f) \equiv \varphi(\| \, f - \tilde{f} \, \|_{\mathcal{G}})$$

attains its minimum on Γ. Further, if Γ is convex and φ is strictly convex, J has a unique minimum on Γ.

Proof The proof is based on identical arguments. •

One can easily observe that a similar result holds for the dual space $\mathcal{G}^*$. This is stated as follows,

Corollary 4.3.6 *Let $\Gamma \subset \mathcal{G}^*$ be a weak star compact set, $\tilde{f} \in \mathcal{G}^*$, and $\varphi : [0, \infty] \longrightarrow [0, \infty]$ be a nondecreasing, continuous, and extended real valued function. Then the functional J given by*

$$J(f) \equiv \varphi(\| \, f - \tilde{f} \, \|_{\mathcal{G}^*})$$

attains its minimum on Γ. Further, if Γ is convex and φ is strictly convex, J has a unique minimum on Γ.

4.4   Applications to SDE

Here we present some applications of the results given in the preceding sections. In particular, we use these results to study approximation theory for input-output maps corresponding to stochastic differential equations driven by Brownian motion. These results are also used in the study of inverse problems known as system identification (problems) in physical and engineering sciences.

(E1): Consider the stochastic differential equation on R^{ℓ}

$$dx = b(t,x)dt + \sigma(t,x)dw, x(0) = x_0, t \in I \equiv (0,T] \qquad (4.1)$$

where w is an R^d-valued standard Brownian motion. The parameters $\{b, \sigma\}$ are vector valued functions $b : I \times R^{\ell} \longrightarrow R^{\ell}$, $\sigma : I \times R^{\ell} \longrightarrow \mathcal{L}(R^d, R^{\ell})$. They are measurable in $t \in I$, locally Lipschitz continuous in x and there exists a constant $\eta > 0$ such that

$$|b(t,x)|^2_{R^{\ell}} + Tr(\sigma(t,x)\sigma(t,x)^*) \leq \eta\big(1 + |x|^2_{R^{\ell}}\big)$$

for all $(t, x) \in I \times R^{\ell}$. Further, we assume that $E|x_0|^2_{R^{\ell}} < \infty$.

It is well known [Skorohod (1965)], [Ahmed (1998)], [Shigekawa (1998/2004)], [Ikeda (1989)] that, under the given assumptions, the system (4.1) has a unique strong solution x which belongs to $C(I, R^{\ell})$ with probability one and further

$$\sup\{E|x(t)|^2_{R^{\ell}}, t \in [0,T]\} < \infty.$$

Clearly, it follows from this result that $x_T \equiv x(T) \in L_2(\Omega, \mathcal{B}, \mu; R^{\ell}) \cong \mathcal{G}$. Since Φ is an isomorphism between the Hilbert spaces G and $\mathcal{G}$, there exists a unique $K \in G$ such that

$$x_T = \Phi(K) = \sum_{s=0}^{\infty} f_s(K_s). \qquad (4.2)$$

In practical applications, often it is desirable to approximate a system, equivalently, its input-output representation, by a finite Wiener-Itô series. In general, one may wish to find an element from a closed linear subspace M_0 of $\mathcal{G}$. In this case one can formulate the problem as: find $f \in M_0$ that minimizes the functional Z given by

$$Z(f) \equiv \| x_T - f \|_{\mathcal{G}} . \qquad (4.3)$$

Clearly, this functional is radially unbounded, that is, $\lim_{\|f\|_{\mathcal{G}} \to \infty} Z(f) = \infty$. The reader can easily verify that it is weakly lower semicontinuous on

$\mathcal{G}$ and that $m_0 \equiv \inf\{Z(f), f \in M_0\} \geq 0$, and it is also finite. Thus, there exists a minimizing sequence $\{f^n\} \subset M_0$ such that $\lim_{n\to\infty} Z(f^n) = m_0$. We show that there exists an $f^o \in M_0$ at which the infimum is attained. Since Z is radially unbounded, and $Z(f^n) \to m_0$, the sequence $\{f^n\}$ is contained in a bounded subset of $\mathcal{G}$. A bounded set in a Hilbert space is conditionally weakly compact. Hence, there exists a subsequence of the sequence $\{f^n\}$, relabeled as the original sequence, and an f^o in $\mathcal{G}$ such that $f^n \xrightarrow{w} f^o$ in $\mathcal{G}$. By weak lower semicontinuity of Z we have

$$Z(f^o) \leq \liminf_{n\to\infty} Z(f^n) \leq \lim_{n\to\infty} Z(f^n) = m_0.$$

Now we must verify that $f^o \in M_0$. Since the sequence $\{f^n\} \subset M_0$ and f^n converges weakly to f^o and M_0 is closed linear subspace of $\mathcal{G}$, it follows from Hahn-Banach theorem that f^o must be in M_0. Hence $Z(f^o) \geq m_0$ and the conclusion now follows from the above opposite inequalities.•

Remark 4.4.1 The result presented above also holds for stochastic functional differential equations of the form

$$dx = b(t, x_t)dt + \sigma(t, x_t)dw,$$
$$x(s) = \zeta(s), -r \leq s < 0, x(0) = \xi, t \in I \equiv (0, T], \qquad (4.4)$$

where $x_t \equiv x(t+s), -r \leq s \leq 0$. The basic assumptions are similar [Ahmed (1970)].

In fact these results also hold for differential inclusions [Ahmed (1992)] of the form

$$dx \in B(t, x)dt + \Sigma(t, x)dw, x(0) = x_0$$

where both B and Σ are measurable multifunctions with closed convex values in R^ℓ and $\mathcal{L}(R^d, R^\ell)$ respectively and x_0 is an R^ℓ-valued second order random variable. The standard assumption is that both B and Σ satisfy the following Lipschitz and growth conditions with respect to the Hausdorff metric ρ_H as stated below,

$$\rho_H(B(t, x), B(t, y)) \leq K|x - y|_{R^\ell}, \quad \rho_H(\Sigma(t, x), \Sigma(t, y)) \leq K|x - y|_{R^\ell}$$
$$\sup\{|z|_{R^\ell}, z \in B(t, x)\} \leq K(1 + |x|_{R^\ell}),$$
$$\sup\{|\sigma|_{\mathcal{L}(R^d, R^\ell)}, \sigma \in \Sigma(t, x)\} \leq K(1 + |x|_{R^\ell}).$$

Remark 4.4.2 An example of a linear subspace M_0 of significant practical importance is

$$M_0 \equiv \{f \in \mathcal{G} : f = \sum_{n=0}^{m} f_n(K_n), K_n \in \hat{L}_2(I^n, M_n)\}$$

where m is a finite positive integer. In other words, one wishes to approximate the given functional by a finite Wiener-Itô series. This is certainly important in practical applications.

(E2) Inverse Problem A: Inverse problems are known as identification problems in Physical sciences and engineering. Often, Engineers want to identify their system through an input-output representation. Mathematically, a convenient model is the Volterra series given by

$$y(t) \equiv F_t(x) \equiv \sum_{n=1}^{\infty} \int_0^t \cdots \int_0^t K_n(\tau_i, \cdots, \tau_n) x(\tau_1) \cdots x(\tau_n) d\tau_1 \cdots d\tau_n, t \in I,$$

$$(4.5)$$

where x is the input and y the output. Given the input-output pair $\{x, y\}$ over a given period of time, say, $I \equiv [0, T]$, the problem is to identify the kernels $\{K_n\}$ so that the model output approximates the actual response as closely as possible. Recall the Hilbert space G_α with the norm topology given by the expression (2.8). Suppose $K \equiv \{K_n\} \in G_\alpha$. Then one can readily verify that, for every $x \in L_2(I)$, and $t \in I$, we have

$$|y(t)| \leq \sum_{n=1}^{\infty} \| K_n \|_{L_2(I^n)} \| x \|_{L_2(I)}^n \leq \| K \|_{G_\alpha} \exp\{(1/2) \| x \|_{L_2(I)}^2\}. (4.6)$$

Thus, the operator F, as defined above, is a bounded nonlinear operator from $L_2(I)$ to $L_\infty(I)$ and it is also continuous and locally Lipschitz. For mathematical convenience, we assume that the kernels $\{K_n, n \geq 1\}$ are symmetrized. Now, consider the map $F_T(\dot{w}) = f(w)$ where w is the standard Brownian motion. As in Chapter 2, this is defined as the infinite series of Wiener-Itô stochastic integrals. To identify the kernels, we consider the sequence of orthogonal functionals $\{g_n(L_n)\}$, for $L \equiv \{L_n, n \geq 1\} \in G_\alpha$. It is clear that

$$< f, g_n(L_n) > = n!(K_n, L_n)_{L_2(I^n)} \quad \text{for every } n \in N,$$

provided the n-th order kernel of the system being identified is K_n. Define the functional

$$L_n \longrightarrow \ell(L_n) \equiv (1/n!) < f, g_n(L_n) >$$

where the angle bracket means integration with respect to Wiener measure. Clearly, this is a continuous linear functional on $\hat{L}_2(I^n)$. We can determine K_n by maximizing this functional on the unit ball of $\hat{L}_2(I^n)$. It is easily shown that a point at which ℓ attains its maximum on the unit ball is given by $L_n^o \equiv K_n/ \parallel K_n \parallel$ which is a point on the boundary of the unit ball. Since an L_2 space is strictly convex (that is, its unit ball is strictly convex), and ℓ is linear, the maximizer is unique. Thus the maximizer is uniquely related to the unknown kernel. The unknown kernel K_n is equivalent to the maximizer L_n^o modulo a multiplicative constant β_n with

$$\beta_n = \frac{< f, g_n(L_n^o) >}{n!}.$$

In other words, $K_n^o = \beta_n L_n^o$. Thus, in principle, we have identified the system F in the form of an infinite Volterra series with kernels given by $K^o \equiv \{K_n^o, n \geq 1\}$. Note that the norm of f is given by

$$\parallel f \parallel_{\mathcal{G}_\alpha} = \parallel K^o \parallel_{G_\alpha} = \left(\sum n! \beta_n^2 \right)^{1/2}. \tag{4.7}$$

(E3) Inverse Problem B: For the inverse problem A, f was assumed to be an element of the regular space $\mathcal{G}_\alpha$ which is a proper subspace of $\mathcal{G}_\alpha^*$. However, the same procedure applies even if f is in the larger space. We construct the linear functional exactly as before giving

$$\ell(L_n) = (1/n!) < f, g_n(L_n) > .$$

Maximizing this functional on the unit ball of $\hat{L}_2(I^n)$, we obtain L_n^o and then construct $K_n^o \equiv \beta_n L_n^o$ for each $n \in N$. The normalizing factors $\{\beta_n\}$ are still given by the same expression as in the regular case. Since in this case $f \in \mathcal{G}_\alpha^*$, it follows from the isomorphism [see Theorem 3.2.2] that $K^o \equiv \{K_n^o, n \in N\}$ is an element of G_α^*. Thus

$$\parallel f \parallel_{\mathcal{G}_\alpha^*} = \parallel K^o \parallel_{G_\alpha^*} = \left(\sum_{n=1}^{\infty} (1/n!) \parallel K_n^o \parallel_{L_2(I^n)}^2 \right)^{1/2}$$

$$= \left(\sum_{n=1}^{\infty} (1/n!) \beta_n^2 \right)^{1/2}. \tag{4.8}$$

Note the distinction between the two expressions (4.7) and (4.8). In the case of regular functionals (Wiener-Itô class) the normalizing factors decay very rapidly with respect to the index $\{n\}$ while in the case of generalized functionals it is not so and even may grow unbounded. Clearly, this shows the power of the generalized functionals which are elements of spaces $\mathcal{H}$ or more generally, $\mathcal{G}^*$.

4.5 Vector Measures

Here we consider vector measures with values in any of the spaces $\{\mathcal{G}_\alpha, \mathcal{G}_\beta, \mathcal{G}_\gamma, \mathcal{G}_\delta, \mathcal{G}_\varpi, \mathcal{G}_\varphi, \mathcal{G}_\Pi\}$ and the corresponding duals. For convenience of presentation, we use one symbol $\mathcal{G}$ to denote any of the members of the family $\{\mathcal{G}_\alpha, \mathcal{G}_\beta, \mathcal{G}_\gamma, \mathcal{G}_\delta, \mathcal{G}_\varpi, \mathcal{G}_\varphi, \mathcal{G}_\Pi\}$ and $\mathcal{G}^*$ the corresponding dual. First, we present the following definition.

Definition 4.5.1 *Let $(S, \mathcal{B}, \nu)$ be a measure space. A Banach space X is said to have the Radon-Nikodym property (RNP) with respect to $(S, \mathcal{B}, \nu)$ if, for every ν-continuous X valued vector measure μ, there exists an $f \in L_1(\nu, X) \equiv L_1((S, \mathcal{B}, \nu), X)$ such that $d\mu = f d\nu$, that is, for every $\sigma \in \mathcal{B}$,*

$$\mu(\sigma) = \int_\sigma f(s)\nu(ds).$$

A Banach space X is said to have the RNP if it has the Radon-Nikodym property with respect to every finite measure space.

It is well-known [Diestel and Uhl.Jr., [Diestel (1977)]] that reflexive Banach spaces and separable dual spaces satisfy RNP. Hence all Hilbert spaces have RNP.

First we consider the vector space G_β and its isometric image $\mathcal{G}_\beta = \Phi_\beta(G_\beta)$.

Theorem 4.5.2 *Let $(S, \mathcal{B}, \nu)$ be a finite measure space and $\mu : \mathcal{B} \longrightarrow \mathcal{G}_\beta$ be a countably additive vector measure of bounded variation and suppose it is ν-continuous. Then, there exists a map*

$$L = \{L_n, n \in N\} : \mathcal{B} \longrightarrow G_\beta$$

such that

$$\mu(\sigma) = \sum_{n=0}^{\infty} f_n(L_n(\sigma)) \equiv \sum_{n=0}^{\infty} \int_{I^n} L_n(\sigma; \tau_1, \cdots, \tau_n) dw(\tau_1) \otimes \cdots \otimes dw(\tau_n)$$

(4.9)

for every $\sigma \in \mathcal{B}$.

Proof Since every Hilbert space has the Radon-Nikodym property, and μ is ν-continuous countably additive $\mathcal{G}_\beta$ valued vector measure having bounded total variation, there exists a $z \in L_1(\nu, \mathcal{G}_\beta)$ such that

$$\mu(\sigma) = \int_\sigma z(s)\nu(ds).$$

Thus $z(s) \in \mathcal{G}_\beta$ ν-almost all $s \in S$. Then, it follows from the isometric isomorphism Φ_β between $\mathcal{G}_\beta$ and G_β that, for ν-almost all $s \in S$, there exists a $K(s) \equiv \{K_n(s), n \in N\} \in G_\beta$ such that

$$z(s) = \sum_{n=0}^{\infty} f_n(K_n(s)) \tag{4.10}$$

where

$$f_n(K_n(s)) \equiv \int_{I^n} K_n(s; \tau_1, \tau_2 \cdots \tau_n) dw(\tau_1) \otimes dw(\tau_2) \cdots \otimes dw(\tau_n).$$

Since z is Bochner integrable with respect to the measure ν, it follows from Fubini's theorem that for almost all $(\tau_1, \tau_2, \cdots, \tau_n) \in I^n$, each of the kernels K_n is ν integrable. Thus $L_n(\sigma), \sigma \in \mathcal{B}$, defined by

$$L_n(\sigma; \tau_1, \tau_2, \cdots, \tau_n) \equiv \int_\sigma K_n(s; \tau_1, \tau_2, \cdots, \tau_n) \, \nu(ds),$$

is a vector measure with values in $\hat{L}_2(I^n)$. This is true for every $n \in N_0 \equiv \{0, 1, 2, \cdots\}$ and hence one can verify that L given by $L(\cdot) \equiv \{L_n(\cdot), n \in N\}$ is a G_β valued vector measure and satisfies the statement of the theorem. This completes the proof. •

Similar results hold for the spaces $\mathcal{G}_\alpha, \mathcal{G}_\gamma, \mathcal{G}_\delta, \mathcal{G}_\varpi, \mathcal{G}_\varphi, \mathcal{G}_\Pi$. We state this for the vector spaces G_γ and its isometric image $\mathcal{G}_\gamma = \Phi_\gamma(G_\gamma)$.

Theorem 4.5.3 *Let $(S, \mathcal{B}, \nu)$ be a finite measure space and $\mu : \mathcal{B} \longrightarrow \mathcal{G}_\gamma$ be a countably additive vector measure of bounded variation and suppose it* *is ν-continuous. Then there exists a map $L : \mathcal{B} \longrightarrow G_\gamma$ such that*

$$\mu(\sigma) = \sum_{n=0}^{\infty} e_n(L_n(\sigma)) \equiv \sum_{n=0}^{\infty} \int_{D^n} L_n(\sigma; \xi_1, \xi_2, \cdots, \xi_n) W(d\xi_1) W(d\xi_2) \cdots W(d\xi_n)$$

(4.11)

for every $\sigma \in \mathcal{B}$.

Proof Proof is identical to that of Theorem 4.5.2.

Remark 4.5.4 Results similar to those of Theorem 4.5.2 (Theorem 4.5.3) also hold for all the dual spaces $\{\mathcal{G}_\alpha^*, \mathcal{G}_\beta^*, \mathcal{G}_\gamma^*, \mathcal{G}_\delta^*, \mathcal{G}_\varpi^*, \mathcal{G}_\varphi^*, \mathcal{G}_\Pi^*\}$.

In the study of optimization problems involving vector measures questions on weak compactness are very important. Let $\mathcal{M}_{cabv}(\mathcal{B}, \mathcal{G})$ denote the Banach space of countably additive $\mathcal{G}$-valued vector measures having bounded total variation. Then we have the following result.

Theorem 4.5.5 *A set $M_0 \subset \mathcal{M}_{cabv}(\mathcal{B}, \mathcal{G})$ is relatively weakly compact if and only if the following conditions hold:*

(i): M_0 is bounded,

(ii): there exists a nonnegative countably additive bounded measure ν such that $\lim_{\nu(\sigma) \to 0} |\mu|(\sigma) = 0$ uniformly with respect to $\mu \in M_0$.

Proof This is a special case of the Bartle-Dunford-Schwartz theorem [Theorem 1.6.12]. Since all the spaces $\{\mathcal{G}_\alpha, \mathcal{G}_\beta, \mathcal{G}_\gamma, \mathcal{G}_\delta, \mathcal{G}_\varpi, \mathcal{G}_\varphi, \mathcal{G}_\Pi\}$ including their duals are Hilbert spaces, they satisfy RNP and further every bounded set in any of them is relatively weakly compact. Thus the proof follows directly from Theorem 1.6.12. •

Corollary 4.5.6 *A set $M_0 \subset \mathcal{M}_{cabv}(\mathcal{B}, \mathcal{G}^*)$ is relatively weak star compact if and only if the following conditions hold:*

(i): M_0 is bounded,

(ii): there exists a nonnegative countably additive bounded measure ν such that $\lim_{\nu(\sigma) \to 0} |\mu|(\sigma) = 0$ uniformly with respect to $\mu \in M_0$.

Proof Proof is identical. Since $\mathcal{G} \in \{\mathcal{G}_\alpha, \mathcal{G}_\beta, \mathcal{G}_\gamma, \mathcal{G}_\delta, \mathcal{G}_\varpi, \mathcal{G}_\varphi, \mathcal{G}_\Pi\}$ and its dual $\mathcal{G}^*$ are all Hilbert spaces, weak and weak star topologies are equivalent. So relative weak or weak star compactness in the statement of the theorem is immaterial. •

The compactness results presented above hold for σ-algebra $\mathcal{B}$ of subsets of S and countably additive vector measures. In fact these results also hold for finitely additive measures which is a generalization by Brooks and Dinculeanu [Brooks (1974)], [[Diestel (1977)], IV.2.6, Diestel, p106] of the countably additive case. This is stated in the following Theorem.

Theorem 4.5.7 *Let $\mathcal{B}$ denote any field (algebra) of subsets of S. Let $\mathcal{M}_{fabv}(\mathcal{B}, \mathcal{G}^*)$ denote the class of finitely additive $\mathcal{G}^*$-valued vector measures. Then a set $M_0 \subset \mathcal{M}_{fabv}(\mathcal{B}, \mathcal{G}^*)$ is relatively weakly (equivalently weak star)compact if and only if the following conditions hold:*

(i): M_0 is bounded (in variation norm),

(ii): there exists a finitely additive nonnegative measure ν on $\mathcal{B}$ such that $\lim_{\nu(\sigma) \to 0} |\mu|(\sigma) = 0$ uniformly with respect to $\mu \in M_0$.

Before we complete this section, we consider an optimization problem on the space of $\mathcal{G}^*$-valued measures. Let S be any (not necessarily compact) Hausdorff space and $\mathcal{B}$ the sigma algebra of Borel subsets of the set S. We consider the Banach space $\mathcal{M}_{cabv}(\mathcal{B}, \mathcal{G}^*)$ endowed with the topology induced by the total variation norm.

Theorem 4.5.8 *Suppose M_0 is a weak star (equivalently weakly) compact subset of the space $\mathcal{M}_{cabv}(\mathcal{B}, \mathcal{G}^*)$ and $f : R^n \longrightarrow R$ is a lower semi continuous function satisfying*

$$f(\xi) + \alpha|\xi|^q_{R^n} + \beta \geq 0, \ \forall \ \xi \in R^n$$

for any $\alpha, \beta \in R$ and $q \in [1, \infty)$. Let $\{\varphi_i\}_{1 \leq i \leq n} \subset B_\infty(S, \mathcal{G})$ be any given set and consider the cylindrical functional F on $\mathcal{M}_{cabv}(\mathcal{B}, \mathcal{G}^*)$ given by

$$F(\mu) \equiv f(< \varphi_1, \mu >, < \varphi_2, \mu >, \cdots, < \varphi_n, \mu >)$$

where $< \varphi_i, \mu > \equiv \int_S < \varphi_i(s), \mu(ds) >_{\mathcal{G}, \mathcal{G}^*}$. Then, F attains its minimum on M_0. Further, if f is strictly convex, F has a unique minimizer.

Proof Since M_0 is weak star compact, it is bounded and hence

$$\sup\{| < \varphi_i, \mu > |, 1 \leq i \leq n, \mu \in M_0\} < \infty.$$

Thus, it follows from the assumption on f that the functional F is bounded away from $-\infty$. Further, it follows from the lower semi continuity of f on

R^n that F is weak star lower semi continuous on M_0. Hence, weak star compactness of M_0, and weak star lower semi continuity of F, imply that F attains its minimum on M_0. If f is strictly convex on R^n, then F is also strictly convex on M_0, and hence, the minimizer is unique. This completes the proof. ●

Remark 4.5.9 The results presented in this section also hold for the spaces $\{\mathcal{G}_\varphi, \mathcal{G}_\varphi^*\}$ and $\{\mathcal{G}_\pi, \mathcal{G}_\pi^*\}$ related to fractional Brownian motion and the Lévy process respectively. The reader may carry out the details.

The results presented here are applied in the next two sections where we study nonlinear filtering problems.

4.6   Application to Nonlinear Filtering

In this section we consider applications of the vector measure theory of the previous section to nonlinear filtering. We have seen in Chapter 1 that the unnormalized conditional probability measure is governed by the Zakai equation in its weak form,

$$d\mu_t(\varphi) = \mu_t(A\varphi)dt + \mu_t(\varphi\Gamma h) \cdot d\hat{v}(t)$$
$$\mu_0(\varphi) = \nu_0(\varphi), t \in I \equiv [0, T], T < \infty, \qquad (4.12)$$

where φ is any test function such as C^∞ function with compact supports, $x \cdot y \equiv (x, y)$ and the operator Γ is given by $\Gamma(h) \equiv (\sigma_0 R)^{-1}(h)$ and $\hat{v}$ is a d-dimensional Brownian motion (innovation process) with covariance R which is the same as that of the original measurement noise v. See Chapter 1 for details. Without any loss of generality, we will assume that $\hat{v}$ has been normalized so that it is again a standard R^d-valued Brownian motion. We denote the corresponding Wiener measure on $\Omega \equiv C_0(I, R^d)$ by μ^W. Using the Markov semigroup $S(t), t \geq 0$, corresponding to the infinitesimal generator A, equation (4.12) can be written as an integral equation in the weak form

$$\mu_t(\varphi) = \nu_0(S(t)\varphi) + \int_0^t \mu_\tau((S(t-\tau)\varphi)(\Gamma h)) \cdot d\hat{v}(\tau), t \in I, \varphi \in C_0^\infty(R^n).$$
$$(4.13)$$

Let $\hat{\mathcal{F}}_t$ denote the filtration induced by the innovation process $\{\hat{v}(s), 0 \leq s \leq t\}$. It is evident that the measure valued process μ_t is $\hat{\mathcal{F}}_t$ adapted. We show in the following Lemma that for each $t \in I$, it's variation has bounded second moment.

Lemma 4.6.1 *Consider the filter equation (4.12) in the weak form and suppose that*

$$\| \Gamma(h) \|_\infty \equiv \sup\{|\Gamma(h)(x)|_{R^d}, x \in R^n\} < \infty.$$

Then the measure valued process $\{\mu_t\}$ is $\hat{\mathcal{F}}_t$ adapted having finite second moment in the sense that its variation norm $\| \mu_t \|$ has finite second moment, that is,

$$\mathbb{E}\{\| \mu_t \|^2\} < \infty \ \forall \ t \in I.$$

Further, $t \to \mu_t$ is weakly continuous.

Proof We use the integral equation (4.13). It is evident from this equation that μ_t is $\hat{\mathcal{F}}_t$ adapted in the weak sense. Taking the expectation of the square of the absolute value on either side of this equation we obtain the following inequality

$$\mathbb{E}|\mu_t(\varphi)|^2 \leq 2|\nu_0(S(t)\varphi)|^2 + 2 \int_0^t \mathbb{E}|\mu_\tau((S(t-\tau)\varphi)\Gamma(h))|^2 d\tau. \quad (4.14)$$

Now using the fact that $S(t), t \geq 0$, is a contraction semigroup, it follows from the above inequality that

$$\mathbb{E}|\mu_t(\varphi)|^2 \leq 2 \| \nu_0 \|^2 \| \varphi \|_\infty^2 + 2 \| \Gamma(h) \|_\infty^2 \| \varphi \|_\infty^2 \int_0^t \mathbb{E} \| \mu_\tau \|^2 d\tau. \quad (4.15)$$

Hence,

$$\mathbb{E}|\mu_t(\varphi/ \| \varphi \|_\infty)|^2 \leq 2 \| \nu_0 \|^2 + 2 \| \Gamma(h) \|_\infty^2 \int_0^t \mathbb{E} \| \mu_\tau \|^2 d\tau, \ t \geq 0, \quad (4.16)$$

for all $\varphi \in C_0^\infty(R^n)$. Taking the supremum over all $\varphi(\neq 0) \in C_0^\infty(R^n)$, it follows from this inequality that

$$\mathbb{E} \| \mu_t \|^2 \leq 2 \| \nu_0 \|^2 + 2 \| \Gamma(h) \|_\infty^2 \int_0^t \mathbb{E} \| \mu_\tau \|^2 d\tau, \ t \geq 0. \quad (4.17)$$

By use of Gronwall inequality applied to (4.17), we obtain

$$\mathbb{E} \parallel \mu_t \parallel^2 \ \leq 2 \parallel \nu_0 \parallel^2 \exp\{2 \parallel \Gamma(h) \parallel^2_\infty t\}, \ t \in I.$$

Thus $\mathbb{E} \parallel \mu_t \parallel^2 < \infty$, $\forall \ t \in I$. In other words, for each $t \in I$, $\mu_t \in M_{cabv}(\mathcal{B}_n, \mathcal{G})$ where $\mathcal{B}_n$ denotes the Borel algebra of subsets of the space R^n. Weak continuity simply follows from the strong continuity of the semigroup $S(t), t \geq 0$, on the Banach space $BUC(R^n)$ of bounded uniformly continuous functions on R^n, and the integral relation (4.13). This completes the proof. ●

For some applications, we are interested in the terminal value of the measure valued process $\{\mu_t, t \in I\}$. In order to avoid new notations, we use $\mathcal{G}_\beta$ again to denote the space of scalar valued Wiener-Itô functionals of R^d valued Brownian motions. That is, in this case the range space R^ℓ of section 2.3 is replaced by R. Since μ_T is (weakly) $\hat{\mathcal{F}}_T$ measurable, it follows from the above result that

$$\mu_T : \mathcal{B}_n \longrightarrow \mathcal{G}_\beta$$

and so for each $\triangle \in \mathcal{B}_n$, $\mu_T(\triangle) \in \mathcal{G}_\beta$. Hence, by virtue of the isometric isomorphism (see Corollary 2.3.5), there exists a vector measure $K : \mathcal{B}_n \longrightarrow G_\beta$ such that

$$\mu_T(\triangle) = \Phi_\beta(K(\triangle)) = \sum_{m=0}^{\infty} f_m(K_m(\triangle)) \qquad (4.18)$$

where $\{f_m\}$ is the family of orthogonal functionals (homogeneous chaos) introduced in section 2.3 of chapter 2 and $K_m(\triangle) \in \hat{L}_2(I^m, M_m)$ giving the homogeneous functionals

$$f_m(K_m(\triangle)) \equiv \int_{I^m} K_m(\triangle; \tau_1, \tau_2, \cdots, \tau_m) d\hat{v}(\tau_1) \otimes d\hat{v}(\tau_2) \otimes \cdots \otimes d\hat{v}(\tau_m), m \in N.$$

Note that $\{K_m\}$ is also a family of vector measures taking values from the Hilbert space $\hat{L}_2(I^m, M_m)$. Under certain smoothness assumptions on the coefficients of the infinitesimal generator A and the observation operator h,(sufficient differentiability with respect to the spatial variables), it follows from Malliavin calculus [[Bell (1987)], Theorem 7.1, Theorem 7.2], (see also Chapter 8, section 8.7), that the measure μ_T is absolutely continuous with respect to the Lebesgue measure. In this situation μ_T has a density, say, ρ_T and the kernels $\{K_m\}$, which are generally vector valued measures, are given by Lebesgue measurable (tensor valued) functions. In that case the expression (4.18) can be written as

$$\rho_T(x) = \Phi_\beta(K(x)) = \sum_{\ell=0}^{\infty} f_\ell(K_\ell(x)) \tag{4.19}$$

for almost all $x \in R^n$.

In general the measure μ_T may not have density with respect to Lebesgue measure. Absolute continuity of the measure μ_T with respect to Lebesgue measure is usually proved by use of integration by parts formula based on the Malliavin calculus; see Bell [Bell (1987)] for details and also chapter 8. This requires strong assumptions demanding smoothness and uniform boundedness of the drift and diffusion parameters. We will have occasion to discuss this further in more details in Chapter 8.

Now returning to the filtering problem discussed in Chapter 1, it is evident that the performance of the UMV (unbiased minimum variance) filter depends very much on the choice of the sensor-measurement strategy determined by the SDE,

$$dy = h(\xi)dt + \sigma_0 dv, t \geq 0,$$

where ξ is the Markov process with A being its infinitesimal generator. In particular, the map $h : R^n \longrightarrow R^d$ is important; its choice is crucial in the performance of the optimal filter. In many applications, [Ahmed and Charalambos, [Ahmed (2006)], [Ahmed (1988)]], h can be chosen from an admissible set of bounded functions, say,

$$H_{ad} \subset H_r \equiv \{h \in C_b(R^n, R^d) : \sup\{|h(x)|_{R^d}, x \in R^n\} \leq r\}$$

where r is any finite positive number. It is important to indicate this freedom of choice and dependence on h through the filtration $\hat{\mathcal{F}}_t^h$ by using the superscript h. As seen in Chapter 1, given the sensor policy h, for each $\phi \in C_b(R^n)$ it follows from the definition of unbiased minimum variance filter that

$$\mathbb{E}\{\phi(\xi_T)|\hat{\mathcal{F}}_T^h\} \equiv \int_{R^n} \phi(x)Q_T^h(dx),$$

where the conditional probability measure valued process $Q_t^h, t \geq 0$, and the un-normalized measure valued process $\mu_t^h, t \geq 0$, are both $\hat{\mathcal{F}}_t$ adapted and are related through the expression $\mu_t^h(\cdot) = \mu_t^h(R^n)Q_t^h(\cdot)$ for all $t \in I \equiv [0, T]$. An interesting problem is to determine the best measurement strategy, that is, $h \in H_{ad}$, that gives the best unbiased minimum variance filter. Let

$$\psi(T) \equiv \mathbb{E}\{\phi(\xi_T)|\hat{\mathcal{F}}_T^h\}$$

denote the unbiased minimum variance estimate of $\phi(\xi_T)$. The performance of this filter can be measured in terms of the mean square error as follows

$$
\begin{aligned}
J_e(h) &\equiv \mathbb{E}\{(\psi(T) - \phi(\xi_T))^2|\hat{\mathcal{F}}_T^h\} \\
&= \mathbb{E}\{\phi^2(\xi_T)|\hat{\mathcal{F}}_T^h\} - \left(\mathbb{E}\{\phi(\xi_T)|\hat{\mathcal{F}}_T^h\}\right)^2 \\
&= \int_{R^n} \phi^2(x)Q_T^h(dx) - \left(\int_{R^n} \phi(x)Q_T^h(dx)\right)^2. \quad (4.20)
\end{aligned}
$$

Using the un-normalized measure, this can be written as

$$
\begin{aligned}
\tilde{J}_e(\mu^h) &\equiv J_e(h) \\
&= \frac{1}{(\mu_T^h(R^n))^2}\left\{\mu_T^h(R^n)\int_{R^n} \phi^2(x)\mu_T^h(dx) - \left(\int_{R^n} \phi(x)\mu_T^h(dx)\right)^2\right\}.
\end{aligned}
$$

$$(4.21)$$

Clearly, this functional is nonnegative and our objective now is to choose an $h \in H_{ad}$ that minimizes this functional. Strictly speaking, this is a path wise optimization process since μ^h is dependent on the innovation process $\hat{v}$ appearing in the filter equation (4.13). Thus we may consider the average performance rather than the path wise one. Hence, the appropriate performance measure is the average of the above functional on the Wiener measure space $\Omega \equiv C_0(I, R^d)$ furnished with the Wiener measure $W(\cdot)$ induced by the innovation process $\hat{v}$. This is given by

$$C(h) \equiv \int_\Omega \tilde{J}_e(\mu^h)(v)\, W(dv), \quad (4.22)$$

with $\tilde{J}_e(\mu^h)(\cdot)$ denoting the path wise value of $\tilde{J}_e(\mu^h)$. This is the functional that we should consider as the true objective functional and find the observer h that minimizes it.

Lemma 4.6.2 *For each* $\phi \in C_b(R^n)$, *the functional* $\mu \longrightarrow \tilde{J}_e(\mu)$ *given by the expression (4.21) is weakly continuous on the space of bounded positive Borel measures* $\mathcal{M}_b(R^n)$.

Proof Suppose $\{\mu^k\}$ is a sequence of bounded positive measures converging weakly to μ^o. Since ϕ is a bounded continuous function on R^n, it is evident that ϕ^2 is also a bounded continuous function. Thus

$$\lim_{k \to \infty} \int_{R^n} \phi^\gamma(x)\mu^k(dx) \longrightarrow \int_{R^n} \phi^\gamma(x)\mu^o(dx)$$

for $\gamma = 1, 2$. Similarly, $\mu^k(R^n) \equiv \int_{R^n} I(x)\mu^k(dx) \longrightarrow \int_{R^n} I(x)\mu^o(dx)$, where $I(x) \equiv 1$ for all $x \in R^n$. Hence, we conclude that the functional $\mu \longrightarrow \tilde{J}_e(\mu)$ given by the expression (4.21) is weakly continuous. $\bullet$

Now we are prepared to consider the question of existence of an optimal observer that minimizes the filtering error.

Theorem 4.6.3 *Suppose the admissible set* H_{ad} *is compact in the topology* τ_{uck} *of uniform convergence on compact subsets of* R^n. *Then the functional* $h \longrightarrow C(h)$ *is* τ_{uck} *continuous and there exists an* h^o *at which* $C(h)$ *given by (4.22) attains its minimum.*

Proof Let $h_n \xrightarrow{\tau_{uck}} h_o$ and let μ^n and μ^o be the corresponding weak solutions of the Zakai equation (4.12). Using the integral equation,

$$\mu_t(\varphi) = \nu(S(t)\varphi) + \int_0^t \mu_r(S(t-r)\Gamma h\varphi) \cdot dv, t \in I,$$

corresponding to $h = h_n$ and $h = h_o$ respectively, and choosing any test function $\varphi \in C_0(R^n)$ (bounded continuous functions with compact support), it is easy to verify that $\mu^n \xrightarrow{w} \mu^o$. Then, it follows from Lemma 4.6.2 that $\tilde{J}_e(\mu^n) \longrightarrow \tilde{J}_e(\mu^o)$ almost surely with respect to the Wiener measure W. Using the expression (4.21), the reader can easily verify that

$$\tilde{J}_e(\mu)(\cdot) \leq |\phi|^2_\infty \quad W - a.s.$$

Thus, it follows from dominated convergence theorem that

$$\lim_{n \to \infty} C(h_n) = \lim_{n \to \infty} \int_\Omega \tilde{J}_e(\mu^n)(v) \, W(dv) = \int_\Omega \tilde{J}_e(\mu^o)(v) \, W(dv) = C(h_o).$$

Therefore the functional $h \longrightarrow C(h)$ is τ_{uck} continuous on $C_b(R^n, R^d)$. Since $H_{ad} \subset C_b(R^n, R^d)$ is τ_{uck} compact, we conclude that $C(h)$ attains its minimum on H_{ad}. This completes the proof. $\bullet$

We have assumed that the set H_{ad} is compact in the topology of uniform convergence on compacts. For necessary and sufficient conditions for this see [[Willard (1970)], Theorem 42.3, Theorem 43.15].

4.7 Application to Infinite Dimensional Systems

The results presented above can be easily extended to infinite dimensional stochastic systems of the form

$$dx = Axdt + B(x)dt + C(x)dW(t), x(0) = x_0, \tag{4.23}$$

where A is the infinitesimal generator of a C_0-semigroup $S(t), t \geq 0$, on a Hilbert space H, and $B : H \longrightarrow H$ is locally Lipschitz having at most linear growth. The operator $C : H \longrightarrow \mathcal{L}_2(U, H)$ is locally Lipschitz with uniformly bounded Hilbert-Schmidt norm. This is equivalent to the statement that the map $C^*C : H \longrightarrow \mathcal{L}_1(U)$ is nuclear having uniformly bounded nuclear norm. The process W is the cylindrical Brownian motion with values in another separable Hilbert space U. As seen in Chapter 1, section 1.3, under the above assumptions, the process $x \in L_\infty(I, L_2^a(H))$, that is, $x(t)$ is $\mathcal{F}_t$ adapted and $\sup_{t \in I}\{\mathbb{E}|x(t)|_H^2\} < \infty$. Further, $x \in C(I, H)$ with probability one. Thus $x(t)$ is defined for all $t \in I$. In view of this result, for each $t \in I$, $x(t)$ has Wiener-Itô expansion given by $x(t) \equiv \sum_{n \geq 0} J_n(t, K_n)$ where the homogeneous chaos J_n is given by

$$
\begin{aligned}
&J_n(t, K_n(t, \cdot))) \\
&\equiv \int_{I_t^n} K_n(t; \tau_1, \tau_2, \cdots; \tau_n) \otimes dW(\tau_1) \otimes dW(\tau_2) \cdots \otimes dW(\tau_n) \tag{4.24}
\end{aligned}
$$

with Hilbert-Schmidt kernels $K_n(t, \cdot) \in \hat{L}_2(I^n, \mathcal{L}_2(U^{\otimes n}, H))$. That is, for each $t \in I$,

$$\| K_n(t, \cdot)) \|_{HS}^2 \equiv \int_{I_t^n} \| K_n(t; \tau_1, \cdots, \tau_n) \|_{\mathcal{L}_2(U^{\otimes n}, H)}^2 \, d\tau_1 \cdots d\tau_n < \infty.$$

Here, the vector space $\mathcal{K}_{\varpi}$ is given by $\mathcal{K} \equiv \prod_{n \geq 0} \hat{L}_2(I^n, \mathcal{L}_2(U^{\otimes n}, H))$ which may be furnished with a family of seminorms as in section 2.6 (see equation (2.6)). We are interested in the Hilbert space G_{ϖ} as follows:

$$G_{\varpi} \equiv \{K \in \mathcal{K}_{\varpi} : \| K \|_G^2 \equiv \sum_{n \geq 0} n! \| K_n \|_{H.S}^2 < \infty\},$$

with the natural inner product given by

$$(K, L)_{G_\varpi} \equiv \sum_{n \geq 0} (\sqrt{n!} K_n, \sqrt{n!} L_n)_{L_2(I^n, \mathcal{L}_2(U^{\otimes n}, H))}$$

for $K, L \in G_\varpi$. Clearly,

$$G_\varpi = \sum_{n \geq 0} \oplus \sqrt{n!} \; \hat{L}_2(I^n, \mathcal{L}_2(U^{\otimes n}, H)).$$

Again, using the classical Wiener measure space $(C_0, \mathcal{B}_0, \mu^W)$ and the Hilbert space over it $L_2(C_0, \mathcal{B}_0, \mu^W; H) \equiv L_2(\mu^W, H) \equiv \mathcal{G}_\varpi$, one can prove, exactly as in the finite dimensional case, that $G_\varpi \cong \mathcal{G}_\varpi$ which we have denoted by $G_\varpi \leftrightarrow \mathcal{G}_\varpi$. Using the isometric isomorphism Φ_ϖ, we can write

$$x(t) = \Phi_\varpi(K(t)), t \in I.$$

Since $x \in C(I, H)$ almost surely, there exists a $K \in C(I, G_\varpi)$ satisfying the above identity.

The subject of continuous dependence of solutions on certain parameters is important in applications. We have seen in Chapter 1, that if the drift vector is continuously dependent on a parameter $\vartheta \in \Lambda$, where Λ is any Hausdorff space, then, under certain assumptions, the solution $\vartheta \longrightarrow x^\vartheta$ is also continuous. The implication of this result in the orthogonal decomposition is that the representing kernel $K^\vartheta \in G_\varpi$ is also continuous in $\vartheta \in S$. More precisely we have the following result.

Theorem 4.7.1 *Consider the system*

$$dx = Axdt + B(\vartheta, x)dt + C(x)dW(t), x(0) = x_0 \qquad (4.25)$$

and suppose A is the infinitesimal generator of a C_0-semigroup $S(t), t \geq 0$, on a Hilbert space H, $B : \Lambda \times H \longrightarrow H$ is continuous, locally Lipschitz in the second argument having at most linear growth uniformly with respect to $\vartheta \in \Lambda$. The operator $C : H \longrightarrow \mathcal{L}_2(U, H)$ is locally Lipschitz, with uniformly bounded Hilbert-Schmidt norm and the initial state has finite second moment. Then, for each $\vartheta \in \Lambda$ the representing kernel $K^\vartheta \in G_\varpi$, and further $\vartheta \longrightarrow K^\vartheta$ is continuous with respect to the Hausdorff topology on Λ and metric topology on G_ϖ or equivalently, for each $n \in N$, $\vartheta \longrightarrow K_n^\vartheta$ is continuous from Λ to $L_2(I^n, \mathcal{L}_2(U^{\otimes n}, H))$.

Proof By Corollary 1.4.3 of Chapter 1, we have seen that under the given assumptions, if $\vartheta_n \to \vartheta_0$ in Λ, then $x^{\vartheta_n} \to x^{\vartheta_0}$ in $B_\infty^a(I, L_2(\Omega, H))$. Thus it follows from the isomorphism $\mathcal{G}_\varpi \leftrightarrow G_\varpi$, that

$$\| K^{\vartheta_n} - K^{\vartheta_0} \|_{G_\varpi} \to 0$$

as $n \to \infty$. This completes the proof. •

Remark 4.7.2 If Λ is a compact Hausdorff space, the set $\{K^\vartheta, \vartheta \in \Lambda\}$ is a compact subset of G_ϖ and hence under the isomorphism, the set of solutions of the system (4.25) is a compact subset of $\mathcal{G}_\varpi$.

Without going into details, we mention that identical problems on system approximation and inverse problems as presented in sections 4.4 and 4.5 can be formulated for infinite dimensional SDE's like (4.23) without any difficulty. Results are similar.

We present here a result on infinite dimensional Filtering as discussed in Chapter 1, section 1.5. Recall that the filtering equation is given by (1.45) which is reproduced below for convenience of the reader. For each $t \in I$,

$$d\mu_t(\varphi) = \mu_t(\mathcal{A}\varphi)dt < \mu_t(\varphi R_2^{-1}h), dy >, \mu_0(\varphi) = \nu(\varphi), \varphi \in D_0, \quad (4.26)$$

where D_0 is the domain of the operator $\mathcal{A}$. See section 1.5. Using similar arguments, involving change of measure as in the finite dimensional case (see (1.38)), we can rewrite this equation in terms of R^d valued standard Brownian motion called innovation process giving

$$d\mu_t(\varphi) = \mu_t(\mathcal{A}\varphi)dt < \mu_t(\Gamma\varphi), dv >, \mu_0(\varphi) = \nu(\varphi), \varphi \in D_0, \quad (4.27)$$

where the operator Γ is given by the multiplication operator $\Gamma\varphi \equiv R_2^{-1/2}h\varphi$ and $v \equiv R_2^{-1/2}y$ [[Ahmed (1997)], Lemma 2.5, p187]. In general Γ : $B_b(H) \longrightarrow B_b(H, R^d)$ where $B_b(H)$ is the Banach space of Bounded Borel measurable functions on H furnished with the sup norm topology. In Ahmed, Fuhrman and Zabczyk [Ahmed (1997)], an invariant measure μ^o was used to reformulate this measure evolution equation as an evolution equation in the Hilbert space $L_2(\mu^o, H)$. This admits an extension of the associated Markov semigroup to a C_0 semigroup on $L_2(\mu^0, H)$. Then, under certain assumptions, one can take the solution space to be the space of random processes taking values from the Hilbert space $L_2(\mu^0, H)$ and consider the evolution of density in this space governed by equation

$$d\rho(t) = \mathcal{A}^*\rho(t)dt + < \Gamma\rho(t), dv >, \rho(0) = \rho_0, t \in I. \qquad (4.28)$$

The initial state ρ_0 is the Radon-Nikodym derivative of ν with respect to the invariant measure μ^0 and it is assumed that $\rho_0 \in L_2(\mu^0, H)$. Under this assumption, existence (in the weak sense) and uniqueness of a density valued solution was proved in [[Ahmed (1997)], Theorem 3.2, p189].

The natural space for solutions, however, is the space of measure valued processes as considered in [Ahmed (1996)]. We follow this later path. Consider the complete filtered probability space $(\Omega, \mathcal{F}, \mathcal{F}_t, t \geq 0, P)$ which supports the Brownian motion v as introduced above. Let $\mathcal{Z} \equiv L_2(\Omega, P)$ and consider the space of regular bounded finitely additive measures $M_{rba}(\Sigma_H, \mathcal{Z})$ on the Borel sigma algebra Σ_H of the Hilbert space H. Existence of measure valued solutions was proved in [Ahmed (1996)] which we present here without detailed proof. Let $B^a_\infty(I, M_{rba})$ denote the function space of weakly $\mathcal{F}_t$-adapted essentially bounded vector measures.

Theorem 4.7.3 *Suppose* $\Gamma \in \mathcal{L}(B(\Sigma_H), B(\Sigma_H, R^d))$ *and* $S(t), t \geq 0$, *the Markov semigroup associated with the Kolmogorov operator* $\mathcal{A}$. *Then, for each* $\nu \in M_{rba} \equiv M_{rba}(\Sigma_H, \mathcal{Z})$ *equation (4.27) has a unique mild solution* $\mu \in B^a_\infty(I, M_{rba})$. *Further, the solution is continuously dependent on the data* $\{\nu, \Gamma\}$.

Proof For detailed proof see Ahmed [[Ahmed (1996)] Theorem 4.1, p260]. We present a brief outline. Write equation (4.27) as an integral equation in the weak form on the space of measure valued stochastic processes $B_\infty(I, M_{rba})$. This is given by

$$\mu_t(\varphi) = \nu(S(t)\varphi) + \int_0^t < \mu_s(S(t-s)\Gamma\varphi), dv(s) >, t \in I, \varphi \in B(\Sigma_H),$$

$$(4.29)$$

where $S(t), t \geq 0$, is the Markov semigroup with infinitesimal generator $\mathcal{A}$. We formulate this as a fixed point problem in the Banach space $B_\infty(I, M_{rba}(\Sigma_H, \mathcal{Z}))$ and use contraction principle to arrive at the conclusion. This completes our outline. •

Remark 4.7.4 For each $t \geq 0$, define $\mathcal{G}_\beta^t \equiv \mathcal{G}_\beta|_{[0,t]}$ where $\mathcal{G}_\beta^T = \mathcal{G}_\beta$ is the Hilbert space introduced in Chapter 2, section 2.3 with the range space R^ℓ replaced by R. Then, it follows from the above result that for each $t \in I$, $\mu_t \in M_{rba}(\Sigma_H, \mathcal{G}_\beta^t)$. Thus, for the terminal time T, $\mu_T \in M_{rba}(\Sigma_H, \mathcal{G}_\beta)$.

Remark 4.7.5 Note that, in the above theorem we have assumed that Γ is a bounded operator. We can relax this by admitting measure solutions taking values from the space of generalized functionals $\mathcal{G}_\beta^*$. In this case $\mu_T \in M_{rba}(\Sigma_H, \mathcal{G}_\beta^*)$. We discuss this further in Chapter 6.

In Chapter 7, section 7.7, we return to the nonlinear filtering in infinite dimensional space and consider the optimization problem of the observation operator as discussed in section 4.6 for the finite dimensional case.

4.8 Lévy Optimization Problem

In this brief section we consider an optimization problem related to the Lévy (jump) process (centered Poisson random measure $\{q\}$ with the associated Lévy measure Π). This, in turn, is related to the Hilbert space G_Π (see section 2.8 and Theorem 3.7.4) and the corresponding Hilbert space $\mathcal{G}_\Pi$ of square integrable (nonlinear) Lévy functionals.

Let $M_B(Z)$ denote the linear space of bounded regular signed Borel measures on the sigma algebra $\mathcal{B}(Z)$ of subsets of the set Z. Let $LM(Z) \subset M_B(Z)$ denote the class of nonnegative measures which we call Lévy measures on the state space Z.

We introduce the notion of dominating measures in the space $M_B(Z)$.

Definition 4.8.1 *A measure* $\mu_1 \in LM(Z)$ *is said to be dominated by a measure* $\mu_2 \in LM(Z)$ *if for every* $E \in \mathcal{B}(Z)$, $\mu_1(E) \leq \mu_2(E)$. *This is denoted by* $\mu_1 \leq \mu_2$.

It is clear from this definition that

(1) if $\mu \in LM(Z)$ and $\nu \in LM(Z)$, then $\mu + \nu \in LM(Z)$;
(2) if $\mu \in LM(Z)$, then $a\mu \in LM(Z)$ for $a \geq 0$;
(3) if $\mu_1 \leq \mu_2$ and $\mu_2 \leq \mu_3$ then $\mu_1 \leq \mu_3$;
(4) if $\mu_1 \leq \mu_2$ and $\mu_2 \leq \mu_1$ then $\mu_1 = \mu_2$.

Clearly this introduces a partial ordering on the space $M_B(Z)$ turning it into a partially ordered lattice. Clearly, associated with every Lévy measure $\mu \in LM(Z)$ there exists a Hilbert space G_μ characterized by the expression (2.61). The reader can easily verify that for $\mu_1, \mu_2 \in LM(Z)$ satisfying $\mu_1 \leq \mu_2$, we have $G_{\mu_1} \supset G_{\mu_2}$. Thus, this induces a partial ordering in the family of Hilbert spaces $\{G_\mu, \mu \in LM(Z)\}$.

We summarize these results in the following Proposition.

Proposition 4.8.2 *Let $\{\mu_i, 1 \leq i \leq k\} \in LM(Z)$ satisfying the order relation $\mu_1 \leq \mu_2 \leq \cdots \leq \mu_k$. Let G_{μ_i} denote the Hilbert space characterized by the expression (2.61) with the measure Π replaced by μ_i, and $\mathcal{G}_{\mu_i}$ the Hilbert space of square integrable functionals of the Lévy process corresponding to the Lévy measure μ_i. Then the following diagram holds where $\downarrow$ denotes the correspondence and $\updownarrow$ denotes the isometric isomorphism;*

$$
\begin{array}{ccccc}
\mu_1 & \leq & \mu_2 & \leq & \cdots \mu_k \\
\downarrow & & \downarrow & & \cdots \downarrow \\
G_{\mu_1} & \supset & G_{\mu_2} & \supset & \cdots G_{\mu_k} \\
\updownarrow & & \updownarrow & & \cdots \updownarrow \\
\mathcal{G}_{\mu_1} & \supset & \mathcal{G}_{\mu_2} & \supset & \cdots \mathcal{G}_{\mu_k}.
\end{array}
\qquad (4.30)
$$

Proof The proof follows from Theorem 2.8.2, Corollary 2.8.3 and Theorem 3.7.4. ●

Remark 4.8.3 It is well known that for finite measures, if $\mu_1(E) \leq \mu_2(E)$ for every $E \in \mathcal{B}$, then $\mu_1 \prec \mu_2$, that is, μ_1 is absolutely continuous with respect to μ_2 and there exists a $g \in L_1^+(Z, \mu_2)$, the Radon-Nikodym derivative of μ_1 with respect to μ_2, such that $d\mu_1 = g d\mu_2$. Note that $\mu_1 \prec \mu_2$ is not sufficient to guarantee the order relation $G_{\mu_1} \supset G_{\mu_2}$.

Problem Statement (An Inverse Problem): Let $\mathcal{G}$ denote the Hilbert space of square integrable functionals of an unknown Lévy process corresponding to a certain Lévy measure in $LM(Z)$. Let M_0 be any countable subset of $LM(Z)$ and $\{\mathcal{G}_\mu, \mu \in M_0\}$ be the corresponding (countable) family of Hilbert spaces of functionals of the centered Poisson random measure

$\{q^\mu, \mu \in M_0\}$ with mean zero and variance $\ell \times \mu$. Note that ℓ denotes the Lebesgue measure on I, and μ the Lévy measure on Z corresponding to the Lê'vy process q^μ. It is known to the statistician that

$$\bigcup \{\mathcal{G}_\mu, \mu \in M_0\} \equiv \mathcal{G}^o \subset \mathcal{G}.$$

The problem is, given an element $f^o \in \mathcal{G}$, find an $f^* \in \mathcal{G}^o$ that minimizes the functional

$$J(f) \equiv \| f - f^o \|_{\mathcal{G}} . \tag{4.31}$$

In other words, given the (choice) set M_0, find the Lévy measure $\mu \in M_0$ such that the corresponding Hilbert space $\mathcal{G}_\mu$ contains an element f^* closest to the given $f^o \in \mathcal{G}$. This is truly an approximate inverse or identification problem and we call this (AIP).

Theorem 4.8.4 *Consider the identification problem (AIP) as stated above and suppose $\mathcal{G}$ is a Hilbert space of functionals of certain unknown Lévy process and M_0 is any countable subset of the class of Lévy measures $LM(Z)$ satisfying*

$$\bigcup \{\mathcal{G}_\mu, \mu \in M_0\} \equiv \mathcal{G}^0 \subset \mathcal{G}.$$

Then the problem (AIP) has a solution.

Proof Take any $\mu \in M_0$ and consider the corresponding Hilbert space $\mathcal{G}_\mu$. Let $\mathcal{G}_\mu^o$ denote the strong closure of $\mathcal{G}_\mu$ with respect to the norm topology of $\mathcal{G}$, that is ,

$$\mathcal{G}_\mu^o \equiv \overline{\mathcal{G}}_\mu = s - cl \ \mathcal{G}_\mu.$$

Now considering the functional J given by (4.31) on the closed linear subspace $\mathcal{G}_\mu^o$ of $\mathcal{G}$ and denoting it by $J_\mu(\cdot)$ we prove that it attains its infimum on $\mathcal{G}_\mu^o$. Let $\{f_n\} \in \mathcal{G}_\mu$ be a minimizing sequence. Clearly, since J_μ is non-negative and radially unbounded, the minimizing sequence is contained in a bounded subset of $\mathcal{G}$. Since $\mathcal{G}$ is a Hilbert space, there exists a subsequence of the sequence $\{f_n\}$, relabeled as the original sequence, and an element $f^\mu \in \mathcal{G}$ such that $f_n \xrightarrow{w} f^\mu$. Since $\mathcal{G}_\mu^o$ is a closed subspace and so a convex subset of $\mathcal{G}$, by Mazur's theorem it is weakly closed and therefore $f^\mu \in \mathcal{G}_\mu^o$. Thus

$$0 \leq m_\mu \equiv J_\mu(f^\mu) \leq J_\mu(f) \ \forall \ f \in \mathcal{G}_\mu^o. \tag{4.32}$$

Since M_0 is countable it is clear that $m \equiv \inf\{m_\mu, \mu \in M_0\}$ exists and hence there exists a $\mu^o \in M_0$ and $f^* \in \mathcal{G}^o_{\mu^o}$ such that

$$m = \inf\{J_{\mu^o}(f), f \in \mathcal{G}^o_{\mu^o}\} = J_{\mu^o}(f^*) \le J_\mu(f^\mu), \mu \in M_0.$$

This proves that the problem (AIP) has a solution $\mu^o \in M_0$. •

Remark 4.8.5 The reader may have noticed that $f^* = f^{\mu^o} \in \mathcal{G}^o_{\mu^o}$ the strong closure of $\mathcal{G}_{\mu^o}$. In the other words f^* may not be an element of $\mathcal{G}_{\mu^o}$. But since $\mathcal{G}_{\mu^o}$ is dense in $\mathcal{G}^o_{\mu^o}$, it can be approximated by an element from $\mathcal{G}_{\mu^o}$ to any degree of accuracy. This is what is expected from the solution of the approximation problem (AIP).

4.9   Some Problems for Exercise

P1: Compare Theorem 4.2.1 with the classical result on conditional compactness for bounded sets in the space $L_2(R^n, \lambda)$. Hints: Consider the set $A \subset L_2(R^n, \lambda)$ where λ is the standard Lebesgue measure. The set A is conditionally compact if, and only if, (1): A is bounded, (2): $\lim_{h \to 0} \int_{R^n} |f(x+h) - f(x)|^2 \lambda(dx) \longrightarrow 0$ uniformly with respect to $f \in A$, (3): $\lim_{r \to \infty} \int_{B'_r(R^n)} |f(x)|^2 \lambda(dx) \longrightarrow 0$ uniformly with respect to $f \in A$, where $B'_r(R^n)$ denotes the complement of the closed ball of radius r in R^n.

P2: Consider the normalized duality map $J : G_\alpha \longrightarrow G^*_\alpha$ defined by

$$J(K) \equiv \{L \in G^*_\alpha :< L, K >_{G^*_\alpha, G_\alpha} = \| K \|^2_{G_\alpha} = \| L \|^2_{G^*_\alpha}\}.$$

(a): Show that L, given by $L \equiv \{L_n, n \in N\} \equiv \{n!K_n, n \in N\}$, is in $J(K)$ where $K = \{K_n, n \in N\} \in G_\alpha$.

(b): Define $\varphi(K) = (1/2) \| K \|^2_{G_\alpha}$. Verify that $\partial\varphi(K) = J(K)$ and that the duality map $J(K)$ is single valued.

P3: Characterize the bidual of G_α or equivalently the dual of G^*_α (the space of continuous linear functionals on G^*_α).

P4: Refer to example **(E1)** of section 4.4 and consider the approximation problem (4.3) on the closed ball $B_r(\mathcal{G}) \subset \mathcal{G}$ of radius $r > 0$. Prove the existence and uniqueness of a minimizer.

P5: Consider the example **(E2)** giving the input-output map F. Prove the inequality given by (4.6).

P6: Consider the ∞-dimensional filter equation (generalized Zakai equation) (4.27) in its integral form given by equation (4.29). Let $\Xi \subset \mathcal{L}(B(\Sigma_H), B(\Sigma_H, R^d))$. Find the weakest topology on $\mathcal{L}(B(\Sigma_H), B(\Sigma_H, R^d))$ and sufficient conditions on the set Ξ (endowed with the relative topology) that guarantees almost sure weak compactness of the family of vector measures $\{\mu_T^\Gamma, \Gamma \in \Xi\}$. Hints: Here we make use of Theorem 4.5.7.

P7: Refer to problem **(P6)** and let $f \in C_b(R^m)$ be bounded away from $-\infty$. Prove the existence of an operator $\Gamma^o \in \Xi$ such that

$$J(\Gamma) \equiv \mathbb{E}\left\{ f(\mu_T^\Gamma(\varphi_1), \mu_T^\Gamma(\varphi_2), \cdots, \mu_T^\Gamma(\varphi_m)) \right\}$$

attains its minimum at Γ^o, where $\{\varphi_i\} \in BC(H)$.

P8: Let $\Psi : [0, \infty] \longrightarrow [0, \infty]$ be a continuous monotone non decreasing function satisfying $\Psi(0) = 0$ and $\lim_{s \to \infty} \Psi(s) = \infty$. For a given $f^o \in \mathcal{G}$ consider the functional $J(f) \equiv \Psi(\| f - f^o \|_{\mathcal{G}})$ on $\mathcal{G}^0$. Prove that for every $\varepsilon > 0$ there exists an $f^* \in \mathcal{G}^0$ such that

$$J(f^*) \leq \inf\{J(f), f \in \mathcal{G}^0\} + \varepsilon.$$

Hint: Follow the notations and the approach used in the proof of Theorem 4.8.4 related to the inverse problem (AIP).

P9: Let $(\Omega, \mathcal{F}, P)$ be a probability space and suppose $\{q_k, q_o\}$ is a sequence of centered Poisson random measures on $\mathcal{B}(I \times Z)$ with mean measure given by the product of Lebesgue measure and the sequence of Lévy measures $\{\pi_k, \pi_o\}$ respectively. Suppose $q_k \to q_0$ in the vague topology with probability one in the sense that for every bounded measurable function η on $I \times Z$,

$$\int_{I \times Z} \eta(t, z) q_k(dt \times dz) \to \int_{I \times Z} \eta(t, z) q_o(dt \times dz)$$

with probability one. Define the Poisson chaos $G_n(q)$ by

$$G_n(q) \equiv \int_{(I \times Z)^n} K_n(t_1, z_1; \cdots; t_n, z_n) q(dt_1 \times dz_1) \cdots q(dt_n \times dz_n).$$

Develop sufficient conditions for the kernel K_n such that

$$G_n(q_k) \to G_n(q_o)$$

in the mean square sense as $k \to \infty$.

Chapter 5

L_2-Based Generalized Functionals of White Noise II

5.1 Introduction

In this Chapter, we present some of the pioneering results of Hida [Hida (1978)], [Hida (1978, 1980)] on generalized functionals (or equivalently distributions) of Brownian motion. It is known that the derivative of the Wiener process does not exist in the classical sense, but in the sense of distribution it has derivatives of all orders. Let $\mathcal{D}$ denote the space of C^{∞} functions on I with compact support and $\mathcal{D}^*$ the space of Schwartz distributions on I. Then we can consider the white noise or the derivative of the Wiener process $\dot{w}$ as a random distribution in the sense that it's action on $\mathcal{D}$ is given by

$$\dot{w}(\phi) = (-1) \int_I w(t)\dot{\phi}(t)dt \quad \text{for all} \ \ \phi \in \mathcal{D}. \tag{5.1}$$

Note that $\dot{w}(\phi)$ is a well defined Gaussian random variable, parameterized by $\phi \in \mathcal{D}$, with mean zero and covariance $\| \phi \|^2 \equiv \int_I |\phi(t)|^2 dt$. In fact the n-th derivative of w, denoted by $w^{(n)}$, is also a random distribution and

$$w^{(n)}(\phi) = (-1)^n \int_I w(t)\phi^{(n)}(t)dt \quad \text{for all} \ \ \phi \in \mathcal{D}. \tag{5.2}$$

Clearly, this is also a Gaussian random variable, parameterized by $\phi \in \mathcal{D}$, with mean zero and variance $\int_I |\phi^{(n-1)}(t)|^2 dt$. Thus, the derivatives of the Wiener process are actually stochastic distributions in the sense of Schwartz and we may call them generalized random processes. It is easy to see that for any fixed $s \in (0, T)$, $f_s(w) \equiv \dot{w}(s)$ is an unbounded linear functional of Brownian motion and not an element of $L_2(\Omega, \mathcal{B}, \mu)$, where $(\Omega, \mathcal{B}, \mu)$ is any classical Wiener measure space. Hence, it is not a regular functional. Similarly, $F_s(w) \equiv (\dot{w}(s))^n$ is a homogeneous functional of white noise of degree n but it is not an element of $L_2(\Omega, \mathcal{B}, \mu)$. This certainly justifies the need of generalized functionals and we shall see later that the functionals mentioned above are well defined generalized functionals.

5.2   Characteristic Function of White Noise

According to standard definition, the characteristic function of the white noise is given by

$$\mathbb{E}\{\exp(i\ \dot{w}(\phi))\} \equiv \exp -\{(1/2)\int_I |\phi(t)|^2 dt\}. \tag{5.3}$$

Define the functional

$$C(\phi) \equiv \exp -\{(1/2)\int_I |\phi(t)|^2 dt\}.$$

It is clear that the function C is also well defined on the Hilbert space $X \equiv L_2(I)$, that is, it is continuous and positive definite on X with $C(0) = 1$. However, there exists no countably additive measure μ on $\mathcal{B}(X)$ that satisfies the relation,

$$C(\phi) = \int_H e^{i\langle x,\phi\rangle} d\mu(x), \phi \in X. \tag{5.4}$$

A necessary and sufficient condition for countable additivity is provided by Minlos-Sazanov theorem (generalizing Bochner's theorem) [Gihman (1971)](p342), which states that the measure is countably additive if, and only if, the corresponding covariance operator is nuclear. Thus, if we consider the functional C on the Sobolev space H_0^1 then the white noise has a countably additive measure on the dual $(H_0^1)^* = H^{-1}$. In this case

$$C(\phi) \equiv \int_{H^{-1}} \exp\{i < x, \phi >\} d\mu(x), \tag{5.5}$$

where $< x, \phi > \equiv < x, \phi >_{H^{-1}, H_0^1}$ denotes the duality pairing between the elements of H^{-1} and H_0^1. We can justify this as follows. First, note that

$$\mathbb{E}(< \dot{w}, \phi >)^2 = \int_{I \times I} (t \wedge \tau)\dot{\phi}(t)\dot{\phi}(\tau) dt d\tau \equiv < R\phi, \phi > . \tag{5.6}$$

Clearly, the kernel of this integral operator given by $K(t, \tau) \equiv t \wedge \tau$ is symmetric and positive. Since $\int_I K(t, t)dt = T^2/2 < \infty$, the corresponding linear operator Q given by $Q\xi(t) \equiv \int_I K(t, \tau)\xi(\tau)d\tau$ is a nuclear operator in the Hilbert space $X = L_2(I)$. Indeed, let $\{e_i\}$ be any complete orthonormal basis of the Hilbert space X. Then, for any $n \in N$, it follows from simple computation using Fubini's theorem that

$$\sum_{i=1}^{n}(Qe_i, e_i)_X = \sum_{i=1}^{n}\int_{I^2} K(t, s)e_i(t)e_i(s)dtds$$

$$= \sum_{i=1}^{n}\mathbb{E}(w, e_i)^2 = \mathbb{E}\{\sum_{i=1}^{n}(w, e_i)^2\}.$$

Since, for a finite interval I, $Prob.\{w \in X\} = 1$ it follows from the above identity that for every $n \in N$, we have

$$\sum_{i=1}^{n}(Qe_i, e_i)_X = \mathbb{E}\{\sum_{i=1}^{n}(w, e_i)^2\} \leq \mathbb{E} \parallel w \parallel_X^2 = T^2/2.$$

Hence, Q is a nuclear operator in X and consequently the covariance operator R, as defined by (5.6), is a nuclear operator from H_0^1 to H^{-1}. Similarly, one can verify that the m-th order distributional derivative $w^{(m)}$ of the Wiener process w induces a countably additive Gaussian measure on H^{-m} which is the dual of H_0^m. Here, we are interested only in the white noise since the Wiener's multiple integrals are functionals of the white noise. Thus, to capture the support of the white noise measure, we must choose a larger space, larger than X, for example the dual S^* of a nuclear space $S \subseteq H_0^1$ so that

$$\mathcal{D} \hookrightarrow S \hookrightarrow X \hookrightarrow S^* \hookrightarrow \mathcal{D}^*.$$

Let $(S^*, \mathcal{B}(S^*), \mu)$ denote the measure space with μ denoting the white noise measure on $\mathcal{B}(S^*)$ and let $M(S^*, \mathcal{B}(S^*), \mu) \equiv M(S^*, \mu)$ denote the class of μ-measurable functions on S^* and $L_2(S^*, \mu)$ the Hilbert space of μ-measurable functions on S^* which are square-integrable with respect to the measure μ. For a detailed treatment of this topic see the excellent books by Hida [Hida (1978)]-[Hida (1980)] and, Hida and Si [Hida (2008)].

We shall use $F(S^*, \mu)$ to denote the universal space of generalized functionals of white noise. That is, these functionals are not necessarily $\mathcal{B}(S^*)$ measurable, nor are they defined point wise. They are generally distributions and are characterized by their actions on suitable test functions.

5.3   Multiple Wiener-Itô Integrals

In view of the above discussion, the multiple Wiener-Itô integrals can also be written as

$$g_n(\zeta) = \int_{I^n} K_n(\tau_1, \cdots, \tau_n)\zeta(\tau_1), \cdots, \zeta(\tau_n)d\tau_1 \cdots d\tau_n, \quad \zeta \in S^*. \quad (5.7)$$

In other words, these are the multiple white noise integrals considered as measurable functions on S^*. As stated in the introduction, Hida constructed a large class of generalized functionals by introducing a Fourier like transform on the space $L_2(S^*, \mu)$ as follows:

$$(\mathcal{F}f)(\xi) \equiv \int_{S^*} f(\zeta) \exp\{i < \zeta, \xi >\}d\mu(\zeta), \xi \in S, \quad (5.8)$$

where the duality pairing in (5.8) is between S^* and S. For convenience of notation, we shall denote this transform by $\hat{f}(\xi) \equiv (\mathcal{F}f)(\xi)$. It is clear from this definition that for $f \in L_2(S^*, \mu)$, $\hat{f}(\xi)$ is a bounded analytic function on S. Indeed for $f = g_n$, it is not difficult to verify that

$$\hat{g}_n(\xi) = (i)^n C(\xi) \int_{I^n} K_n(\tau_1, \cdots, \tau_n)\xi(\tau_1), \cdots, \xi(\tau_n)d\tau_1 \cdots d\tau_n, \quad (5.9)$$

where $\| \xi \|$ is the $L_2(I)$ norm of ξ. The proof is easily obtained for simple functions K_n by keeping in mind the fact that the increments of Brownian motion on disjoint intervals are independent Gaussian random variables with mean zero and variance equal to the lengths of the intervals. For general $K_n \in L_2(I^n)$, one then uses the density argument and the fact that simple functions are dense in L_2. A much simpler technique is to use the characteristic functional of the white noise measure μ given by

$$C(\xi) \equiv \hat{\mu}(\xi) \equiv \int_{S^*} e^{i<z,\xi>}\mu(dz) = e^{-(1/2)\|\xi\|^2} \quad (5.10)$$

and the fact that

$$\int_{S^*} z \; e^{i<z,\xi>}\mu(dz) = (-i)D_\xi C(\xi),$$

where D_ξ denotes the Frèchet derivative of $C(\xi)$. Clearly it follows from (5.9) that

$$\hat{g}_n(\xi) \equiv (\mathcal{F}g_n)(\xi) = g_n(i\xi)C(\xi). \quad (5.11)$$

Thus, for $\phi(x) \equiv \sum_{n=0}^{\infty} g_n(x)$, we have

$$\hat{\phi}(\xi) = C(\xi) \sum_{n=0}^{\infty} g_n(i\xi) = C(\xi)\phi(i\xi). \tag{5.12}$$

It is well known that the standard Fourier transform is an isometric isomorphism between any two L_2 spaces over finite dimensional spaces like R^d for $d < \infty$. Since S^* is different from S and also they are infinite dimensional, the Fourier-Hida transform is not an isometry. However, it follows from the above relations that it is an isomorphism. Indeed if $\hat{g}_n(\xi) = 0$ for all $\xi \in S$ then $g_n(i\xi) = 0$ for all $\xi \in S$ and hence $K_n \equiv 0$. Thus, for any $\phi \in L_2(S^*, \mu)$, $\hat{\phi}(\xi) = 0$ for all $\xi \in S$ implies that $\phi = 0$ or equivalently $K(\in G) = 0$. The converse is obvious.

5.4   Generalized Hida-Functionals

Now we are prepared to introduce the generalized functionals constructed by Hida [Hida (1978)], [Hida (1978, 1980)], [1978-1980]. In this section we take $I^n = R^n$. Define

$$\mathcal{H}_n \equiv \{g_n(K_n) \equiv g_n(K_n, \cdot), K_n \in \hat{L}_2(I^n)\} \tag{5.13}$$

where, for any fixed $K_n \in \hat{L}_2(I^n)$, $g_n \in L_2(S^*, \mu)$ is a homogeneous functional of white noise of degree n as defined by the expression (5.3.7). Then define

$$\mathcal{F}_n \equiv \mathcal{F}(\mathcal{H}_n). \tag{5.14}$$

This is a linear vector space isomorphic to $\mathcal{H}_n$. For any nonnegative integer $n \in N$, we use $\alpha \equiv (\alpha_1, \alpha_2, \cdots, \alpha_n)$ to denote the multi index with $\alpha_i (\geq 0) \in N$ and norm $|\alpha| = \sum \alpha_i$. Let $D^\alpha \equiv D^{\alpha_1} D^{\alpha_2} \cdots D^{\alpha_n}$ with $D^{\alpha_i} \equiv \partial^{\alpha_i}/\partial t_i^{\alpha_i}$ denote the differential operator of degree $|\alpha|$. For any integer $m \geq 0$, let

$$H^m \equiv \{\phi \in \hat{L}_2(R^n) : D^\alpha \phi \in L_2(R^n), |\alpha| \leq m\}$$

denote the Sobolev space of real valued symmetric functions on R^n with the usual norm topology

$$\| \phi \|_{H^m} \equiv \Big(\sum_{|\alpha| \leq m} \| D^\alpha \phi \|_{L_2(R^n)}^2 \Big)^{1/2}.$$

As usual, we let $\hookrightarrow$ denote the injection (embedding) and $\updownarrow$ denote isomorphism (not isometry, only for this chapter). For $m = m(n) \equiv ((n+1)/2)$, it is known from Sobolev embedding theorem [Adams (1975)] that the injection $H^{m(n)} \hookrightarrow C(R^n)$ is continuous. This is the correct number for construction of generalized functionals as seen below. Since the dual of $H^{m(n)}$ is $H^{-m(n)}$ it is clear that

$$H^{m(n)} \hookrightarrow \hat{L}_2(R^n) \hookrightarrow H^{-m(n)} \tag{5.15}$$

where the injections are continuous and dense. Define

$$\mathcal{H}_n^{m(n)} \equiv \{g_n(K_n, \cdot) \in L_2(S^*, \mu) : K_n \in H^{m(n)}\},$$
$$\mathcal{F}_n^{m(n)} = \mathcal{F}(\mathcal{H}_n^{m(n)}) \text{ and}$$
$$\mathcal{F}_n \equiv \mathcal{F}(\mathcal{H}_n), \tag{5.16}$$

where $\mathcal{H}_n$ is given by (5.13). Identifying $\mathcal{F}_n$ with its topological dual $\mathcal{F}_n^*$, it follows from (5.15) that

$$\mathcal{F}_n^{m(n)} \hookrightarrow \mathcal{F}_n \hookrightarrow \mathcal{F}_n^{-m(n)}. \tag{5.17}$$

The following result is fundamental and it was due to Hida [Hida (1978)].

Theorem 5.4.1 (Hida) *Under the above assumptions, for each $n \in N$, the following diagram holds:*

$$\begin{array}{ccc}
\mathcal{F}_n^{m(n)} \hookrightarrow \mathcal{F}_n \hookrightarrow \mathcal{F}_n^{-m(n)} \\
\updownarrow \qquad \updownarrow \qquad \updownarrow \\
\mathcal{H}_n^{m(n)} \hookrightarrow \mathcal{H}_n \hookrightarrow \mathcal{H}_n^{-m(n)}.
\end{array} \tag{5.18}$$

Proof The Fourier-Hida transform is an isomorphism (not an isometry). It's inverse is well defined on both $\mathcal{F}_n^{m(n)}$ and $\mathcal{F}_n$. Since the embeddings in (5.15) are continuous and dense, again by the principle of extension by continuity, the inverse of the Fourier-Hida transform can be extended to $\mathcal{F}_n^{-m(n)}$. Thus by application of the inverse transform one obtains the last column of (5.18) giving the result as indicated by the diagram.$\bullet$

Recall that $\mathcal{H}_n$ is the space of regular functionals of degree n of white noise (S^*, μ) while the elements of $\mathcal{H}_n^{-m(n)}$ are generalized functionals thereof.

For example, consider the functional $\hat{\Psi}$ on S given by

$$\tilde{\Psi}(\xi) \equiv \int_{R^s} K_s(\vartheta_1,\cdots,\vartheta_s)(\xi(\vartheta_1))^{n_1}(\xi(\vartheta_2))^{n_2}\cdots(\xi(\vartheta_s))^{n_s}d\vartheta_1\cdots d\vartheta_s;$$

(5.19)

where $\sum_{i=1}^{s} n_i = n$. Clearly, $\tilde{\Psi}$ is a homogeneous functional of degree n on S but it is not an element of $\mathcal{F}_n$. It is however an element of $\mathcal{F}_n^{-m(n)}$. Thus by the Fourier-Hida inverse transform we have $\Psi \equiv \mathcal{F}^{-1}\big((i)^n C\tilde{\Psi}\big) \in \mathcal{H}_n^{-m(n)}$, where C is the characteristic function of white noise as introduced earlier given by the expression (5.10).

Hence, the functionals in the class $\mathcal{H}_n^{-m(n)}$ have the same integral form as (5.19) with the polynomials in $\xi \in S$ replaced by suitable polynomials in $x \in S^*$ but of the same degrees. Consider the homogeneous generalized functional of white noise of degree n given by

$$g(x) \equiv \int_{R^s} K_s(\vartheta_1,\cdots,\vartheta_s)(x(\vartheta_1))^{n_1}(x(\vartheta_2))^{n_2}\cdots(x(\vartheta_s))^{n_s}d\vartheta_1\cdots d\vartheta_s.$$

(5.20)

Clearly, this is not a regular Wiener-Itô functional as considered in the previous chapters. In fact, it belongs to the class $\mathcal{H}_n^{-m(n)}$. It is not difficult to show that its Fourier-Hida (FH) transform is given by

$$\hat{g}(\xi) = \kappa_s(\xi)\int_{I^s} K_n(t_1,t_2,\cdots,t_s)\left(\prod_{k=1}^{s} H_{n_k}(\xi(t_k))\right)dt_1 dt_2\cdots dt_s$$

(5.21)

where $\kappa_s(\xi) = (i)^n\left(\Pi_{k=1}^s(n_k)!\right)C(\xi)$ and $H_n(\zeta), n \geq 0$, is the classical Hermite polynomial (of degree n) given by

$$H_n(\zeta) = \frac{(-1)^n}{n!}e^{\zeta^2/2}\left(\frac{d^n}{d\zeta^n}e^{-\zeta^2/2}\right).$$

To verify the identity (5.21), use the expression for the characteristic functional $C(\xi)$ of the white noise measure given by the equation (5.10), differentiate it n_i times with respect to $\{\xi(t_i)\}$ on either side and multiply the resulting expressions with the kernel $K_n(t_1,t_2,\cdots,t_s)$ and then integrate. It is clear from the expression (5.21) that whenever g is a homogeneous functional of degree n on the dual space S^*, its Fourier-Hida transform is a homogeneous functional of degree n on the predual S.

Now we wish to extend the above result to infinitely many variables. Recall that the embedding $H^{m(n)} \hookrightarrow L_2(R^n)$ is continuous and the embedding constant is ≤ 1. We introduce a norm topology on the vector space $\mathbf{H}^+ = \sum \bigoplus_{n=0}^{\infty} \mathcal{H}_n^{m(n)}$ as follows. For $\phi \in \mathbf{H}^+$ we write

$$\| \phi \|_{\mathbf{H}^+}^2 \equiv \sum_{n=0}^{\infty} n! \, \| K_n \|_{H^{m(n)}}^2 .$$

Similarly, for $\mathbf{H} \equiv \sum \bigoplus_{n=0}^{\infty} \mathcal{H}_n$, we have already the norm topology derived from

$$\| \phi \|_{\mathbf{H}}^2 \equiv \sum_{n=0}^{\infty} n! \, \| K_n \|_{L_2(R^n)}^2 .$$

Letting $\mathbf{H}^-$ denote the dual of $\mathbf{H}^+$ we have the following result.

Theorem 5.4.2 *Under the assumptions of Theorem 5.4.1, and the norm topologies as introduced above for the spaces $\mathbf{H}^+, \mathbf{H}$, and $\mathbf{H}^-$ and identifying $\mathbf{H}$ with it's own dual, we have the following embeddings*

$$\mathbf{H}^+ \hookrightarrow \mathbf{H} \hookrightarrow \mathbf{H}^- \tag{5.22}$$

which are continuous and dense. •

Define

$$\mathcal{F}^+ \equiv \sum_{n=0}^{\infty} \bigoplus \mathcal{F}_n^{m(n)}, \mathcal{F} \equiv \sum_{n=0}^{\infty} \bigoplus \mathcal{F}_n, \mathcal{F}^- \equiv \sum_{n=0}^{\infty} \bigoplus \mathcal{F}_n^{-m(n)}.$$

Then it follows from Fourier-Hida transform and Theorem 5.4.2 that the following result holds.

Corollary 5.4.3 *The following diagram holds with continuous and dense embeddings*

$$\mathcal{F}^+ \hookrightarrow \mathcal{F} \hookrightarrow \mathcal{F}^-$$

$$\updownarrow \quad \updownarrow \quad \updownarrow$$

$$\mathbf{H}^+ \hookrightarrow \mathbf{H} \hookrightarrow \mathbf{H}^-, \tag{5.23}$$

where $\updownarrow$ indicates isomorphism (not isometry)

Remark 5.4.4 For more on Hida distributions see the book [Si (2011)].

5.5 Application to Quantum Mechanics

In this section we present a brief account of application of white noise measure. One of the most interesting applications of Brownian motion is found in quantum mechanics. For example, the Feynman-Kac formula solves the heat equation with a potential (cooling). Consider the Brownian motion $w(t), t \geq 0$, with values in R^d starting from $w(0) = x \in R^d$. Let $u_0, V \in C_b(R^d)$ and consider the function given by

$$u(t,x) \equiv \mathbb{E}\left\{ u_0(w(t))e^{-\int_0^t V(w(s))ds} \big| w(0) = x \right\}, t \geq 0. \qquad (5.24)$$

Using Itô formula applied to the process $r \longrightarrow u(t, w(r)), r \in [0, t]$, it is easy to verify that u solves the heat equation

$$\partial u/\partial t = (1/2)\Delta u - Vu,$$
$$u(0, x) = u_0(x), t \geq 0, x \in R^d. \qquad (5.25)$$

In general, the Laplacian can be replaced by any elliptic differential operator and the Brownian motion by the corresponding Markov process.

The interesting point to learn from this is that the solution of the classical heat equation can be obtained by averaging a suitable functional over the Brownian path space. More precisely, let $C_x([0, t], R^d)$ denote the space of continuous functions with values in R^d starting from $x \in R^d$. Define the functional

$$F(t, \xi) \equiv u_0(\xi(t))e^{-\int_0^t V(\xi(s))ds}$$

on the path space $C_x([0, t], R^d)$. Let μ^w denote the Wiener measure on C_x. Then

$$u(t, x) = \int_{C_x} F(t, \xi)\mu^w(d\xi). \qquad (5.26)$$

Thus, in principle, the solution of the heat equation can be obtained by Monte Carlo simulation which essentially computes the average over a very large number of sample paths using fast computers. Particularly, for $V = 0$, this gives the spatial distribution of temperature as it evolves with time starting from the initial distribution u_0.

Informally, the Schrödinger equation can be obtained by simply chang-
ing the real time to imaginary time, $t \longrightarrow it$. Consider the Schrödinger
equation for $m = 1, \hbar = 1$ given by

$$i\partial\psi/\partial t = -(1/2)\Delta\psi + V\psi$$
$$\psi(0, x) = \psi_0(x). \tag{5.27}$$

Then the solution ψ is given by

$$\psi(t, x) = E\left\{\psi_0(\sqrt{i}\ w(t))e^{-i\int_0^t V(\sqrt{i}\ w(s))ds}|w(0) = x\right\}$$
$$= \int_{C_x} F(t, \sqrt{i}\xi)\mu^w(d\xi). \tag{5.28}$$

Historically, these results came after the great breakthrough in quantum
mechanics due to Feynman [Feynman (1948)]. In essence, Feynman demonstrated that the solution of the Schrödinger's equation is a result of the integral of a suitable action functional over the path space. It is well known
that every mechanical system possesses an action functional, generally the
integral of the difference between the kinetic and the potential energies.
And its motion is determined by De Alembert's principle of least action.
The action functional is given by

$$S(x) \equiv \int_I L(x(s), \dot{x}(s))ds, I \equiv [t_0, t], t \geq t_0, \tag{5.29}$$

where L is the Lagrangian. Feynman's postulate is that the probability
amplitudes for all trajectories are equal in modulus, but their phases are
determined by the action in units of $\hbar$. The total probability amplitude
$a(t, x)$, for a particle to move from (t_0, x_0) to (t, x) in the phase space, is
given by the sum (!) over all possible paths connecting (t_0, x_0) to (t, x).
Hence

$$a(t, x) \equiv \int_{\mathcal{P}(t,x)} e^{(i/\hbar)S(\zeta)}D\zeta \tag{5.30}$$

where $\mathcal{P}(t, x)$ denotes all possible continuous paths in the phase space connecting (t_0, x_0) to (t, x) and $D\zeta$ denotes the infinitesimal volume element
in the path space (Feynman measure). This is what is known as the Feynman's path integral, also called the flat integral. The Feynman measure
appears to be treated as Lebesgue measure which is translation invariant.

Unfortunately, there is no such (translation invariant) measure in infinite dimensional spaces. However, heuristic arguments used by physicists show [Sudbery (1986)] that $a(t, x) = \psi(t, x)$ giving the solution of Schrödinger equation.

In recent years, Hida and his school [Hida (1993)] have proposed to justify the Feynman flat measure by use of the white noise measure. As we have seen in section 5.2, the correct space that supports white noise measure is $\mathcal{S}^*$, the dual of the nuclear space $\mathcal{S}$. Thus, the action functional S should be a functional on $\mathcal{S}^*$ and may be taken as

$$S(\dot{x}) \equiv \int_I L\left(\dot{x}(s), \int_{t_0}^s \dot{x}(r)dr\right)ds. \tag{5.31}$$

Let $Q(t, x)$ denote the closure of the set $\{\zeta \in L_2(I) : x = x_0 + \int_{t_0}^t \zeta(\tau)d\tau\}$ with respect to the $\mathcal{S}^*$-topology on $L_2(I)$. Then the Feynman integral for the probability amplitude should be given by

$$a(t, x) = \int_{Q(t,x) \subset \mathcal{S}^*} e^{-(i/\hbar)S(\zeta)} \mu(d\zeta) \tag{5.32}$$

with μ being the white noise measure and $S(\zeta)$ being a generalized functional defined on $\mathcal{S}^*$. Since the white noise $\dot{w}(t), t \geq 0$, is an indexed family of independent generalized Gaussian random variables for all $t \geq 0$, this seems to provide some justification for the Feynman measure, not on the path space but rather, on the space of velocity fields. Hida and his school [Hida (1993)] propose to use (5.31) and (5.32) along with the Fourier-Hida transform given by (5.8). Consider the Feynman's integrand $F(\zeta) \equiv e^{-(i/\hbar)S(\zeta)}$ defined on $\mathcal{S}^*$. This is a generalized functional and its Fourier transform is computed as

$$\hat{F}(\xi) \equiv \int_{Q(t,x)} F(\zeta)e^{i<\zeta,\xi>_{\mathcal{S}^*,\mathcal{S}}} \mu(d\zeta).$$

The Feynman's path integral is then given by $a(t, x) = \hat{F}(0)$. This is the essence of the technique used by Hida and his school. The emphasis here is the space $\mathcal{S}^*$ and the white noise measure on it. Clearly, for the existence of $\hat{F}$ it is necessary that $F \in L_1(\mathcal{S}^*, \mu)$ for the given action functional $S(\zeta), \zeta \in \mathcal{S}^*$. This puts significant restrictions on the class of admissible action functionals. For a detailed discussion on this topic and some illustrative examples the interested reader is referred to [Hida (2008)], [Hida (1993)].

5.6   Some Problems for Exercise

P1: Use the expression for the characteristic functional given by (5.10) to show that the Fourier-Hida (FH) transform of the functional $\phi_\tau(x) \equiv (x(\tau))^m, x \in S^*$, is given by

$$\hat{\phi}(\xi) = (i)^m m! H_m(\xi(\tau)) C(\xi), \xi \in S,$$

where H_m is the Hermite polynomial of degree m.

P2: Consider the white noise functional

$$\varphi(x) = \exp\{\int_I \beta(t) x(t) dt\}$$

on the space (S^*, μ) where $\beta \in L_2(I)$. Show that

$$\mathbb{E}\{\varphi\} \equiv \int_{S^*} \varphi(x) \mu(dx) = \exp(1/2) \int_I |\beta(t)|^2 dt.$$

Hints: Show that $\hat{\varphi}(\xi) = C(\xi - i\beta)$ and then set $\xi = 0$.

P3: Consider the functional $\varphi(x) = e^{\alpha x(\tau)}, \tau \in I$, on (S^*, μ) and show that this functional does not posses Fourier-Hida transform. Hints: For any $\varepsilon > 0$, define the step function $\chi_\varepsilon(t) = (1/\varepsilon)$ for $t \in [\tau - \varepsilon/2, \tau + \varepsilon/2]$ and 0 otherwise and approximate φ by

$$\varphi_\varepsilon(x) \equiv \exp \alpha\{\int_I \chi_\varepsilon(t) x(t) dt\}.$$

Its FH transform is given by $\hat{\varphi}_\varepsilon(\xi) = C(\xi - i\alpha\chi_\varepsilon)$. From this verify that $\hat{\varphi}$ does not exist and that $\mathbb{E}\varphi = \infty$.

P4: Verify that the FH transform of the multiple white noise integral $g(x)$ given by the expression (5.20) is $\hat{g}(\xi)$ given by (5.21).

P5: Refer to the expression (5.31) for the action functional and suppose it is given by

$$S(x) \equiv \int_I K(t) x^2(t) dt$$

for $x \in S^*$. Compute the FH transform of the functional

$$\varphi(x) \equiv \exp\{(i/\hbar) \int_I K(t) x^2(t) dt\}$$

and verify the validity of the result:

$$\hat{\varphi}(\xi) = C(\xi) \exp\{(i/\hbar) \int_I K(t)(\xi^2(t) - 1) dt.$$

(**P6**): Consider the action functional

$$S(x) \equiv \int_I K_1(t) x(t) + \int_I K_2(t) x^2(t) dt$$

for suitable $\{K_1, K_2\}$ and compute the path integral

$$J(S) = \int_{S^*} \exp\{(i/\hbar) S(x)\} \mu(dx).$$

Hints: Compute the FH transform of the function $\varphi(x) \equiv \exp\{(i/\hbar) S(x)\}$ giving $\hat{\varphi}(\xi)$ on S and then set $\xi = 0$.

Chapter 6

L_p-Based Generalized Functionals of White Noise III

6.1 Introduction

So far we have considered regular and generalized functionals of Brownian motion based on Hilbertian structures. In the following two chapters we present a much more broader class of generalized functionals (of white noise) as introduced by the author [[Ahmed (1970)]]. In this chapter we introduce the class of generalized functionals based on L_p spaces and their weighted versions including the inductive and projective limits of such spaces. Also considered here are the generalized functionals of random fields, and vector measures with values in the space of generalized functionals of random fields. Applications of these results to physical sciences and nonlinear filtering and identification problems are considered.

These functionals cover those already introduced in the preceding chapters including the Hida class and much more and apply to a larger class of physical problems where the preceding classes fail.

6.2 Homogeneous Functionals of Degree n

For each $p \in [1, \infty]$, and each positive integer n, let $\hat{L}_p(I^n)$ denote the equivalence classes of symmetric functions from the class $L_p(I^n)$ where the Lebesgue measure $\ell(I)$ of I is finite. Introduce the linear space,

$$G^{p,n} \equiv \{K_n \in \hat{L}_p(I^n) : \| K_n \|_{G^{p,n}} \equiv (n!)^{1/p} \| K_n \|_{L_p(I^n)} < \infty\},$$

endowed with the norm topology as indicated. Clearly, $G^{p,n}$ is a Banach space. For each $L_n \in G^{p,n}$, define the functional

$$\phi_n(L_n) \equiv \int_{I^n} L_n(\tau_1, \tau_2, \cdots, \tau_n) \dot{w}(\tau_1) \dot{w}(\tau_2) \cdots \dot{w}(\tau_n) d\tau_1 d\tau_2 \cdots d\tau_n, \quad (6.1)$$

where $\dot{w} \in S^*$ with the white noise measure μ. We consider this functional as a μ measurable function on S^*. We shall use $M(S^*, \mu)$ to denote all μ measurable functions on S^*. Clearly, for each $p \geq 2$ and $K_n \in G^{p,n}$, $z \equiv \phi_n(K_n)$ (see Chapter 2 where $\phi_n = g_n$) defines a regular homogeneous chaos of degree n of Wiener-Itô class and belongs to $M(S^*, \mu)$. This expression is equivalent to the multiple Wiener-Itô integral discussed in Chapter 2. Define

$$\mathcal{G}^{p,n} \equiv \{z \in M(S^*, \mu) : z = \phi_n(K_n), K_n \in G^{p,n}\}. \quad (6.2)$$

This is a linear vector space. For $z \in \mathcal{G}^{p,n}$, corresponding to the kernel $K_n \in G^{p,n}$, define

$$\| z \|_{\mathcal{G}^{p,n}} \equiv \| K_n \|_{G^{p,n}} . \quad (6.3)$$

Lemma 6.2.1 *For $2 \leq p < \infty$, $\mathcal{G}^{p,n}$ is a Banach space with respect to the norm topology given by (6.3) and it is isometrically isomorphic to $G^{p,n}$.*

Proof Since I is a finite interval, for each $n \in N$ and $p \geq 2$, there exists a finite positive number $c = c(p, n)$ such that $\| L_n \|_{G^{2,n}} \leq c \| L_n \|_{G^{p,n}}$ for all $L_n \in G^{p,n}$. Hence, for each z given by $z \equiv \phi_n(L_n)$ with $L_n \in G^{p,n}$, we have

$$\| z \|_{\mathcal{G}^{2,n}} \leq c \| z \|_{\mathcal{G}^{p,n}} .$$

It follows from this that (6.3) defines a norm on $\mathcal{G}^{p,n}$ and since $G^{p,n}$ is a Banach space so also is $\mathcal{G}^{p,n}$. Consider the map ϕ_n given by (6.1). Clearly $\phi_n : G^{p,n} \longrightarrow \mathcal{G}^{p,n}$ is linear and $\phi_n(L_n) = 0$ $\mu-a.s$ if and only if $L_n = 0$ a.e. Thus ϕ_n is an isomorphism between $G^{p,n}$ and $\mathcal{G}^{p,n}$, and the isometry follows from the definition of the norm. $\bullet$

The functionals belonging to the class $\mathcal{G}^{p,n}, p \geq 2$, are regular functionals defined in the Wiener-Itô sense. For $1 \leq q < 2$ and $L_n \in G^{q,n}$, the random element $z^* \equiv \phi_n(L_n)$ is not defined in the classical Wiener-Itô sense. However, we can define them in a weak sense as generalized random elements as introduced in [Ahmed (1970)]. This is given in Lemma 6.2.2 as presented below.

Recall that $F(S^*, \mu)$ denotes the (universal) space of generalized functionals of the white noise (S^*, μ), generally not defined point wise. Let $\{p, q\}$ denote the conjugate pair, $(1/p) + (1/q) = 1$, satisfying $1 \leq q \leq 2 \leq p \leq \infty$.

Lemma 6.2.2 *For each $L_n \in G^{q,n}, 1 \leq q \leq 2$, the functional $z^* \equiv \phi_n(L_n)$ is a well defined element of $F(S^*, \mu)$, and that z^* is a continuous linear functional on $\mathcal{G}^{p,n}$.*

Proof For detailed proof see [[Ahmed (1970)], Lemma 3.2, p143]. We will present an outline of the proof. Since $\hat{L}_2(I^n)$ is dense in $\hat{L}_q(I^n)$, $1 \leq q \leq 2$, there exists a sequence $\{L_{n,s}, s = 1, 2, 3, \cdots\} \in \hat{L}_2(I^n)$ such that $L_{n,s} \xrightarrow{s} L_n$ (strongly) in $\hat{L}_q(I^n)$ as $s \to \infty$. Then $\{z_s^* \equiv \phi_n(L_{n,s}), s = 1, 2, \cdots\}$ is a sequence of regular functionals belonging to $L_2(\Omega, \mathcal{B}, \mu)$. Hence, for any $z \in \mathcal{G}^{p,n}$ which, by the previous lemma, has the representation $z = \phi_n(K_n)$ for some $K_n \in G^{p,n}$, we have

$$\begin{aligned}
< z_s^*, z >_{\mathcal{G}^{q,n}, \mathcal{G}^{p,n}} &\equiv \mathbb{E}\{(z_s^* \cdot z)\} = n!(L_{n,s}, K_n)_{L_2(I^n), L_2(I^n)} \\
&= n!(L_{n,s}, K_n)_{L_q(I^n), L_p(I^n)} \\
&= (L_{n,s}, K_n)_{G^{q,n}, G^{p,n}}.
\end{aligned} \tag{6.4}$$

Since $L_{n,s} \xrightarrow{s} L_n$ in $L_q(I^n)$, it also converges weakly to the same element. Thus the limit of the right hand expression of equation (6.4) exists for all $K_n \in G^{p,n}$. Hence, the limit of the expression on the left hand side is well defined for every $z \in \mathcal{G}^{p,n}$. We call the limit of z_s^*, in the above sense, the generalized random variable z^* given by $z^* \equiv \phi_n(L_n)$. Hence, z^* is well defined as a continuous linear functional on $\mathcal{G}^{p,n}$, that is, $z^* \in (\mathcal{G}^{p,n})^*$ and, therefore, it belongs to $F(S^*, \mu)$. ●

On the basis of the above result we can introduce a duality pairing between the spaces $\mathcal{G}^{q,n}$ and $\mathcal{G}^{p,n}$, with $1 \leq q \leq 2 \leq p < \infty$ and $p^{-1} + q^{-1} = 1$, as follows:

$$< z^*, z >_{\mathcal{G}^{q,n}, \mathcal{G}^{p,n}} \equiv \mathbb{E}\{(z^* z)\} = (L_n, K_n)_{G^{q,n}, G^{p,n}}, \tag{6.5}$$

where $z^* \equiv \phi_n(L_n)$ for $L_n \in G^{q,n}$ and $z \equiv \phi_n(K_n)$, for $K_n \in G^{p,n}$. In view of this result, we can introduce a norm topology on $\mathcal{G}^{q,n}$ by setting

$$\| z^* \|_{\mathcal{G}^{q,n}} \equiv \sup\{| < z^*, z > |, z \in \mathcal{G}^{p,n}, \| z \|_{\mathcal{G}^{p,n}} = 1\}.$$

Now we can present the following fundamental result.

Theorem 6.2.3 *For each nonnegative integer n, and $1 < q \leq 2 \leq p < \infty$ with p being conjugate of q, we have*

(1):
$$\mathcal{G}^{p,n} \hookrightarrow \mathcal{G}^{2,n} \hookrightarrow (\mathcal{G}^{p,n})^* = \mathcal{G}^{q,n}.$$

(2):
$$\begin{array}{ccc}
G^{p,n} \hookrightarrow G^{2,n} \hookrightarrow G^{q,n} \\
\updownarrow \quad\quad \updownarrow \quad\quad \updownarrow \\
\mathcal{G}^{p,n} \hookrightarrow \mathcal{G}^{2,n} \hookrightarrow \mathcal{G}^{q,n}.
\end{array} \qquad (6.6)$$

where all the injections are continuous and dense.

Proof We follow the proof given in [[Ahmed (1970)],Theorem 3.1]. The first inclusion follows from the inequality

$$\| z \|_{\mathcal{G}^{2,n}} \leq c \| z \|_{\mathcal{G}^{p,n}}, \text{ for all } p \geq 2, \text{ and } z \in \mathcal{G}^{p,n}$$

which was seen in course of the proof of Lemma 6.2.1. That the embedding is dense follows from the fact that for $p \geq 2$, $\hat{L}_p(I^n)$ is dense in $\hat{L}_2(I^n)$ implying that $\mathcal{G}^{p,n}$ is dense in $\mathcal{G}^{2,n}$. Now identifying the dual $(\mathcal{G}^{2,n})^*$ of $\mathcal{G}^{2,n}$ with itself we obtain $\mathcal{G}^{p,n} \hookrightarrow \mathcal{G}^{2,n} = (\mathcal{G}^{2,n})^* \hookrightarrow (\mathcal{G}^{p,n})^*$. In order to complete the proof of (1), we must justify that the dual of $\mathcal{G}^{p,n}$ is $\mathcal{G}^{q,n}$. By virtue of Lemma 6.2.2, in particular the duality pairing (6.5), it is clear that each element $\tilde{z} \in \mathcal{G}^{q,n}$ defines a continuous linear functional on $\mathcal{G}^{p,n}$ and hence $\mathcal{G}^{q,n} \subset (\mathcal{G}^{p,n})^*$. So it suffices to prove the reverse inclusion. Let $f \in (\mathcal{G}^{p,n})^*$; then by definition there exists a finite positive number c such that

$$|f(z)| \leq c \| z \|_{\mathcal{G}^{p,n}} \quad \forall \ z \in \mathcal{G}^{p,n}.$$

By the isometric isomorphism between $\mathcal{G}^{p,n}$ and $G^{p,n}$, denoted by $\mathcal{G}^{p,n} \leftrightarrow G^{p,n}$, for each $z \in \mathcal{G}^{p,n}$ there exists a unique $L_n \in \hat{L}_p(I^n)$ such that $z = \phi_n(L_n)$. Hence,

$$|f(z)| = |f(\phi_n(L_n))| = |(f \circ \phi_n)(L_n)| \leq c \| L_n \|_{G^{p,n}}$$

and consequently the composition map $f \circ \phi_n$ is a continuous linear functional on the Banach space $G^{p,n}$. Thus there exists a unique $K_n \in G^{q,n}$ such that

$$f(z) = (f \circ \phi_n)(L_n) = (K_n, L_n)_{G^{q,n}, G^{p,n}} \qquad (6.7)$$

for all $L_n \in G^{p,n}$ with $z = \phi_n(L_n)$. Therefore, it follows from (6.5), defining the duality pairing between $\mathcal{G}^{q,n}$ and $\mathcal{G}^{p,n}$, and the expression (6.7) and the definition of the vector space $\mathcal{G}^{q,n}$ that there exists a $z^* \in \mathcal{G}^{q,n}$ such that

$$f(z) = <z^*, z>_{\mathcal{G}^{q,n}, \mathcal{G}^{p,n}} \ \forall \ z \in \mathcal{G}^{p,n}.$$

Identifying f with z^*, we have $f \in \mathcal{G}^{q,n}$ proving the reverse inclusion $(\mathcal{G}^{p,n})^* \subset \mathcal{G}^{q,n}$ and hence the identity. This completes the proof of (1). For (2), the isometric isomorphism between $\mathcal{G}^{q,n}$ and $G^{q,n}$ follows easily from the identity (6.5). Thus the diagram (2) follows from (1) and the fact that the embeddings $G^{p,n} \hookrightarrow G^{2,n} \hookrightarrow G^{q,n}$ are continuous and dense. •

We note that, for $1 \leq q < 2$, $\mathcal{G}^{q,n}$ is a system of generalized functionals of the white noise (S^*, μ). This system is more general containing the class $\mathcal{G}^{2,n}$.

Remark 6.2.4. The diagram (2) also holds for $q = 1$ and $p = \infty$. The proof is direct. Clearly, the spaces $\{\mathcal{G}^{r,n}, 1 < r < \infty\}$ are reflexive Banach spaces.

Remark 6.2.5. The fact that $z^* \in \mathcal{G}^{q,n}$ does not, in any sense, imply that $\mathbb{E}\{|z^*|^q\} < \infty$. In other words, $\mathcal{G}^{q,n} \not\subset L_q(S^*, \mu)$.

6.3 Nonhomogeneous Functionals

Now we consider nonhomogeneous generalized functionals of degree $N \in N_+$. Define, for $r \geq 1, N < \infty$, the vector space

$$G_{r,N} \equiv \{L : L = (L_0, L_1, L_2, \cdots, L_N), L_n \in G^{r,n}, 0 \leq n \leq N\} \quad (6.8)$$

which, furnished with the norm topology

$$\| L \|_{G_{r,N}} \equiv \Big(\sum_{n=0}^{N} n! \| L_n \|_{L_r(I^n)}^r \Big)^{1/r}, \qquad (6.9)$$

is a Banach space. Here L_0 is a scalar. Since the Lebesgue measure $\ell(I) < \infty$, it is easy to verify that, for $1 \leq q \leq 2 \leq p < \infty$, $G_{p,N} \hookrightarrow G_{q,N}$ and that there exists a positive constant $C_{p,N}$ such that for each $L \in G_{p,N}$ we have

$$\| L \|_{G_{q,N}} \leq C_{p,N} \| L \|_{G_{p,N}}, \tag{6.10}$$

for every finite positive integer N. Thus the embedding $G_{p,N} \hookrightarrow G_{q,N}$ is continuous, and it follows from the density of the embedding $L_p \hookrightarrow L_q$ that it is also dense. The upper bound of the embedding constant can be shown to be

$$C_{p,N} \leq \left(\sum_{n=0}^{N} n!(\ell(I))^n \right)^{(p-1)/p}.$$

Now, we introduce the map Φ_N as

$$\Phi_N(K) \equiv \sum_{n=0}^{N} \int_{I^n} K_n(\tau_1, \cdots, \tau_n) dw(\tau_1) \cdots dw(\tau_n)$$

$$\equiv \sum_{n=0}^{N} \int_{I^n} K_n(\tau_1, \cdots, \tau_n) \dot{w}(\tau_1) \cdots \dot{w}(\tau_n) d\tau_1 \cdots d\tau_n, \tag{6.11}$$

defined for any $K \in G_{r,N}, r \geq 1$. The second identity allows one to consider this functional on the white noise space (S^*, μ). For any $r \geq 1$, this defines a larger class of functionals on (S^*, μ) given by

$$\mathcal{G}_{r,N} \equiv \{ z \in F(S^*, \mu) : z = \Phi_N(K), \quad \text{for some } K \in G_{r,N} \}. \tag{6.12}$$

We can introduce a norm topology on $\mathcal{G}_{r,N}$ by defining

$$\| z \|_{\mathcal{G}_{r,N}} \equiv \| K \|_{G_{r,N}}, \quad \text{for } z = \Phi_N(K), K \in G_{r,N}. \tag{6.13}$$

This makes Φ_N an isometric isomorphism between $G_{r,N}$ and $\mathcal{G}_{r,N}$. This is stated more precisely in the following theorem.

Theorem 6.3.1. *Given the topologies as defined above, for $1 \leq q \leq 2 \leq p < \infty$, the following diagram holds*

$$G_{p,N} \hookrightarrow G_{2,N} \hookrightarrow G_{q,N}$$

$$\updownarrow \qquad \updownarrow \qquad \updownarrow \tag{6.14}$$

$$\mathcal{G}_{p,N} \hookrightarrow \mathcal{G}_{2,N} \hookrightarrow \mathcal{G}_{q,N},$$

with continuous and dense embeddings where $\updownarrow$ denotes isometric isomorphism.

Proof The proof is similar to that of Theorem 6.2.3. For details see [[Ahmed (1970)]].

Again, the elements of $\mathcal{G}_{q,N}, 1 \leq q < 2$, are generalized functionals of white noise, that is, $\mathcal{G}_{q,N} \subset F(S^*, \mu)$. The next question is, can the above result be extended to $N = \infty$. The estimate on the embedding constant $C_{p,N}$ as given above is not suitable for this. In fact it diverges to infinity and hence the embeddings may not extend to $N = \infty$. However one can prove a partial result. It is easy to see that as $N \to \infty$, the limits of the Banach spaces $G_{p,N}, G_{q,N}$ and their duals are well defined with norms derived from the limits of the expressions (6.9) giving

$$\| L \|_{G_r} \equiv \left(\sum_{n=0}^{\infty} n! \| L_n \|_{L_r(I^n)}^r \right)^{1/r}, \tag{6.15}$$

provided they are finite. For each $r \in [1, \infty)$, define $\mathcal{K}_r \equiv \prod_{n=0}^{\infty} \hat{L}_r(I^n)$. Equipped with the family of seminorms, $Q_r \equiv \{q_{r,n}, n \in N\}$ with $q_{r,n}(K_n) \equiv \| K_n \|_{L_r(I^n)}$, $\mathcal{K}_r$ is a Fréchet space. Thus, for each $r \in [1, \infty)$, the normed space G_r given by

$$G_r \equiv \{L \in \mathcal{K}_r : \quad \| L \|_{G_r} < \infty \},$$

is a Banach space. For each $L \in G_r$, we can define an element $\Phi(L)$ given by

$$\Phi(L) \equiv \sum_{n=0}^{\infty} \phi_n(L_n).$$

This is a well defined element of $F(S^*, \mu)$ and we denote this class of functionals of Brownian motion by $\mathcal{G}_r$. More precisely we have the following result.

Theorem 6.3.2 *For $1 < q \leq 2 \leq p < \infty$, (i): $\mathcal{G}_p = \lim_{N \to \infty} \mathcal{G}_{p,N}, \mathcal{G}_q = \lim_{N \to \infty} \mathcal{G}_{q,N}$; (ii) $G_p, G_q, \mathcal{G}_p, \mathcal{G}_q$ are reflexive Banach spaces; (iii) $\mathcal{G}_p \leftrightarrow G_p, \mathcal{G}_q \leftrightarrow G_q$ and $G_p^* = G_q, \mathcal{G}_p^* = \mathcal{G}_q$.*

Remark 6.3.3 It is interesting to note that we do not have any inclusion relation between the Banach spaces G_p, G_q and hence no inclusion relation exists between the corresponding Banach spaces of Brownian functionals $\mathcal{G}_p, \mathcal{G}_q$. This is the basic difference between the results of Theorem 6.3.1 and Theorem 6.3.2.

Let $\{p, q\}$ be the conjugate pair satisfying $1 \leq q \leq 2 \leq p < \infty$, and $\{G_q, G_p\}$ the dual pair of Banach spaces introduced above. It is easy to verify that G_p separates points of G_q and hence, by the isomorphism, $\mathcal{G}_p$ separates points of $\mathcal{G}_q$. We show that G_p separates elements of G_q. In other words, it suffices to verify that

$$(K, L)_{G_q, G_p} = 0 \; \forall \; L \in G_p$$

implies $K \equiv 0$. Suppose this is false, then there exists $K^o(\neq 0) \in G_q$ such that $(K^o, L)_{G_q, G_p} = 0$ for all $L \in G_p$. Choose $L = L^o \equiv \{L_n^o, n \in N\}$ with L_n^o given by

$$L_n^o \equiv \frac{|K_n^o|^{q-1}\text{sign}(K_n^o)}{\| K^o \|_{G_q}^{q-1}}.$$

Then, it is easy to verify that $\| L^o \|_{G_p} = 1$ and that the scalar product $(K^o, L^o)_{G_q, G_p} = \| K^o \|_{G_q}$. But since $L^o \in G_p$, $(K^o, L^o) = \| K^o \|_{G_q} = 0$ which contradicts the nontriviality of K^o. Hence, G_p separates points of G_q and consequently $\mathcal{G}_p$ separates points of $\mathcal{G}_q$. For details see [Ahmed (1970)].

Remark 6.3.4 This separation property is not only mathematically interesting; it is also practically useful. For example, if our interest is only in the class of generalized functionals which coincide on the closed linear subspace $M \subset \mathcal{G}_p$, then for all practical purposes we may consider two elements $z_1^* \neq z_2^*$ as identical, $z_1^* \cong z_2^*$, if $(z_1^* - z_2^*)(z) = 0$ for all $z \in M$. In other words, we have the quotient space $\mathcal{G}_q/M^\perp$ where $M^\perp$ denotes the annihilator of M. For approximation, it is the quotient space that matters.

Remark 6.3.5 By setting $p = q = 2$, we obtain from the above results the classical Wiener-Itô functionals.

Remark 6.3.6 For $1 < q \leq 2 \leq p < \infty$, the dual pair of Banach spaces $\{\mathcal{G}_p, \mathcal{G}_q\}$ are reflexive and hence bounded subsets therein are relatively weakly and weak-star compact respectively. But in reflexive Banach spaces weak and weak star topologies are identical. It would be interesting to characterize weak and weak star compactness in $\mathcal{G}_1$ space. The question of compactness is important in optimization problems as seen in Chapter 4.

6.4 Weighted Generalized Functionals

In the spirit of Theorem 3.2.2, we can construct new generalized functionals from those of Theorem 6.3.2 by introducing suitable weighting functions other than just the functions of $n \in N$, like $a(n) \equiv n!$. With this end in view, we introduce the following definition.

Definition 6.4.1. *Let N_+ denote the set of nonnegative integers, and R_+ the set of nonnegative real numbers. A function $a : N_+ \times R \longrightarrow R_+ \cup \{+\infty\}$ is said to be an admissible weight if it is a monotone increasing function of both the variables and satisfies the following properties:*

$$(i) : a(n,s)a(n,-s) = 1 \ \forall \ n \in N_+ \text{ and } s \geq 0, (ii) : a(n,0) = 1 \ \forall \ n \in N_+,$$
$$(iii) : \lim_{s \to \infty} a(n,s) = \infty, n > 1, (iv) : \lim_{n \to \infty} a(n,s) = \infty, s > 0.$$

$$(6.16)$$

Clearly, it follows from (i) and (iii) that $\lim_{s \to \infty} a(n,-s) = 0$.

For $p \in [1, \infty]$, define

$$\Xi_p \equiv \{(K_0, K_1, K_2, \cdots K_n \cdots) : K_n \in \hat{L}_p(I^n), n \in N_+\} \equiv \prod_{n=0}^{\infty} \hat{L}_p(I^n).$$

Clearly, this is only a linear vector space (with the product topology) but without any norm topology. Corresponding to Ξ_p, we can introduce the vector space of white noise functionals

$$\hat{\Xi}_p \equiv \{(\phi_0(K_0), \phi_1(K_1), \phi_2(K_2), \cdots \phi_n(K_n) \cdots), K \in \Xi_p\},$$

where $\{\phi_n\}$ are the elementary homogeneous functionals of white noise as defined in Lemma 6.2.1 and Lemma 6.2.2. Being the cartesian product of infinitely many Banach spaces, these spaces have only the linear structure with the product topology. Thus the map

$$\{K_n, n \in N_+\} \longrightarrow \{\phi_n(K_n), n \in N_+\}$$

from Ξ_p to $\hat{\Xi}_p$, denoted here by Ψ, is merely an algebraic isomorphism.

Now we shall introduce various norm topologies and construct normed spaces so that Ψ becomes an isometry: For $p \in [1, \infty), s \geq 0$, define

$$Z_{p,s} \equiv \left\{ K \in \Xi_p : \left(\sum_{n=0}^{\infty} n!(a(n,s))^p \parallel K_n \parallel_{L_p(I^n)}^p \right)^{1/p} < \infty \right\}. \quad (6.17)$$

Furnished with this norm topology, $Z_{p,s}$ is a Banach space. And for a fixed p, $Z_{p,s} \subset Z_{p,r}$, for $s > r$. For $1 < q \leq 2 \leq p < \infty$, $\{p, q\}$ conjugate pair, $(Z_{p,s})^* = Z_{q,-s}$. Now we can introduce the generalized functionals of white noise through the isomorphism Ψ as follows:

$$\mathcal{Z}_{p,s} \equiv \Psi(Z_{p,s}), \mathcal{Z}_{q,-s} \equiv \Psi(Z_{q,-s}). \quad (6.18)$$

Based on the results of section 6.2 and section 6.3, we have the following result.

Theorem 6.4.2. *For $1 < q \leq 2 \leq p < \infty$ and $s \in R_+$, the following results hold:*

(i): $Z_{p,s}$, $Z_{q,-s}$, $\mathcal{Z}_{p,s}$, and $\mathcal{Z}_{q,-s}$ are reflexive Banach spaces.

(ii): $Z_{p,s} \leftrightarrow \mathcal{Z}_{p,s}$, $Z_{q,-s} \leftrightarrow \mathcal{Z}_{q,-s}$ and $Z_{p,s}^* = Z_{q,-s}$, and $\mathcal{Z}_{p,s}^* = \mathcal{Z}_{q,-s}$

(iii): $\mathcal{Z}_{q,-s} \subset F(S^*, \mu)$, are generalized functionals of white noise for $s \geq 0$.

Remark 6.4.3 For $s = 0$, we obtain the results of Theorem 6.3.2; and, in particular, $\mathcal{Z}_{q,-0} = \mathcal{G}_q$. The family of Banach spaces $\{\mathcal{Z}_{q,-s}, s \in R_+\}$ is an expanding family of generalized functionals of white noise more general than those of Theorem 6.3.2, and they are the duals of the Banach spaces $\{\mathcal{Z}_{p,s}, s \in R_+\}$.

The ultimate generalization of this section is obtained by use of inductive and projective limit topologies. For example, the following theorem generalizes the results of the previous sections.

Theorem 6.4.4 Let $1 < q \leq 2 \leq p < \infty$. Then, (1): For $0 \leq r \leq s$, the identity map $\mathcal{Z}_{q,-r} \hookrightarrow \mathcal{Z}_{q,-s}$ is continuous. (2): $\mathcal{Z}_q \equiv \bigcup_{r \geq 0} \mathcal{Z}_{q,-r}$ is the inductive limit of the spaces $\mathcal{Z}_{q,-r}$ which has the finest locally convex *topology for which the embedding $\mathcal{Z}_{q,-r} \hookrightarrow \mathcal{Z}_q$ is continuous. This is the inductive limit topology. (3): $\mathcal{Z}_q$ is the topological dual of $\mathcal{Z}_p \equiv \bigcap_{r \geq 0} \mathcal{Z}_{p,r}$; with $\mathcal{Z}_p$ having the projective limit topology.*

The space $\mathcal{Z}_q, 1 \leq q \leq 2 \leq p$, is the space of generalized functionals of Brownian motion and the space $\mathcal{Z}_p$ is the space of test functionals.

Remark 6.4.5 Note that the generalized functionals presented in this section cover those of Nualart and Zakai [Nualart (2009)], and all those presented in Chapters 2 to 4.

Remark 6.4.6 Most of the results in the preceding sections also hold for functionals of fractional Brownian motion and Lévy process. In other words one can construct generalized functionals of fractional Brownian motion and Lévy process following exactly the same procedure.

6.5 Some Examples Related to Section 6.4

(E1) Integral Operator: Consider the random variable Z given by,

$$Z(w) \equiv \int_0^T U(t, w(t))dw(t)$$

with U satisfying

$$|U(t, x)| \leq k_0(t) + k_1 \exp\left(rx^2\right)$$

for some $r > 0$ and $k_0 \in L_q$ for $1 < q \leq 2$. We show that there exists an $r > 0$ such that $Z \in L_q(\Omega)$ where Ω is the Wiener measure space. Define the retract

$$R_n(x) = \begin{cases} x, & \text{for } |x| \leq n \\ nx/|x|, & \text{for } |x| > n, \end{cases}$$

and approximate Z by

$$Z_n \equiv \int_0^T R_n(U(t, w(t)))\, dw(t).$$

Let $\mathcal{F}_t, t \geq 0$, denote the current of sigma algebras generated by the Brownian motion w. Clearly, R_n is $\mathcal{F}_t$ adapted and bounded and so square integrable. Thus Z_n belongs to $L_2(\Omega)$ for every finite integer n. Let p be the conjugate of q. For test function, take $Y = \int_0^T h(t)dw(t), h \in L_p^a$ where L_p^a denotes the class of $\mathcal{F}_t$ adapted processes so that

$$\mathbb{E} \int_0^T |h(t)|^p dt < \infty.$$

Clearly $L_p^a(I) \subset L_2^a(I)$. Using Fubini's theorem and the natural duality pairing we have

$$\mathbb{E}\{ZY\} = \lim_{n\to\infty} \mathbb{E}\{Z_n Y\} = \lim_{n\to\infty} \int_0^T \mathbb{E}\{R_n(U(t, w(t)))h(t)\}dt.$$

By Holder's inequality, we have

$$\left| \int_0^T \mathbb{E}\{R_n(U(t, w(t)))h(t)\}dt \right|$$
$$\leq \left(\int_0^T \mathbb{E}|R_n(U(t, w(t)))|^q dt \right)^{1/q} \left(\int_0^T \mathbb{E}|h(t)|^p dt \right)^{1/p}.$$

Now, note that there exists a constant $C_q > 0$ independent of n such that

$$\int_0^T \mathbb{E}|R_n|^q dt \leq C_q \left\{ \int_0^T |k_0(t)|^q dt + \int_0^T \mathbb{E}\{\exp\{rq|w(t)|^2\}\}dt \right\}.$$

Since $w(t)$ is Gaussian with mean zero and variance t, the reader can easily check that

$$\mathbb{E}\exp\{rq|w(t)|^2\} = (1/\sqrt{2\pi t}) \int_R \exp-\{(1/2t) - rq\}x^2 dx = (1/\sqrt{1 - 2trq}).$$

Clearly, it follows from this that for $r < (1/2Tq)$, this function has finite Lebesgue integral over the interval $[0, T]$. Thus for sufficiently small $r > 0$, $Z \in L_q(\Omega) \supset L_2(\Omega)$.

On the other hand, suppose $r > 0$ is given and we are to find a space $L_q(\Omega), q \geq 1$, so that Z is a member of this space. This can be done if $(1/2Tr) > 1$ because in that case we can choose q so that $1 \leq q < (1/2Tr)$.

(E2) Urysohn Operator: Many nonlinear systems give rise to Urysohn's integral operators. Here we present an example which involves $\mathcal{G}^*$. For illustration, consider the scalar Brownian motion w and the Brownian functional given by the Urysohn's integral operator,

$$F_t(w) \equiv \int_0^t U(t, s, w(s))\, dw(s), t \in [0, T]. \tag{6.19}$$

We are concerned with the terminal value $f(w) \equiv F_T(w)$. Suppose the Urysohn kernel satisfies the polynomial growth

$$|U(T, t, x)| \leq C_1 + C_2|x|^r, r \in [0, \infty).$$

Then one can easily verify that there exist nonnegative constants $\{\alpha_1, \alpha_2\}$ depending on $\{C_1, C_2\}$ and r such that

$$E|F_T|^2 \leq \alpha_1 + \alpha_2 T^{r+1}.$$

This shows that $F_T \in \mathcal{G}_\alpha$ for any U satisfying a polynomial growth as indicated. In fact this remains true even for exponential growth of the form

$$|U(T, t, x)| \leq C_1 + C_2 e^{rx}, r \in [0, \infty).$$

In this case

$$E|F_T|^2 \leq \alpha_1(T) + \alpha_2(T) \exp(2r^2 T)$$

with $\alpha_i(T)$ being finite for every finite T. Thus $F_T \in \mathcal{G}_\alpha$. As seen in the previous example, a case where $f \equiv F_T$ may not belong to $\mathcal{G}_\alpha$ is obtained under the assumption that U satisfies the exponential growth of the form

$$|U(T, t, x)| \leq C_1 + C_2 \exp(rx^2), r \in [0, \infty).$$

For $0 \leq r < (1/4T)$, $f(\equiv F_T) \in \mathcal{G}_\alpha$ otherwise f may escape $\mathcal{G}_\alpha$. In any case one can always compute the Fourier-Wiener kernels $L \equiv \{L_n\}$ corresponding to f. Clearly, $f \notin \mathcal{G}_\alpha$ if we find that $\| L \|_{G_\alpha} = \infty$. It is possible that $L \in G_\alpha^* \setminus G_\alpha$ and hence f is a generalized functional. Recalling the norm topologies of G_α^* and G_α, it is clear that the infinite series

$$\| L \|_{G_\alpha^*}^2 \equiv \sum (1/n!) \| L_n \|_{L_2(I^n)}^2$$

has a better chance of convergence compared with the series $\| L \|_{G_\alpha}^2 \equiv \sum (n!) \| L_n \|_{L_2(I^n)}^2$. If this also fails, that is $\| L \|_{G_\alpha^*} = \infty$, then we may consider the (larger) Banach spaces $\mathcal{Z}_{p,s}$ and their duals $\mathcal{Z}_{q,-s}$ given in Theorem 6.4.2. The Fourier-Wiener kernel L may belong to $\mathcal{Z}_{q,-s}$ for some $q \in [1, 2]$ and $s \in (0, \infty)$. In that case, $f \in \mathcal{Z}_{q,-s}$ the dual of the Banach space $\mathcal{Z}_{p,s}$. If these also fail, we use the Banach spaces $\mathcal{Y}_{q,-s}$ introduced in the next chapter.

In view of the example (E2), we may conclude that the generalized functionals, introduced so far, provide a much broader scope for inclusion of functionals not covered either by the well known classical Wiener-Itô class or the Hida class.

6.6   Generalized Functionals of Random Fields Applied

In the study of distributed signal processing or equivalently processing of signals which are functions of both space and time, it is natural to consider noise distributed both in space and time. Solutions of partial differential equations, deterministic as well as stochastic, present a great variety of distributed signals. Typical examples are heat equations, wave equations, reaction diffusion equations, Navier-Stokes equation, Schrödinger equation, population equations etc. subject to perturbation by random fields. In particular, 2-D and 3-D image processing is one such area where substantial work has been done in mathematical modeling and processing such as smoothing, segmentation, registration etc [Chan (2002)], [Sigurd (2006)]. See also the extensive references therein. One important problem in image processing is reconstruction (or restoration,inpainting) of an entire image from its available parts. The problem is to find the processing operator that transforms the partial information into an approximate replica of the true image. In recent years, using suitable Lagrangians dictated by the desired properties of images, Lagrange principle and variational calculus have been used to develop empirical models for generation of images. These are generally nonlinear degenerate diffusion equations governed by second order partial differential equations of parabolic type (PDE's). Engineers and doctors are often interested in input-output models so that from the input data they can readily obtain the desired information from the corresponding output. In the study of structural mechanics, vibration of beams, plates and complex structures like suspension bridges, tall buildings etc. are of major concern to engineers. Here, they are interested in the displacement and its rate at certain strategic locations for evaluation of structural integrity. These systems are all described by nonlinear partial differential equations which may be subject to stochastic forces.

A challenging problem is to construct input output models for such systems. For example, a nonlinear Euler beam equation with clamped boundary condition is given by

$$\partial^2 y/\partial t^2 + c_1 D_x^4 y + c_2 N(y) D_x^2 y + c_3 \partial y/\partial t = \dot{W}(t,x), t > 0, x \in (0,L)$$

$$y(t,0) = y(t,L) = 0, \ D_x y(t,0) = D_x y(t,L) = 0, \ y(0,\cdot) = y_0, y_t(0,\cdot) = y_1,$$

where $D_x \equiv \partial/\partial x$, the nonlinear operator N is given by

$$N(y) = \int_0^L (D_x y)^2 dx$$

which represents the membrane force and $\dot{W}$ is the space time white noise.

This model is a good approximation for moderately large vibrations [Srinivasan (1971)], [Ahmed (1994)]. In the absence of noise, for any given initial state $y_0 \in H_0^1(D), y_1 \in L_2(D)$, this equation has a well defined solution with $y \in C(I, H_0^1), y_t \in C(I, L_2(D))$. In the presence of noise $\dot{W}$, the solution $y(t,x), t > 0, x \in D \equiv (0,L)$ is a random field. One may be interested in displacement at the mid point of the structure like $y(t, L/2)$ which is considered as the output corresponding to the noisy input. By Sobolev embedding $H_0(D) \subset C(D)$ and so the point value at $x = L/2$ is well defined.

For image analysis, the typical equations are of the form

$$(\partial/\partial t)\mathcal{I} = G(I, \bigtriangledown\mathcal{I}, \triangle\mathcal{I}) + \dot{W}(t,x), (t,x) \in (0,T] \times D, \mathcal{I}(0,x) = \mathcal{I}_0.$$

The solution of all such equations can be approximated by the canonical expansion involving multiple integrals with respect to the Gaussian random measure.

Let $D \subset R^n$ ($n = 1, 2, 3$ for applications) be an open bounded set, $I \equiv [0, T]$ a finite time interval and ℓ the Lebesgue measure on the sigma algebra of subsets of the set $I \times D$ and $D_t \equiv [0, t] \times D$. The most general input-output model for distributed signal processing is given by

$$y(t, x) = F(\varphi)(t, x), (t, x) \in D_T, \tag{6.20}$$

where the input-output operator F is given by

$$F(\varphi)(t, x)$$
$$\equiv \sum_{n=1}^{\infty} \int_{D_t^n} L_n(t, x, z_1, z_2, \cdots, z_n)\varphi(z_1)\varphi(z_2)\cdots\varphi(z_n)\ell(dz_1)\cdots\ell(dz_n).$$

$$\tag{6.21}$$

The input signal is denoted by $\varphi(z) \equiv \varphi(\tau, \xi), z \equiv (\tau, \xi) \in D_t$, which represents the space-time distribution of the input. The output $y(t, x)$ describes the response at time $t \in I$ at the location $x \in D$. For convenience of notation we write

$$L_n(t, x) \equiv L_n(t, x; z_1, z_2, \cdots, z_n) \quad \text{for} \quad z_i \in D_t$$

and consider them as elements of suitable function spaces. Assuming that for each integer $n \in N$ and $(t, x) \in D_T$, the kernel $L_n(t, x)$ belongs to the Lebesgue space $L_q(D_T^n)$ and the input $\varphi \in L_p(D_T)$ for $1 \leq q \leq 2 \leq p < \infty$ with $(1/p + 1/q = 1)$, it is easy to verify that

$$|y(t, x)| \equiv |F(\varphi)(t, x)| \leq \sum \| L_n(t, x) \|_{L_q(D_t^n)} \| \varphi \|_{L_p(D_t)}^n . \quad (6.22)$$

To avoid using too many new notations, we shall use G_q also for the Banach space with the norm topology given by

$$\| K \|_{G_q} \equiv \left(\sum_{n=1}^{\infty} n! \, \| K_n \|_{L_q(D_T^n)}^q \right)^{1/q},$$

where I^n of the expression (6.15) has been replaced by D_T^n. Assuming that, for each $(t, x) \in D_T$, the G_q-valued kernel L is given by $L(t, x) \equiv \{L_n(t, x), n \in N\} \in G_q$, it is not difficult to verify that

$$|y(t, x)| \leq \| L(t, x) \|_{G_q} \exp\{(1/p) \| \varphi \|_{L_p(D_t)}^p\}. \quad (6.23)$$

This follows from Holder's inequality applied to (6.22). From this analysis we have the following result.

Lemma 6.6.1 *Consider the operator F defined by (6.21) and suppose the kernel $L \in L_\infty(D_T, G_q)$ $(C(D_T, G_q))$ with $\{p, q\}$ satisfying $(1/p) + (1/q) = 1, 1 \leq q \leq 2 \leq p < \infty$. Then the input-output operator F is continuous and bounded from $L_p(D_T)$ into $L_\infty(D_T)(C(D_T))$.*

Remark 6.6.2 Note that if we assume that $L \in L_\infty(D_T, \mathcal{H})$ with $\mathcal{H}$ denoting the Hilbert space with the norm topology given by

$$\| L(t, x) \|_2 \equiv \left(\sum_{n=1}^{\infty} \| L_n(t, x) \|_{L_2(D_t^n)}^2 \right)^{1/2}$$

then the operator F is only defined on the open unit ball in $L_2(D_T)$. More precisely,

$$F : B_1^o(L_2(D_T)) \longrightarrow L_\infty(D_T),$$

where $B_1^o(L_2(D_T))$ denotes the open unit ball in $L_2(D_T)$ centered at the origin.

For identification of the characteristic kernels, we may follow the same principle as in Chapter 4 (section 4.4), in particular, the examples (E2) and (E3). We consider the case $p = q = 2$. Here we use as input, the orthogonal and normalized space time Gaussian random measure as indicated in Remark 2.5.4. Let $z \equiv (\tau, \xi)$ denote the generic points in D_t. The response y is experimentally obtained by using the space-time Gaussian random measure as the input and recording the corresponding output giving $\{y_e(t, x)\}$. Then, as in examples (E1) and (E2) of section 4.4, to identify the n-th degree kernel we choose the test functional as

$$u_n(t, x) = e_n(K_n)(t, x) \equiv \int_{D_t^n} K_n(t, x; z_1, \cdots, z_n) W(dz_1) \cdots W(dz_n)$$

$$(6.24)$$

which is then used to determine the scalar product,

$$\ell(K_n) \equiv (1/n!) < y_e(t, x), u_n(t, x) >_{\mathcal{G}}, \qquad (6.25)$$

giving the linear functional ℓ. In practice, this has to be determined by use of Monte Carlo simulation techniques. The characteristic n-th order kernel L_n is then determined by maximizing the functional ℓ on the unit ball $B_1(L_2(D_T^n))$ giving $L_n = \beta_n K_n^o$, with K_n^o being the maximizer of ℓ on the unit ball as stated above and β_n are suitable scalars as explained below. In principle, this way we can identify the unknown system giving

$$y_e(t, x) \approx y(t, x) = \sum_{n=1}^{\infty} e_n(L_n)(t, x) \qquad (6.26)$$

where $\{L_n\}$ are the true characteristic kernels. It is clear that if $y_e \in L_\infty(D_T, \mathcal{G})$, then we have the exact identity; otherwise y is the best element of $L_\infty(D_T, \mathcal{G})$ approximating the experimentally observed output y_e. The constants β_n^2 determine the energy content of the n-th degree polynomial (mode) chaos.

Remark 6.6.3 In case $1 \le q < 2$, with $L(t, x) \in G_q$, the functional $y(t, x)$ given by (6.26) is a generalized functional of the Gaussian random field W

and belongs to $\mathcal{G}_q$ (see Theorem 6.3.2). In this case the test functionals u_n must be chosen with the kernel $K_n(t, x) \in L_p(D_T^n)$.

In the field of image processing, the input images having textures and sharp boundaries etc., are better described by functions of bounded variation. In this situation, the regular signal process φ is replaced by a signed measure m having bounded variation on D_T. Let $\mathcal{B}(D_T)$ denote the sigma algebra of Borel subsets of the set D_T and $M_{cabv}(\mathcal{B}(D_T))$ the space of countably additive signed Borel measures on $\mathcal{B}(D_T)$ having bounded variation. The model (6.21) is then written as

$$y(t, x) \equiv F(m)(t, x) \equiv \sum_{n=1}^{\infty} \int_{D_t^n} L_n(t, x, z_1, \cdots, z_n) m(dz_1) \cdots m(dz_n)$$

(6.27)

where $m \in M_{cabv}(\mathcal{B}(D_T))$.

For analysis of the above model we need some special Banach spaces. Let $\{E_n\}$ be an infinite family of (real) Banach spaces and $p \in [1, \infty)$ and define the vector space $\Upsilon_p \equiv \ell^p(\prod_{n=1}^{\infty} E_n)$. Furnished with the topology induced by the norm,

$$\| e \|_{\Upsilon_p} \equiv \left(\sum_{n=1}^{\infty} \| e_n \|_{E_n}^p \right)^{1/p}, e \in \Upsilon_p,$$

Υ_p is a Banach space. For the system (6.27), we take $E_n \equiv \sqrt{n!} \hat{L}_\infty(D_T^n)$ and use the vector space

$$\Upsilon_2 = \ell^2(\prod_{n=1}^{\infty} E_n) \equiv \ell^2(\prod_{n=1}^{\infty} \sqrt{n!} \hat{L}_\infty(D_T^n))$$

endowed with the norm topology given by

$$\| K \|_{\Upsilon_2} \equiv \left(\sum_{n=1}^{\infty} n! \| K_n \|_{L_\infty(D_T^n)}^2 \right)^{1/2}.$$

The input-output model (6.27) is then well defined if

$$\| L(t, x) \|_{\Upsilon_2}^2 \equiv \sum_{n=1}^{\infty} n! \| L_n(t, x) \|_{L_\infty(D_T^n)}^2 < \infty \quad \forall \ (t, x) \in D_T. \quad (6.28)$$

In this case we have

$$|y(t, x)| \equiv |F(m)(t, x)| \leq \| L(t, x) \|_{\Upsilon_2} \exp\{(1/2)|m|_v^2\} \quad (6.29)$$

where $|m|_v$ denotes the total variation norm of the signed measure m. Thus we have proved the following result.

Lemma 6.6.4 *Consider the map F given by (6.27). If $L \in L_\infty(D_T, \Upsilon_2)(C(D_T, \Upsilon_2))$ then the operator F is a bounded and continuous map from $M_{cabv}(\mathcal{B}(D_T))$ to $L_\infty(D_T)(C(D_T))$.*

In general, the output y given by the expression (6.26) with e_n as defined by (6.24) is a random field belonging to $L_\infty(D_T, \mathcal{G}_q)$. Even though $\mathcal{G}_q$ is also a Banach space of generalized functionals of the Gaussian random measure W, it can be substantially generalized to admit much larger class of generalized functionals. In order to do this we introduce the following Banach spaces.

For $1 \leq q \leq 2 \leq p < \infty$, with $\{p, q\}$ a conjugate pair, define

$$F_p \equiv \left\{ K \equiv (K_0, K_1, K_2, \cdots) \in \prod_{n=0}^\infty \hat{L}_p(D_T^n) : \right.$$
$$\left. \left(\sum_{n=0}^\infty (n!)^{p-1} \| K_n \|_{L_p(D_T^n)}^p \right)^{1/p} < \infty \right\}, \qquad (6.30)$$

where as usual, $L_p(D_T^0) = R$, the set of real numbers. This is a Banach space with respect to the norm topology defined above. Its dual is given by $\tilde{F}_q$ with the norm topology

$$\tilde{F}_q \equiv \left\{ L \equiv (L_0, L_1, L_2, \cdots) \in \prod_{n=0}^\infty \hat{L}_q(D_T^n) : \right.$$
$$\left. \left(\sum_{n=0}^\infty (1/n!) \| L_n \|_{L_q(D_T^n)}^q \right)^{1/q} < \infty \right\}. \qquad (6.31)$$

This is a much larger space compared with $\mathcal{G}_q$. In fact $\tilde{F}_q \supset \mathcal{G}_q$ for all $q \in [1, 2]$. The isomorphic map Φ assigns to each element $K \in F_p$ the random variable $z = \Phi(K) \equiv \sum_{n=0}^\infty e_n(K_n)$ where e_n denotes the multiple Wiener-Itô integrals with respect to the Gaussian random measure $\{W\}$ as given by the expression (6.24). The image of F_p under the isomorphism Φ is denoted by $\mathcal{F}_p \equiv \Phi(F_p)$. The space $\mathcal{F}_p$ is the linear vector space of regular functionals of Gaussian random field and it is given the norm topology

$$\| z \|_{\mathcal{F}_p} \equiv \| \Phi(K) \|_{\mathcal{F}_p} \equiv \| K \|_{F_p}$$

with respect to which it is a Banach space. This turns Φ into an isometric isomorphism between the spaces F_p and $\mathcal{F}_p$. Using the continuous extension of this isomorphism to $\tilde{F}_q$, again denoted by Φ, we obtain the space $\tilde{\mathcal{F}}_q \equiv \Phi(\tilde{F}_q)$, which is the dual of $\mathcal{F}_p$ and it is the space of generalized functionals. We state this in the following theorem.

Theorem 6.6.5 *For $1 \leq q \leq 2 \leq p < \infty$ with $\{p,q\}$ being the conjugate pair, the space $\tilde{\mathcal{F}}_q$ is the space of generalized functionals of the space-time Gaussian random field. Endowed with the norm topology given by*

$$\| z^* \|_{\tilde{\mathcal{F}}_q} \equiv \sup\{| < z^*, z >_{\tilde{\mathcal{F}}_q, \mathcal{F}_p} | : z \in \mathcal{F}_p, \| z \|_{\mathcal{F}_p} = 1\},$$

$\tilde{\mathcal{F}}_q$ is a Banach space where the duality bracket denotes integration with respect to the Wiener measure induced by the Gaussian random field. The space $\tilde{\mathcal{F}}_q$ containing $\mathcal{G}_q$ (see Theorem 6.3.2) is the dual of $\mathcal{F}_p$ and, for $1 < q \leq 2 \leq p < \infty$, they are reflexive Banach spaces.

The Banach space $\tilde{\mathcal{F}}_q$ offers a much broader class of generalized functionals of Gaussian random measure than those of $\mathcal{G}_q$. In particular, the class of Banach spaces $\{\tilde{\mathcal{F}}_q, 1 \leq q \leq 2\}$ is very rich for approximation of solutions of SPDE's like those mentioned in the introduction of this section and many more.

The necessity of such spaces of generalized functionals is very well demonstrated by Nualart and Rozovskii in their recent paper [Nualart (1997)] on bi-linear SPDE's of the form

$$\partial u / \partial t = Au + u\dot{W}, u(0) = u_0, (t,x) \in I \times R^d, \tag{6.32}$$

where A is a linear second order elliptic partial differential operator and $\dot{W}$ is the space time white noise. It is known that for $d > 1$, this equation has no solution in $L_2(\Omega)$. According to Nualart-Rozovskii's result [[Nualart (1997)], Theorem 3.7], the solution $u(t,x)$ takes values in a weighted Wiener space $L_2^Q(\Omega)$ where Q is a nonnegative self adjoint operator on the Hilbert space $H \equiv L_2(R^d)$ with a decreasing sequence of nonnegative eigenvalues $\{q_i\}$ which provide the weight. Under some standard assumptions on the coefficients (Hölder continuity) of the operator A, the initial data u_0, and the operator Q (for example Hilbert-Schmidt), they prove the existence of a unique weak (not martingale) solution $u \in L_\infty^{loc}(I \times R^d, L_2^Q(\Omega))$ which the authors call Q-soft solution. Further, using Feynman-Kac formula a closed form expression for the solution is given. For more details see [Nualart (1997)]. Using the dual $\tilde{\mathcal{F}}_q$, for $q = 2$, we claim that $u \in L^{loc}_\infty(I \times R^d, \tilde{\mathcal{F}}_2)$. See also Remark 2.5.4 and Corollary 2.6.3.

In the following section we study vector measures taking values in the space of generalized functionals of Brownian motion. This is also illustrated by an application to nonlinear filtering.

6.7 $\tilde{\mathcal{F}}_q$-Valued Vector Measures with Application

For some applications it may be useful to consider the output y as a vector valued measure. There are many physical processes such as the solutions of porous media equations describing the mass density of fluid contents in the medium, the measure solutions of stochastic Navier-Stokes equations in the presence of turbulence [Ahmed (1998)], solutions of stochastic Maxwell-Vlasov equation describing the particle density, solutions of Kushner-Zakai equations arising from finite and infinite dimensional nonlinear filtering [Ahmed (1998)], [Ahmed (1997)]. These solutions are better considered as stochastic vector valued measures. Another interesting physical example is the geographically structured population process considered by Wulfsohn [Wulfsohn (1977)]. This can be considered as a nonnegative measure valued stochastic process. Study of measure valued stochastic processes was pioneered by Dawson [Dawson (1975)] where one can find a stimulating discussion on this topic.

Here, we are interested in the space of $\tilde{\mathcal{F}}_q$-valued vector measures which covers a very large class of measures (not necessarily positive) with values in the space of generalized functionals of random fields. Let $D \subset R^n$, not necessarily bounded, and $\mathcal{B}_D$ the sigma algebra of subsets of the set D. Considering the output y as a vector measure valued function, it is preferably written as $y \equiv \mu_t(\cdot), t \geq 0$. Clearly, this is defined on the sigma algebra $\mathcal{B}_D$ and it takes values possibly from the Banach spaces like $\mathcal{G}_q$ (Theorem 6.3.2) or $\tilde{\mathcal{F}}_q$ (Theorem 6.6.5). Let $\mathcal{M}_{fabv}(\mathcal{B}_D, \tilde{\mathcal{F}}_q)$ denote the space of finitely additive vector measures furnished with the total variation norm. This is a Banach space and $\mathcal{M}_{cabv}(\mathcal{B}_D, \tilde{\mathcal{F}}_q)$ is a closed subspace of $\mathcal{M}_{fabv}(\mathcal{B}_D, \tilde{\mathcal{F}}_q)$ and so a Banach space. We are interested in random process like

$$\{\mu_t, t \in I : \text{ for each } t \in I, \mu_t \in \mathcal{M}_{fabv}(\mathcal{B}_D, \tilde{\mathcal{F}}_q)\}.$$

Let $B_\infty(D, \mathcal{F}_p)$ denote the class of bounded $\mathcal{F}_p$ valued $\mathcal{B}_D$ measurable functions which are uniform limits of $\mathcal{B}_D$ measurable simple functions. Furnished with the sup-norm topology,

$$\sup\{\| \zeta(s) \|_{\mathcal{F}_p}, s \in D\}, \zeta \in B_\infty(D, \mathcal{F}_p),$$

this is a Banach space. Now, it is easy to verify that every $\mu \in \mathcal{M}_{fabv}(\mathcal{B}_D, \tilde{\mathcal{F}}_q)$ defines a continuous linear functional on the Banach space $B_\infty(D, \mathcal{F}_p)$. That is, the functional ℓ_μ, given by

$$\ell_\mu(\zeta) \equiv \int_D < \zeta(s), \mu(ds) >_{\mathcal{F}_p, \tilde{\mathcal{F}}_q},$$

is a well defined continuous linear (and hence bounded) functional on $B_\infty(D, \mathcal{F}_p)$ and

$$|\ell_\mu(\zeta)| \leq \| \zeta \|_{B_\infty(D, \mathcal{F}_p)} |\mu|_{v,q}$$

where $|\mu|_{v,q}$ is the variation norm of μ with respect to the norm topology of $\tilde{\mathcal{F}}_q$. This is easily proved by use of simple functions and limiting arguments. In fact the following result shows that every continuous linear functional on $B_\infty(D, \mathcal{F}_p)$ is given by integration with respect to a measure $\mu \in \mathcal{M}_{fabv}(\mathcal{B}_D, \tilde{\mathcal{F}}_q)$.

Theorem 6.7.1 *Let $\{p, q\}$ be the conjugate pair with $1 \leq q \leq 2 \leq p < \infty$ and $\mathcal{F}_p, \tilde{\mathcal{F}}_q$ the dual pair of Banach spaces as introduced above. Then the following representation holds*

$$(B_\infty(D, \mathcal{F}_p))^* \cong \mathcal{M}_{fabv}(\mathcal{B}_D, \tilde{\mathcal{F}}_q).$$

Proof First we show that every $\mu \in \mathcal{M}_{fabv}(\mathcal{B}_D, \tilde{\mathcal{F}}_q)$ defines a continuous linear functional on $B_\infty(D, \mathcal{F}_p)$ through the expression

$$\ell_\mu(f) \equiv \int_D < f(x), \mu(dx) >_{\mathcal{F}_p, \tilde{\mathcal{F}}_q}.$$

Let $S_\infty(D, \mathcal{F}_p) \subset B_\infty(D, \mathcal{F}_p)$ denote the class of $\mathcal{B}_D$ measurable simple functions with values in $\mathcal{F}_p$ and χ_σ the characteristic function of any $\sigma \in \mathcal{B}_D$. Then, for every $f \in S_\infty(D, \mathcal{F}_p)$, there exists a finite family of pairwise disjoint $\mathcal{B}_D$-measurable sets $\{\sigma_i\}_{i=1}^n \subset D$ and $\{e_i\} \subset \mathcal{F}_p$ such that $f(x) \equiv \sum_{i=1}^n \chi_{\sigma_i}(x)e_i$ and

$$\ell_\mu(f) = \sum_{i=1}^n <e_i, \mu(\sigma_i)>_{\mathcal{F}_p, \tilde{\mathcal{F}}_q} .$$

Clearly,

$$\sum_{i=1}^n \| \mu(\sigma_i) \|_{\tilde{\mathcal{F}}_q} \leq \sup_\pi \sum_{\sigma \in \pi} \| \mu(\sigma) \|_{\tilde{\mathcal{F}}_q} \equiv |\mu|_{v,q}$$

where the supremum is taken over all finite disjoint $\mathcal{B}_D$ measurable partitions $\{\pi\}$ of the set D. Thus,

$$|\ell_\mu(f)| \leq \left(\sup_i |e_i|_{\mathcal{F}_p} \right) \left(\sum_i \| (\mu(\sigma_i) \|_{\tilde{\mathcal{F}}_q} \right) \leq c \| f \|_{B_\infty(D, \mathcal{F}_p)},$$

where $c \equiv |\mu|_{v,q}$ denotes the total variation norm of μ. This shows that ℓ_μ is a well defined continuous linear functional on $S_\infty(D, \mathcal{F}_p)$. Since $S_\infty(D, \mathcal{F}_p)$ is dense in $B_\infty(D, \mathcal{F}_p)$, and the above estimate holds for all $f \in B_\infty(D, \mathcal{F}_p)$, it follows from the principle of extension by continuity that ℓ_μ is defined on all of $B_\infty(D, \mathcal{F}_p)$). Hence, we have proved the inclusion

$$\mathcal{M}_{fabv}(\mathcal{B}_D, \tilde{\mathcal{F}}_q) \subset (B_\infty(D, \mathcal{F}_p))^*. \tag{6.33}$$

For the reverse inclusion, we verify that every continuous linear functional on $B_\infty(D, \mathcal{F}_p)$ or equivalently, every $\ell \in (B_\infty(D, \mathcal{F}_p))^*$, has the integral representation,

$$\ell(f) = \int_d <f(s), \mu(ds)>_{\mathcal{F}_p, \tilde{\mathcal{F}}_q},$$

for some $\mu \in \mathcal{M}_{fabv}(\mathcal{B}_D, \tilde{\mathcal{F}}_q)$. Again, we prove this first for simple functions and then use the limiting and density arguments. Let $f \in S_\infty(D, \mathcal{F}_p)$. Then, by definition there exists a finite family of disjoint $\mathcal{B}_D$ measurable sets $\{\sigma_i, i = 1, 2, \cdots, m\}$ covering D and $\{e_i\} \subset \mathcal{F}_p$ such that

$$f(x) = \sum_i \chi_{\sigma_i}(x)e_i.$$

Since ℓ is linear, we have

$$\ell(f) = \ell\left(\sum_i \chi_{\sigma_i} e_i\right) = \sum_i \ell(\chi_{\sigma_i} e_i). \tag{6.34}$$

For each i, define the functional ℓ_i by setting $\ell_i(e) \equiv \ell(\chi_{\sigma_i} e)$. Since $\chi_{\sigma_i}(\cdot) e_i \in B_\infty(D, \mathcal{F}_p)$ and ℓ is a bounded linear functional on it, the functional ℓ_i is a well defined bounded linear functional on $\mathcal{F}_p$. Recalling that $\tilde{\mathcal{F}}_q$ is the dual of $\mathcal{F}_p$, it is clear that for each functional ℓ_i, there exists a unique $\gamma_i \in \tilde{\mathcal{F}}_q$ such that $\ell_i(e) = < \gamma_i, e >_{\tilde{\mathcal{F}}_q, \mathcal{F}_p}$. Thus there exists a set function $\mu^o : \mathcal{B}_D \longrightarrow \tilde{\mathcal{F}}_q$ such that $\mu^o(\sigma_i) = \gamma_i$ for all $i \in \{1, 2, \cdots, m\}$ and $\mu^o(\emptyset) = 0$. By hypothesis, $\ell \in (B_\infty(D, \mathcal{F}_p))^*$ and so there exists a constant c such that

$$|\ell(f)| \leq c \parallel f \parallel_{B_\infty(D, \mathcal{F}_p)} \tag{6.35}$$

for all $f \in B_\infty(D, \mathcal{F}_p)$. Using the set function defined above, it follows from (6.34) that for the simple function f, we have

$$\ell(f) = \sum_i < \mu^o(\sigma_i), e_i > = \int_D < f(x), \mu^o(dx) >_{\mathcal{F}_p, \tilde{\mathcal{F}}_q} . \tag{6.36}$$

From the above expressions including the inequality (6.35), one can easily deduce that

$$\sum_{i=1}^m \parallel \mu^o(\sigma_i) \parallel_{\tilde{\mathcal{F}}_q} \leq c. \tag{6.37}$$

This holds for every $f \in S_\infty(D, \mathcal{F}_p)$ and hence for every finite disjoint $\mathcal{B}_D$ measurable partition of the set D. Thus we may conclude that the set function μ^o as defined above is finitely additive on $\mathcal{B}_D$ and has bounded variation. Further, it is uniquely determined by ℓ alone. Clearly, it follows from (6.35) and (6.37) that

$$|\ell|_* \equiv \sup\{|\ell(f)|, \parallel f \parallel_{B_\infty(D, \mathcal{F}_p)} \leq 1\} \leq |\mu^o|_{v,q} \leq c.$$

Choosing the smallest number c for which the inequality (6.35) holds, we obtain $|\ell|_* = |\mu^o|_{v,q}$. Thus we have proved that

$$(B(D, \mathcal{F}_p))^* \subset \mathcal{M}_{fabv}(\mathcal{B}_D, \tilde{\mathcal{F}}_q) \tag{6.38}$$

and that the embedding is an isometry. Conclusion of the theorem now follows from the inclusions (6.33) and (6.38). This completes the proof. •

For the study of measure valued random process $\{\mu_t, t \in I\}$, suppose that $t \longrightarrow \mu_t$ is weakly measurable in the sense that for any $\varphi \in B_\infty(D, \mathcal{F}_p)$ the function

$$t \longrightarrow \mu_t(\varphi) \equiv \int_D < \varphi(x), \mu_t(dx) >_{\mathcal{F}_p, \tilde{\mathcal{F}}_q}$$

is measurable. We are interested in functionals like

$$\ell(\mu) \equiv \int_{I \times D} < \varphi(t, x), \mu_t(dx) >_{\mathcal{F}_p, \tilde{\mathcal{F}}_q} dt.$$

The first concern is to determine the appropriate dual pairs for which such integrals are well defined. Define

$$\| \mu \| \equiv \inf\{\alpha \geq 0 : ess-sup_{t \in I}|\mu_t(\varphi)| \leq \alpha \, \| \varphi \|_{B_\infty(D, \mathcal{F}_p)}, \varphi \in B_\infty(D, \mathcal{F}_p)\}.$$

Note that, for $\varphi \in B_1(B_\infty(D, \mathcal{F}_p))$, the set $\{t \in I : ess - sup \, |\mu_t(\varphi)| \leq \alpha\}$ may depend on φ. This is the class of weakly measurable functions on I with values in the space of vector measures.

Closely related to Theorem 6.7.1, is the following result for stochastic processes with values in $\mathcal{M}_{fabv}(\mathcal{B}_D, \tilde{\mathcal{F}}_q)$.

Theorem 6.7.2 *The topological dual of the Banach space*

$$L_1(I, B_\infty(D, \mathcal{F}_p))$$

is the space of weakly measurable finitely additive bounded vector measures with values in $\tilde{\mathcal{F}}_q$. This is denoted by $L_\infty^w(I, \mathcal{M}_{fabv}(\mathcal{B}_D, \tilde{\mathcal{F}}_q))$. In other words,

$$\left(L_1(I, B_\infty(D, \mathcal{F}_p)) \right)^* \cong L_\infty^w(I, \mathcal{M}_{fabv}(\mathcal{B}_D, \tilde{\mathcal{F}}_q)).$$

Proof We present a brief outline of the proof. Consider any Banach space X and its dual X^*. It is well known [Diestel (1977)] (see Diestel, Theorem IV.1.1, p.98) that if X^* satisfies the Radon-Nikodym property (RNP), then $(L_1(I, X))^* \cong L_\infty(I, X^*)$. In case they lack this property, it follows from the theory of "lifting" [Tulcea (1969)](see Theorem 7 and its Corollary, p.94) that $(L_1(I, X))^* \cong L_\infty^w(I, X^*)$. The Banach spaces $B_\infty(D, \mathcal{F}_p)$ and its dual $\mathcal{M}_{cabv}(\mathcal{B}_D, \tilde{\mathcal{F}}_q)$ do not have the RNP and hence by use of the theory of lifting we arrive at the duality as stated in the theorem. This completes the outline of our proof. •

Example 6.7.3 Consider the nonlinear filtering problem. In the uncorrelated case, as seen in Chapter 1, the Zakai equation has the same form for both the finite and infinite dimensional filtering. Written in the weak form (see [Ahmed (1998)], [Ahmed (1997)]), this equation is given by

$$d\mu_t(\varphi) = \mu_t(\mathcal{A}\varphi)dt + \mu_t(\mathcal{B}\varphi) \cdot dV, t \geq 0, \mu_0(\varphi) = \nu(\varphi) \qquad (6.39)$$

where V is a finite dimensional R^d valued standard Brownian motion on a different probability space (see Chapter 2) and ν is the initial measure giving the distribution of the initial state. For the operators $\mathcal{A}, \mathcal{B}$ see Chapter 1. The test function $\varphi \in D(\mathcal{A}) \cap B_\infty(E)$ with $E = R^n$ in the case of finite dimensional filtering; and $E = H$, a separable Hilbert space, in the case of filtering in infinite dimensional spaces. By replacing D by E and $\mathcal{B}_D$ by $\mathcal{B}_E$, we can apply the results of Theorem 6.7.1 and Theorem 6.7.2. Recall that the action of the measure μ_t on a test function ψ is given by the integral

$$\mu_t(\psi) \equiv \int_E \psi(x)\mu_t(dx).$$

Clearly, for any such test function φ, $t \longrightarrow \mu_t(\varphi)$ is a random process adapted to the filtration induced by the standard Brownian motion V. Let $(\Omega_0, \mathcal{B}_0, \mu^V)$ denote the canonical Wiener measure space and let $M_n \equiv \mathcal{L}((R^d)^{\otimes n}, R)$, the space of n-linear forms furnished with the standard operator norm. If for each $t \in I$, and $\varphi \in D(\mathcal{A}) \cap B_\infty(E)$, $\mu_t(\varphi) \in \mathcal{G}_\beta \equiv L_2(\Omega_0, \mathcal{B}_0, \mu^V) \equiv L_2(\Omega_0, \mu^V)$, then it follows from Theorem 2.3.4 that there exists an element $K(t, \varphi) \in G_\beta$ such that $\mu_t(\varphi)$ has a unique Wiener-Itô representation giving

$$\mu_t(\varphi) = \sum_{n=0}^{\infty} f_n(K_n(t, \varphi))$$

$$\equiv \sum_{n=0}^{\infty} \int_{I_t^n} K_n(t, \varphi; \tau_1, \cdots, \tau_n)dV(\tau_1) \otimes \cdots \otimes dV(\tau_n). \quad (6.40)$$

This means that for each $t \in I$, $\varphi \longrightarrow K(t, \varphi)$ is a bounded linear map from $B_\infty(E)$ to G_β. And so, for each $t \in I$ and $n \in N$, $\varphi \longrightarrow K_n(t, \varphi)$ is a continuous linear map from $B_\infty(E)$ to $\hat{L}_2(I^n, M_n)$. Further, by virtue of the isometry, we have

$$\mathbb{E}^{\mu^V}\{|\mu_t(\varphi)|^2\} = \sum_{n=0}^{\infty} n! \parallel K_n(t, \varphi) \parallel^2_{L_2(I_t^n, M_n)} < \infty, I_t = [0, t], \quad (6.41)$$

for all $t \in I \equiv [0, T], T < \infty$, and $\varphi \in D(\mathcal{A}) \cap B_\infty(E)$. This is the regular case. But it is conceivable that, under relaxed assumptions on the pair of operators $\{\mathcal{A}, \mathcal{B}\}$, $\mu_t(\varphi)$ may be a generalized functional as discussed above. More precisely, $\mu_t \in \mathcal{M}_{fabv}(\mathcal{B}_E, \tilde{\mathcal{F}}_q)$ or equivalently $\mu_t(\varphi) \in \tilde{\mathcal{F}}_q$ for some $q \in [1, 2]$ and every $\varphi \in D(\mathcal{A}) \cap B_\infty(E)$. In other words, for every such φ and every $f \in \mathcal{F}_p$, we have

$$| < f, \mu_t(\varphi) >_{\mathcal{F}_p, \tilde{\mathcal{F}}_q} | \le \| f \|_{\mathcal{F}_p} \| \mu_t(\varphi) \|_{\tilde{\mathcal{F}}_q}, \tag{6.42}$$

and there exists an element $K(t, \varphi) \in \tilde{F}_q$ such that $\mu_t(\varphi) = \Phi(K(t, \varphi))$ and

$$\| \mu_t(\varphi) \|_{\tilde{\mathcal{F}}_q} = \| K(t, \varphi) \|_{\tilde{F}_q} \equiv \left(\sum_{n=0}^{\infty} (1/n!) \| K_n(t, \varphi) \|_{L_q(I_t^n, M_n)}^q \right)^{1/q}.$$

Clearly, this shows that the space of vector valued measures $\mathcal{M}_{fabv}(\mathcal{B}_E, \tilde{\mathcal{F}}_q)$ offers a much wider scope for the existence of solutions of the filter equation (6.39) corresponding to a larger class of determining operators $\{\mathcal{A}, \mathcal{B}\}$.

In the sequel we will return to this problem and consider optimization of observation operator $\mathcal{B}$.

6.8   Some Problems for Exercise

P1: Refer to Remark 6.2.5. Construct an example verifying this remark.

P2: Consider the Banach spaces $\mathcal{G}_p, \mathcal{G}_q$, for $1 \le q \le 2 \le p < \infty$ satisfying $(1/p) + (1/q) = 1$, and verify that the dual of $\mathcal{G}_p$ is $\mathcal{G}_q$.

P3: In reference to Remark 6.3.3, construct an example verifying the nonexistence of a continuous inclusion between the spaces $\mathcal{G}_p$ and $\mathcal{G}_q$, where $\{p, q\}$ are conjugate pairs with $1 \le q < 2$.

P4: Expand on the Remark 6.3.4 asserting its practical usefulness in the approximation of generalized functionals of Brownian motion. Hint: Suppose M is a weakly dense subset of the closed unit ball $B_1(\mathcal{G}_p)$. Show that $\mathcal{G}_q/M^\perp = \emptyset$.

P5: Let $K_1, K_2 \in L_q(I)$ for $1 \le q < 2$ and consider the random variables $Z_i \equiv \int_I K_i(t) dw(t), i = 1, 2$ where w is a standard Brownian motion. Use the duality argument to verify that $Z_1 = Z_2$ in the sense of generalized random variables if, and only if, $K_1 = K_2$ in the $Lq(I)$ sense.

P6: (a): Consider the input-output relation given by the expression (6.20) and prove the inequality given by (6.23) under the stated assumptions. (b): Consider the measure driven input-output model given by equation (6.27) and prove the inequality (6.29).

P7: Consider the stochastic partial differential equation (6.32) and verify if its solution $u \in L_\infty^{\ell oc}(I \times R^d, \tilde{\mathcal{F}}_2)$.

(P8): Consider the infinite dimensional filtering problem given in example 6.7.3 with measure solution $\mu_t \in \mathcal{M}_{fabv}(\mathcal{B}_E, \tilde{\mathcal{F}}_q)$ for some $q \in [1, 2]$. Verify that each of the representing kernels

$$K_n(t, \cdot) \in M_{fabv}(\mathcal{B}_E, L_q(I_t^n, M_n)).$$

Hint: Read the last part of the discussion following the example.

Chapter 7

$\mathcal{W}^{p,m}$-Based Generalized Functionals of White Noise IV

7.1 Introduction

In this Chapter we introduce several classes of generalized functionals, in the order of increasing generality, that cover the generalized functionals of Hida [Hida (1978)], Nualart and Zakai [Nualart (1989)], Ustunel [Korez (1990)] including those presented in the preceding chapters. We present also vector measures with values in the space of generalized functionals (Wiener-Itô distributions). Further, we present several interesting applications of the abstract results to problems in nonlinear filtering leading to optimal choice of observation operator, stochastic Navier-Stokes equations and optimal control of turbulence etc.

7.2 Homogeneous Functionals

Let $\mathcal{W}^{p,m,n}$ denote the Sobolev space of functions defined on I^n whose distributional derivatives up to order m belong to $L_p(I^n)$ with the usual norm topology. The space $\mathcal{W}_0^{p,m,n}$ is the completion in the topology of $\mathcal{W}^{p,m,n}$ of C^∞ functions having compact supports. The Sobolev space with negative norm (meaning negative exponents) denoted by $\mathcal{W}^{q,-m,n}$, where $(1/p) + (1/q) = 1$, is the dual of $\mathcal{W}_0^{p,m,n}$, and it is a class of distributions on I^n.

Throughout this section, whenever we repeat notations from the preceding chapter, we shall use the hat symbol to distinguish the generalized functionals constructed using Sobolev spaces from those constructed using

L_p spaces.

Let $1 \leq q \leq 2 \leq p < \infty$ satisfying $p^{-1} + q^{-1} = 1$. For each positive integer n, let $m(n)$ be a real number such that $m(n) > (n/p)$ and consider the Sobolev space $\mathcal{W}_0^{p,m(n),n}$ and it's dual $\mathcal{W}^{q,-m(n),n}$. Recall that $\mathcal{W}_0^{p,m,n}$ is the completion in the topology of $\mathcal{W}^{p,m,n}$ of C^∞ functions on I^n with compact supports. By Sobolev embedding theorem [Adams (1975)], $\mathcal{W}_0^{p,m(n),n} \hookrightarrow C^0(I^n)$ and therefore point values of elements of $\mathcal{W}_0^{p,m(n),n}$ are well defined. We shall need the following vector spaces. Recall that by $\hat{L}_p(I^n)$ we mean the space of symmetric kernels from $L_p(I^n)$. Define

$$\hat{G}^{p,m(n),n} \equiv \{L \in \hat{L}_p(I^n) : L \in \mathcal{W}_0^{p,m(n),n}\}$$

$$\hat{G}^{q,-m(n),n} \equiv \left(\hat{G}^{p,m(n),n}\right)^* = \text{ topological dual of } \hat{G}^{p,m(n),n}. \quad (7.1)$$

Furnished with the norm topology,

$$\| L_n \|_{\hat{G}^{p,m(n),n}} \equiv (n!)^{1/p} \| L_n \|_{\mathcal{W}^{p,m(n),n}}, \quad (7.2)$$

the space $\hat{G}^{p,m(n),n}$ is a Banach space. It's dual $\hat{G}^{q,-m(n),n}$ is endowed with it's natural topology given by

$$\| K_n \|_{\hat{G}^{q,-m(n),n}} \equiv (n!)^{1/q} \| K_n \|_{\mathcal{W}^{q,-m(n),n}} \quad (7.3)$$

where the norm of any $f \in \mathcal{W}^{q,-m(n),n}$, is given by

$$\| f \|_{\mathcal{W}^{q,-m(n),n}} \equiv Sup\{|f(\phi)| : \| \phi \|_{\mathcal{W}_0^{p,m(n),n}} \leq 1\}. \quad (7.4)$$

Recall that $\mathcal{S}^*$, the dual of a nuclear space $\mathcal{S}$, supports the measure μ induced by white noise. As introduced in Chapter 5, we use $M(\mathcal{S}^*, \mu)$ to denote the vector space of μ measurable functions defined on $\mathcal{S}^*$. Let ϕ_n denote, as in (6.1), the homogeneous functionals of degree n and define the linear vector space

$$\hat{\mathcal{G}}^{p,m(n),n} \equiv \{z \in M(\mathcal{S}^*, \mu) : z = \phi_n(L_n), L_n \in \hat{G}^{p,m(n),n}\}.$$

Furnished with the norm topology,

$$\| z \|_{\hat{\mathcal{G}}^{p,m(n),n}} \equiv \| L_n \|_{\hat{G}^{p,m(n),n}}, \quad (7.5)$$

this is a Banach space and it is isometrically isomorphic to $\hat{G}^{p,m(n),n}$. Since I is a finite interval, $\hat{G}^{p,m(n),n} \subset \hat{L}_p(I^n) \subset \hat{L}_2(I^n)$ for $p \geq 2$. Thus, these are the regular functionals of white noise or, equivalently, Wiener-Itô functionals. The generalized functionals are the images under ϕ_n (strictly speaking its extension denoted by the same symbol ϕ_n) of the Banach space $\hat{G}^{q,-m(n),n}$. Indeed the following result holds.

Lemma 7.2.1 *For each $K_n \in \hat{G}^{q,-m(n),n}, (1 \leq q \leq 2)$, the functional $z^* \equiv \phi_n(K_n)$ is a well defined generalized random element of $M(S^*, \mu)$.*

Proof The proof is based on Lemma 6.2.2 and the fact that the completion of $L_q(I^n)$ in the topology of $\mathcal{W}^{q,-m(n),n}$ is $\mathcal{W}^{q,-m(n),n}$ itself. For $\psi \in L_q(I^n)$, define

$$\| \psi \|_{\mathcal{W}^{q,-m(n),n}} \equiv \sup\{|\psi(\eta)|, \| \eta \|_{\mathcal{W}^{p,m(n),,n}} \leq 1\}.$$

As stated above, the completion of $L_q(I^n)$ in the topology induced by the above norm is precisely the Sobolev space $\mathcal{W}^{q,-m(n),n}$. Thus, for any $K_n \in \mathcal{W}^{q,-m(n),n}$, there exists a sequence $\{K_{n,s}\}_{s \in N} \in L_q(I^n)$ such that, for every $L_n \in \mathcal{W}_0^{p.m(n),n}$,

$$(K_{n,s}, L_n)_{\mathcal{W}^{q,-m(n),n}, \mathcal{W}^{p,m(n),n}} \longrightarrow (K_n, L_n)_{\mathcal{W}^{q,-m(n),n}, \mathcal{W}^{p,m(n),n}}$$

as $s \to \infty$. Hence, for each $L_n \in \hat{G}^{p,m(n),n}$,

$$(K_{n,s}, L_n)_{\hat{G}^{q,-m(n),n}, \hat{G}^{p,m(n),n}} \longrightarrow (K_n, L_n)_{\hat{G}^{q,-m(n),n}, \hat{G}^{p,m(n),n}}$$

as $s \to \infty$. We have seen in chapter 6 (see Lemma 6.2.2, Lemma 6.2.3) that for such kernels $\{K_{n,s}, s \in N\} \subset G^{q,n} \subset \hat{G}^{q,-m(n),n}$, we can define a sequence of generalized random variables $\{z_s^*\} \subset \mathcal{G}^{q,n} \subset \hat{\mathcal{G}}^{q,-m(n),n}$. Then, it follows from the same arguments as in section 6.2 of chapter 6 that there exists a unique $z^* \in \hat{\mathcal{G}}^{q,-m(n),n}$ such that, as $s \to \infty$,

$$z_s^*(z) \longrightarrow z^*(z)$$

for all $z \in \hat{\mathcal{G}}^{p,m(n),n}$. Hence z^* is a well defined generalized functional. This completes the proof. For details see [[Ahmed (1983)], Lemma 5.1]. •

The class of generalized functionals following from Lemma 7.2.1, is denoted by $\hat{\mathcal{G}}^{q,-m(n),n}$ and is given by $\phi_n(\hat{G}^{q,-m(n),n})$. We can topologize this vector space by a norm. For each $z^* \in \hat{\mathcal{G}}^{q,-m(n),n}$ we define it's norm by

$$\| z^* \|_{\hat{\mathcal{G}}^{q,-m(n),n}} \equiv \mathrm{Sup}\{|z^*(z)|, z \in \hat{\mathcal{G}}^{p,m(n),n}, \| z \|_{\hat{\mathcal{G}}^{p,m(n),n}} \leq 1\}. \quad (7.6)$$

In this case the natural duality pairing between z^* of $\hat{\mathcal{G}}^{q,-m(n),n}$ and z of $\hat{\mathcal{G}}^{p,m(n),n}$ is given by

$$z^*(z) \equiv \; < z^*, z >_{\hat{\mathcal{G}}^{q,-m(n),n}, \hat{\mathcal{G}}^{p,m(n),n}} \; = \; < L_n, K_n >_{\hat{G}^{q,-m(n),n}, \hat{G}^{p,m(n),n}}, \quad (7.7)$$

where $z^* = \phi_n(L_n)$ for some $L_n \in \hat{G}^{q,-m(n),n}$. With this introduction, we can now present the following fundamental result. Again we use ($\hookrightarrow$) for embeddings and $\updownarrow$ for isometric isomorphism respectively.

Theorem 7.2.2 *For each positive integer n, and $1 < q \leq 2 \leq p < \infty$ with $p^{-1} + q^{-1} = 1$, and $m(n) > (n/p)$, the following results hold:*

(i): $\hat{G}^{q,-m(n),n} = \left(\hat{G}^{p,m(n),n}\right)^*$ *and* $\hat{\mathcal{G}}^{q,-m(n),n} = \left(\hat{\mathcal{G}}^{p,m(n),n}\right)^*$.

(ii):

$$
\begin{array}{ccccc}
\hat{G}^{p,m(n),n} & \hookrightarrow & \hat{G}^{2,0,n} & \hookrightarrow & \hat{G}^{q,-m(n),n} \\
\updownarrow & & \updownarrow & & \updownarrow \\
\hat{\mathcal{G}}^{p,m(n),n} & \hookrightarrow & \hat{\mathcal{G}}^{2,0,n} & \hookrightarrow & \hat{\mathcal{G}}^{q,-m(n),n}
\end{array}
\qquad (7.8)
$$

with continuous and dense embeddings.

Proof The proof follows from Lemma 7.2.1 and the above discussion. For details see [Ahmed (1983)].

Remark 7.2.3 Note that for the special case with $p = q = 2$ and $m(n) = (n+1)/2$, we obtain the generalized functionals due to Hida as given by equation (5.18) of Theorem 5.4.1. The example given in the following remark also shows the power of the generalized functionals introduced above.

Remark 7.2.4 Since $m(n) > n/p$, $\mathcal{W}_0^{p,m(n),n} \hookrightarrow C^0(I^n)$ and hence one can easily verify that for any $\tau \in I, z^* \equiv (\dot{w}(\tau))^n \in \hat{\mathcal{G}}^{q,-m(n),n}$. Indeed, for $z_n \in \hat{\mathcal{G}}^{p,m(n),n}$ with kernel $K_n \in \hat{G}^{p,m(n),n}$, we have $< z^*, z_n > = n! K_n(\tau, \tau, \cdots, \tau)$. Since, by the embedding theorem, K_n is continuous it is clear that $| < z^*, z_n > | = n! |K_n(\tau, \tau, \cdots, \tau)| < \infty$. Thus z^* is a well defined bounded linear functional on $\hat{\mathcal{G}}^{p,m(n),n}$, while it is not an element of $L_2(S^*, \mu)$.

7.3 Nonhomogeneous Functionals

Like in the preceding section, for nonhomogeneous functionals we need appropriate Sobolev spaces. We start with the nonhomogeneous generalized functionals of degree $N \in N_+$. Define the vector spaces

$$\hat{G}_{p,N}^+ \equiv \prod_{n=0}^{N} \hat{G}^{p,m(n),n}, \quad 2 \leq p < \infty \tag{7.9}$$

$$\hat{G}_{q,N}^- \equiv \prod_{n=0}^{N} \hat{G}^{q,-m(n),n}, \quad 1 \leq q \leq 2; \tag{7.10}$$

where $m(n) > (n/p)$ for each positive integer n and $1/p + 1/q = 1$. These vector spaces, furnished with the norm topologies:

$$\| L \|_{\hat{G}_{p,N}^+} \equiv \Big(\sum_{n=0}^{N} \| L_n \|_{\hat{G}^{p,m(n),n}}^p \Big)^{1/p} \tag{7.11}$$

and

$$\| K \|_{\hat{G}_{q,N}^-} \equiv \Big(\sum_{n=0}^{N} \| K_n \|_{\hat{G}^{q,-m(n),n}}^q \Big)^{1/q} \tag{7.12}$$

respectively, are Banach spaces (reflexive if $q > 1$) and further $\hat{G}_{q,N}^- = \big(\hat{G}_{p,N}^+ \big)^*$.

Using Φ_N as in the expression (6.11), we introduce the vector spaces

$$\hat{\mathcal{G}}_{p,N}^+ \equiv \{ z \in F(S^*, \mu) : z = \Phi_N(L), L \in \hat{G}_{p,N}^+ \}, \tag{7.13}$$

for $2 \leq p < \infty$. In this case the dual $\hat{\mathcal{G}}_{q,N}^-$ of $\hat{\mathcal{G}}_{p,N}^+$ is a very large class of generalized functionals of white noise covering those given by Theorem 5.4.1 and Theorem 6.3.1. The following theorem is their generalization [[Ahmed (1983)],Theorem 5.2, 1983]. Again, we use $\hookrightarrow$ for embeddings or injections, and $\updownarrow$ and/or $\leftrightarrow$ for isometric isomorphisms.

Theorem 7.3.1 *Given the topologies as defined above, for* $1 < q \leq 2 \leq p < \infty$, *the following results hold:*

(i): $\hat{G}_{q,N}^-$ and $\hat{\mathcal{G}}_{q,N}^-$ are the (topological) duals of $\hat{G}_{p,N}^+$ and $\hat{\mathcal{G}}_{p,N}^+$ respectively.

(ii):

$$\hat{G}_{p,N}^+ \hookrightarrow \hat{G}_{2,N} \hookrightarrow \hat{G}_{q,N}^-$$

$$\updownarrow \qquad \updownarrow \qquad \updownarrow \qquad\qquad (7.14)$$

$$\hat{\mathcal{G}}_{p,N}^+ \hookrightarrow \hat{\mathcal{G}}_{2,N} \hookrightarrow \hat{\mathcal{G}}_{q,N}^-,$$

where the injections are continuous and dense.

Proof For detailed proof see [[Ahmed (1983)], Theorem 5.2]. •

Theorem 7.3.1 can be partially generalized as follows. Let $1 < q \le 2 \le p < \infty$ and introduce the vector spaces

$$\hat{G}_p^+ \equiv \left\{ L \in \prod_{n=0}^{\infty} \hat{G}^{p,m(n),n} : \ \| L \|_{\hat{G}_p^+} \equiv \left(\sum_{n=0}^{\infty} \| L_n \|_{\hat{G}^{p,m(n),n}}^p \right)^{1/p} < \infty \right\},$$

$$\hat{G}_q^- \equiv \left\{ K \in \prod_{n=0}^{\infty} \hat{G}^{q,-m(n),n} : \ \| K \|_{\hat{G}_q^-} \equiv \left(\sum_{n=0}^{\infty} \| K_n \|_{\hat{G}^{q,-m(n),n}}^q \right)^{1/q} < \infty \right\}.$$

Completed with respect to the norm topologies introduced above, these are Banach spaces. Consider Φ as the limit of Φ_N and introduce the vector space

$$\hat{\mathcal{G}}_p^+ \equiv \{ z \in F(S^*, \mu) : z = \Phi(L), L \in \hat{G}_p^+ \}.$$

Again, furnished with the norm topology given by

$$\| z \|_{\hat{\mathcal{G}}_p^+} \equiv \| L \|_{\hat{G}_p^+} \ \text{ for } z = \Phi(L), L \in \hat{G}_p^+,$$

it is easy to see that $\hat{\mathcal{G}}_p^+$ is a Banach space and that it is isometrically isomorphic to the Banach space $\hat{G}_p^+$. Similarly, we introduce the norm topology on

$$\hat{\mathcal{G}}_q^- \equiv \{ z \in F(S^*, \mu) : z^* = \Phi(K), K \in \hat{G}_q^- \},$$

by setting

$$\| z^* \|_{\hat{\mathcal{G}}_q^-} \equiv \sup\{ |z^*(z)|, z \in \hat{\mathcal{G}}_p^+, \| z \|_{\hat{\mathcal{G}}_p^+} \le 1 \}.$$

By virtue of the principle of extension by continuity the map Φ defined on $\hat{G}_p^+$ admits an extension to $\hat{G}_q^-$. We denote this extension by the same symbol. Thus, we have the following result which generalizes Theorem 6.3.2 and it is the limiting version of Theorem 7.3.1.

Theorem 7.3.2 *For* $1 < q \le 2 \le p < \infty$, *(i)* $\hat{G}_p^+$, $\hat{G}_q^-$, $\hat{\mathcal{G}}_p^+$, $\hat{\mathcal{G}}_q^-$ *are reflexive Banach spaces.* *(ii)* $\hat{\mathcal{G}}_p^+ \leftrightarrow \hat{G}_p^+$, $\hat{\mathcal{G}}_q^- \leftrightarrow \hat{G}_q^-$ *and* $(\hat{G}_p^+)^* = \hat{G}_q^-$, $(\hat{\mathcal{G}}_p^+)^* = \hat{\mathcal{G}}_q^-$.

There is no inclusion relation as in (7.14).

Proof For detailed proof see [[Ahmed (1983)], Theorem 5.3].●

For $p \in [2, \infty]$, and $q \in [1, 2]$, $(1/p + 1/q = 1)$, introduce the linear vector spaces $\mathbf{\Sigma}_p^+$ and $\mathbf{\Sigma}_q^-$ as defined by

$$\mathbf{\Sigma}_p^+ \equiv \left\{ K = (K_0, K_1, K_2, \cdots, K_n \cdots) : K_n \in \mathcal{W}_0^{p,m(n),n}, n \in N_+ \right\}$$

$$\equiv \prod_{n=0}^{\infty} \mathcal{W}_0^{p,m(n),n}; \tag{7.15}$$

$$\mathbf{\Sigma}_q^- \equiv \left\{ L = (L_0, L_1, L_2, \cdots, L_n \cdots) : L_n \in \mathcal{W}^{q,-m(n),n}, n \in N_+ \right\}$$

$$\equiv \prod_{n=0}^{\infty} \mathcal{W}^{q,-m(n),n}. \tag{7.16}$$

Note that $\mathbf{\Sigma}_q^-$ is the algebraic dual (not topological) of $\mathbf{\Sigma}_p^+$ for $1 \le q \le p < \infty$. These spaces can be furnished with a family of seminorms $\{q_n^+, q_n^-\}$ as seen in chapter 2 for the space $\mathcal{K}_\alpha$, in particular, the expression (2.6). Here, they are given by

$$\left\{ q_n^+(K) \equiv \parallel K_n \parallel_{\mathcal{W}_0^{p,m(n),n}}, \text{ for } K \in \mathbf{\Sigma}_p^+, n \in N \right\}$$

and

$$\left\{ q_n^-(L) \equiv \parallel L_n \parallel_{\mathcal{W}^{q,-m(n),n}}, \text{ for } L \in \mathbf{\Sigma}_q^-, n \in N \right\}.$$

Furnished with this family of seminorms, both $\mathbf{\Sigma}_p^+$ and $\mathbf{\Sigma}_q^-$ become Fréchet spaces.

As in Chapter 6, we shall use the same symbol Ψ for the algebraic isomorphism between these linear spaces and proper subspaces of the universal space $F(S^*, \mu)$ for the white noise functionals. Thus we have

$$\hat{\mathbf{\Sigma}}_p^+ \equiv \Psi(\mathbf{\Sigma}_p^+) \text{ and } \hat{\mathbf{\Sigma}}_q^- \equiv \Psi(\mathbf{\Sigma}_q^-). \tag{7.17}$$

Now we shall introduce several norm topologies and construct normed spaces so that this algebraic isomorphism becomes an isometry. Let

$\{a(n,s), n \in N_+, s \in R\}$ denote the class of weights introduced in the Definition 6.4.1. For $p \in [2, \infty), s \geq 0$, define

$$Y_{p,s}^+ \equiv \left\{ K \in \Sigma_p^+ : \left(\sum_{n=0}^{\infty} (a(n,s))^p \parallel K_n \parallel_{W_0^{p,m(n),n}}^p \right)^{1/p} < \infty \right\}. \quad (7.18)$$

Similarly, for $q \in (1, 2]$, the conjugate of p, and $s \geq 0$, define

$$Y_{q,-s}^- \equiv \left\{ L \in \Sigma_q^- : \left(\sum_{n=0}^{\infty} (a(n,-s))^q \parallel L_n \parallel_{W^{q,-m(n),n}}^q \right)^{1/q} < \infty \right\}. (7.19)$$

Endowed with the norm topology as defined above, $Y_{p,s}^+$ is a Banach space and for a fixed p, $Y_{p,s}^+ \subset Y_{p,r}^+$, for $s > r$. For $1 < q \leq 2 \leq p < \infty$, the topological dual of $Y_{p,s}^+$ is given by $Y_{q,-s}^-$, that is, $(Y_{p,s}^+)^* = Y_{q,-s}^-$.

Now we can introduce the following spaces of generalized functionals of white noise through the isomorphism Ψ:

$$\mathcal{Y}_{p,s}^+ \equiv \Psi(Y_{p,s}^+), \qquad \mathcal{Y}_{q,-s}^- \equiv \Psi(Y_{q,-s}^-). \quad (7.20)$$

In view of the norm topologies, the algebraic isomorphism Ψ now turns into a topological isomorphism. More precisely, we have the following results.

Theorem 7.3.3 *For $1 < q \leq 2 \leq p < \infty$ and $s \in R_+$, the following results hold:*

(i) $Y_{p,s}^+$, $Y_{q,-s}^-$, $\mathcal{Y}_{p,s}^+$, and $\mathcal{Y}_{q,-s}^-$ are reflexive Banach spaces.

(ii) $\mathcal{Y}_{p,s}^+ \leftrightarrow Y_{p,s}^+$, $\mathcal{Y}_{q,-s}^- \leftrightarrow Y_{q,-s}^-$ and $(Y_{p,s}^+)^* = Y_{q,-s}^-$, and $(\mathcal{Y}_{p,s}^+)^* = \mathcal{Y}_{q,-s}^-$.

Corollary 7.3.4 *For $1 < q \leq 2 \leq p < \infty$, $\{p,q\}$ a conjugate pair, and $s = 0$, results of Theorem 7.3.2 follow from those of Theorem 7.3.3; and, in particular, $\mathcal{Y}_{q,-0}^- = \hat{\mathcal{G}}_q^-$. The family of Banach spaces $\{\mathcal{Y}_{q,-s}^-, s \in R_+\}$ is an expanding family of spaces of generalized functionals of white noise more general than those of Theorem 7.3.2, and they are the duals of the Banach spaces $\{\mathcal{Y}_{p,s}^+, s \in R_+\}$.*

Remark 7.3.5 The results presented in the previous sections including the current one, are also valid for Gaussian random fields with I replaced by any bounded open domain $D \subset R^d$ with smooth boundary ∂D.

Remark 7.3.6 Choose $a(n,s) = (n!)^s$ and $s = 1/q$ where $\{p,q\}$ are the conjugate pairs used above and also take $m(n) = 0$ for all $n \in N$ and replace

I by $I \times D$. With this choice, we recover the results stated in Theorem 6.5.5 from those stated in Theorem 7.3.3.

An Example We present here briefly an example from partial differential equations with nonhomogeneous boundary data that contains white noise. This example will justify the need of generalized functionals of Brownian motion based on Sobolev spaces with negative exponents. Consider the one dimensional heat equation with the Neumann boundary condition as described below

$$\partial u/\partial t = \partial^2 u/\partial x^2, 0 \le x < \infty, t \ge 0$$
$$\partial u(t,0)/\partial x = -g(t), t > 0; \qquad (7.21)$$
$$u(0,x) = 0, 0 \le x < \infty,$$

where the boundary data $g(t) = \beta(\tau(t) - u(t,0))$ for convective heat transfer, and $g(t) = \gamma(\tau^4(t) - u^4(t,0))$ for radiative heat transfer. Here $\beta, \gamma (> 0)$ are the corresponding heat transfer coefficients and τ is the temperature of the flame in which one end of the rod is immersed. Due to chaotic nature of the flame, $\tau(t) = m(t) + n(t)$, where $n = \dot{w}$ is the white noise and m is the mean flame temperature. The temperature of the immersed end of the rod is given by the integral expression

$$u(t,0) = \int_0^t (1/\sqrt{\pi}(t-\tau))g(\tau)d\tau, t > 0. \qquad (7.22)$$

In fact, this is an integral equation since g depends on the unknown $u(t,0), t \in I$. In the case of convective heat transfer, we obtain a linear stochastic integral equation for $u(t,0)$ which contains an additive term of the form

$$S(t) = \beta \int_0^t (1/\sqrt{\pi}(t-\tau))\dot{w}(\tau)d\tau. \qquad (7.23)$$

In the case of radiative heat transfer, the corresponding term is given by

$$S(t) \equiv \int_{I_t^4} (1/\sqrt{\pi}(t-\tau_1))\delta(\tau_1-\tau_2)\delta(\tau_1-\tau_3)\delta(\tau_1-\tau_4) \prod_{i=1}^{i=4} \dot{w}(\tau_i)d\tau_i \quad (7.24)$$

where $I_t^4 \equiv [0,t]^4$. In the linear case it is clear that the kernel $K_1(t,\tau) \equiv (1/\sqrt{\pi}(t-\tau)) \in L_q(0,t)$ only for $1 \le q < 2$ and in the nonlinear case the kernel $K_4 \in \mathcal{W}^{q,-m,4}$ for $m > (4(q-1)/q)$. Clearly, in both the cases $S(t)$ is a generalized functional of white noise and belongs to the class considered in Chapters 6 and 7 but not in the Hida class (see Chapter 5).

7.4   Inductive and Projective Limits

The ultimate generalization of the results of the previous section is given in the following theorem.

Theorem 7.4.1 *Let q, p be the conjugate pair and $1 < q \leq 2 \leq p < \infty$. Then,*

(1): For $0 \leq r \leq s$, the injection (identity map) $\mathcal{Y}_{q,-r}^{-} \hookrightarrow \mathcal{Y}_{q,-s}^{-}$ is continuous.

(2): $\mathcal{Y}_{q}^{-} \equiv \bigcup_{r \geq 0} \mathcal{Y}_{q,-r}^{-}$ is the inductive limit of the spaces $\mathcal{Y}_{q,-r}^{-}$ which has the finest locally convex topology for which the imbedding $\mathcal{Y}_{q,-r}^{-} \hookrightarrow \mathcal{Y}_{q}^{-}$ is continuous for each $r \geq 0$.

(3): $\mathcal{Y}_{q}^{-}$ is the topological dual of the projective limit $\mathcal{Y}_{p}^{+} \equiv \bigcap_{r \geq 0} \mathcal{Y}_{p,r}^{+}$.

For $1 \leq q < 2$, the space $\mathcal{Y}_{q}^{-}$ is the space of generalized functionals of Brownian motion and the space $\mathcal{Y}_{p}^{+}$, for $\infty > p \geq 2$, is the space of test functionals.

The question of regularity of the inductive limit topology is significantly important in applications [see [Mujica (1977)]]. This is also evident from the following definition.

Definition 7.4.2 (D1): *The space $\mathcal{Y}_{q}^{-}, 1 \leq q \leq 2$, is said to be regular, if for every bounded set $B \subset \mathcal{Y}_{q}^{-}$, there exists a finite $r \geq 0$ such that B is a bounded subset of $\mathcal{Y}_{q,-r}^{-}$.*

(D2): *it is said to be compactly regular if every compact subset B of $\mathcal{Y}_{q}^{-}$ is a compact subset of $\mathcal{Y}_{q,-r}^{-}$ for some finite positive real number r.*

(D3): *it is said to be Cauchy-regular if, given a bounded subset B of $\mathcal{Y}_{q}^{-}$, there exists a finite positive number r such that B is a bounded subset of $\mathcal{Y}_{q,-r}^{-}$ and, furthermore, both $\mathcal{Y}_{q}^{-}$ and $\mathcal{Y}_{q,-r}^{-}$ induce the same Cauchy nets.*

Consider the spaces $\{\mathcal{Y}_{q,-r}^{-}, r \geq 0\}$ endowed with the standard weak topology inherited from that of the associated Sobolev spaces as seen above. Thus the phrase compact(Cauchy) used in the above definition (see (D2) and (D3)) means compact(Cauchy) in the weak topology. Recall that a set

B in a topological space is said to be bounded if, for every neighborhood U of the origin (zero vector), there exists a finite positive number α such that $\alpha B \subset U$.

Proposition 7.4.3 *Let* $1 < q \leq 2$ *and consider the space* $\mathcal{Y}_q^-$ *with the inductive limit topology. The space* $\mathcal{Y}_q^-$ *is regular and hence compactly regular.*

Proof We prove this by establishing a contradiction. For every bounded set $B \subset \mathcal{Y}_q^-$, there exists a finite positive number r_o such that $B \subset \mathcal{Y}_{q,-r_o}^-$. If it is false, then for every $r \in [0, \infty)$, there exists an element $f_r \in B$ such that $\| f_r \|_{\mathcal{Y}_{q,-r}^-} \geq r$. This implies that there is no neighborhood U of zero such that $\alpha B \subset U$ for some finite positive number α. This is a contradiction and hence there exists a finite positive number r such that B is a bounded subset of $\mathcal{Y}_{q,-r}^-$. This proves that $\mathcal{Y}_q^-$ is regular. Now we prove that this regularity implies that $\mathcal{Y}_q^-$ is compactly regular. Let B be a compact subset of it. Evidently, B is then a bounded set. Since $\mathcal{Y}_q^-$ is regular there exists a finite positive number r such that $B \subset \mathcal{Y}_{q,-r}^-$ and it is bounded. For $q > 1$, $\mathcal{Y}_{q,-r}^-$ is a reflexive Banach space and therefore B is relatively compact with respect to the weak topology. This completes the proof. ●

Remark 7.4.4 For $1 < q < \infty$, the Banach spaces $\{\mathcal{Y}_{q,-r}^-, r > 0\}$ are reflexive and so weakly sequential complete. Hence $\{\mathcal{Y}_{q,-r}^-, r > 0\}$ is Cauchy-regular.

Example 7.4.5 Here we wish to mention that the Example 6.7.8 on uncorrelated filtering given in chapter 6 can be further generalized by using the topological space $\mathcal{Y}_q^-$ instead of $\tilde{\mathcal{F}}_q$ (see the expression (6.31)). This provides a much broader class of spaces of generalized functionals of Brownian motion in which one may seek for (measure) solutions of a very large class of filtering problems. These spaces may also find applications in the study of nonlinear SPDE's.

7.5   Abstract Generalized Functionals

In this section, we present a class of abstract generalized functionals, in particular, with reference to Gaussian random fields of section 2.5 of chapter 2. Let V_n, furnished with the norm $\| \cdot \|_{V_n}$, be a linear subspace of $G_n \equiv \hat{L}_2(D^n, M_n)$ carrying the structure of a reflexive Banach space with continuous and dense embedding $V_n \hookrightarrow G_n$. Identifying G_n with its dual G_n^*, we obtain the Gelfand triple

$$V_n \hookrightarrow G_n \hookrightarrow V_n^* \tag{7.25}$$

with continuous and dense embeddings. Using the family of maps $\{h_n\}$ as introduced in section 2.5 of Chapter 2, we can construct the vector spaces

$$\mathcal{V}_n \equiv \{h_n(K), K \in V_n\}$$

$$\tag{7.26}$$

$$\mathcal{G}_n \equiv \{h_n(K), K \in G_n\},$$

with norms given by $\| \cdot \|_{\mathcal{V}_n} \equiv \sqrt{n!} \, \| \cdot \|_{V_n}$ and $\| \cdot \|_{\mathcal{G}_n} \equiv \sqrt{n!} \, \| \cdot \|_{G_n}$. The following Lemma characterizes the topological dual $\mathcal{V}_n^*$ of the space $\mathcal{V}_n$.

Lemma 7.5.1 *For each $n \in N$, the map h_n admits a unique continuous extension $\tilde{h}_n$ from V_n to V_n^* and the dual $\mathcal{V}_n^*$ of $\mathcal{V}_n$ is a class of generalized functionals of the Gaussian random field $\{W(\Gamma), \Gamma \in \mathcal{B}_D\}$, characterized by*

$$\mathcal{V}_n^* \equiv \{\tilde{h}_n(L), L \in V_n^*\}$$

with $\mathcal{V}_n$ being the space of test functionals. Further, the injections

$$\mathcal{V}_n \hookrightarrow \mathcal{G}_n \hookrightarrow \mathcal{V}_n^*$$

are continuous and dense.

Proof For detailed proof see [Ahmed (1995)]. Here we give a brief outline. Since G_n is dense in V_n^*, for each $L \in V_n^*$ there exists a sequence $\{L_r\} \in G_n$ such that $L_r \xrightarrow{s} L$ in V_n^* . Thus, for each $K \in V_n$, the following pairings are well defined

$$< h_n(L_r), h_n(K) >_{\mathcal{V}_n^*, \mathcal{V}_n} \equiv (h_n(L_r), h_n(K))_{\mu^w}$$

where $(,)_{\mu^w}$ denotes the expectation with respect to the Wiener measure induced by the Gaussian random field. Since $K \in V_n \subset G_n$, we have

$h_n(K) \in \mathcal{V}_n \subset \mathcal{G}_n$. Thus it follows from the above identity and the embedding $\mathcal{G}_n \hookrightarrow \mathcal{V}_n^*$ that

$$< h_n(L_r), h_n(K) >_{\mathcal{V}_n^*, \mathcal{V}_n} \equiv (h_n(L_r), h_n(K))_{\mu^w}$$
$$= n! < L_r, K >_{\mathcal{G}_n, \mathcal{G}_n}$$
$$= n! < L_r, K >_{V_n^*, V_n} . \qquad (7.27)$$

Clearly, $\lim_{r \to \infty} < L_r, K >_{V_n^*, V_n} = < L, K >_{V_n^*, V_n}$ and hence

$$\lim_{r \to \infty} < h_n(L_r), h_n(K) >_{\mathcal{V}_n^*, \mathcal{V}_n} = n! < L, K >_{V_n^*, V_n} .$$

Since the right hand expression of the above identity defines a natural duality pairing, it is well defined and finite, and so the left hand limit is also well defined and finite. We denote this limit by $\tilde{h}_n(L)$. Thus, for each $L \in V_n^*$, $\tilde{h}_n(L)$ makes sense as an element of $\mathcal{V}_n^*$. The uniqueness follows from the fact that the limit is independent of the choice of the approximating sequence. This also shows that to every $L \in V_n^*$, there corresponds a unique $z^* \in \mathcal{V}_n^*$ through the map $L \longrightarrow \tilde{h}_n(L) \equiv z^*$. Now, we show that every $f \in \mathcal{V}_n^*$ has this representation. Since, by construction, each $\varphi \in \mathcal{V}_n$ has the representation $\varphi = h(L)$ for some $L \in V_n$, it is clear that $f(\varphi) = f(h(L)) = (f \circ h)(L)$ is a continuous linear functional on V_n. Hence, the composition map $f \circ h \in V_n^*$ and there exists a unique $K \in V_n^*$ such that $f(\varphi) = (f \circ h)(L) = n! < K, L >_{V_n^*, V_n} \equiv < \tilde{h}(K), h(L) >_{\mathcal{V}_n^*, \mathcal{V}_n}$. This proves the first part of the statement. The second part follows from the fact that the embeddings $V_n \hookrightarrow G_n \hookrightarrow V_n^*$ are continuous and dense by our construction and these properties are preserved under isomorphism. Thus the embeddings $\mathcal{V}_n \hookrightarrow \mathcal{G}_n \hookrightarrow \mathcal{V}_n^*$ are also continuous and dense. This completes the proof. •

Since the pair of Banach spaces $\{V_n, V_n^*\}$ is assumed to be reflexive, it is clear that the pair of Banach spaces $\{\mathcal{V}_n, \mathcal{V}_n^*\}$ is also reflexive. We shall prove that the norm topology of the Banach space $\mathcal{V}_n^*$, which is compatible with the duality pairings of Lemma 7.5.1, is given by

$$\| \tilde{h}_n(L) \|_{\mathcal{V}_n^*} = \sqrt{n!} \| L \|_{V_n^*}, \text{ for } L \in V_n^*.$$

For convenience of notation, we set $\hat{V}_n \equiv \sqrt{n!} V_n$, $\hat{G}_n \equiv \sqrt{n!} G_n$, and $\hat{V}_n^* \equiv \sqrt{n!} V_n^*$. We need the following result before we can state our main result of this section.

Lemma 7.5.2 *For each $n \in N$, and for the given topologies for the spaces $\{V_n, G_n, V_n^*\}$, the following diagram holds*

$$\hat{V}_n \hookrightarrow \hat{G}_n \hookrightarrow \hat{V}_n^*$$

$$\updownarrow \quad \updownarrow \quad \updownarrow \qquad (7.28)$$

$$\mathcal{V}_n \hookrightarrow \mathcal{G}_n \hookrightarrow \mathcal{V}_n^*,$$

with the embeddings being continuous and dense, and the vertical pairs being isometrically isomorphic. Further, the Banach spaces $\{\mathcal{V}_n, \mathcal{G}_n, \mathcal{V}_n^*\}$ are reflexive.

Proof The continuity and density of the injections follow from Lemma 7.5.1. The isometries in the first two columns follow from the definition of norms following the introduction of the spaces $\{\mathcal{V}_n, \mathcal{G}_n\}$ (see (7.26)). It remains to justify the isometry of the last column. Without loss of generality, we may assume that the spaces V_n and V_n^* are strictly convex. This follows from Asplund's theorem that states that every reflexive Banach can be renormed in such a way that the unit ball is strictly convex and that the new norm is equivalent to the original norm. Let $J_n : V_n^* \setminus \{0\} \longrightarrow V_n$ denote the duality map. By virtue of strict convexity, the map J_n is single valued. Thus by Hahn-Banach theorem, for each $e^*(\not\equiv 0) \in V_n^*$, there exists a unique $e \in V_n$ such that $e = J_n(e^*)$ and

$$< e^*, e >_{V_n^*, V_n} = < e^*, J_n(e^*) > = \| e^* \|_{V_n^*}^2 = \| e \|_{V_n}^2 = \| J_n(e^*) \|_{V_n}^2 .$$

By Lemma 7.5.1, for each $L \in V_n^*$ and $K \in V_n$ we have

$$< \tilde{h}_n(L), \tilde{h}_n(K) >_{\mathcal{V}_n^*, \mathcal{V}_n} = < \tilde{h}_n(L), h_n(K) >_{\mathcal{V}_n^*, \mathcal{V}_n}$$

$$= n! < L, K >_{V_n^*, V_n} . \qquad (7.29)$$

Choosing $K \equiv \left((1/\sqrt{n!}) \, \| L \|_{V_n^*} \right) J_n(L)$ and substituting in (7.29), one can easily verify that

$$< \tilde{h}_n(L), h_n(K) >_{\mathcal{V}_n^*, \mathcal{V}_n} = \sqrt{n!} \, \| L \|_{V_n^*} . \qquad (7.30)$$

Note that for this choice of K, $\| h_n(K) \|_{\mathcal{V}_n} = 1$. On the other hand, for every $K \in \mathcal{V}_n$,

$$| < \tilde{h}_n(L), h_n(K) >_{\mathcal{V}_n^*, \mathcal{V}_n} | \le \left(\sqrt{n!} \, \| L \|_{V_n^*} \right) \| h_n(K) \|_{\mathcal{V}_n} .$$

This shows that

$$\| \tilde{h}_n(L) \|_{\mathcal{V}_n^*} \leq \sqrt{n!} \, \| L \|_{V_n^*} . \tag{7.31}$$

Hence, it follows from (7.30) and the above inequality that

$$\| \tilde{h}_n(L) \|_{\mathcal{V}_n^*} = \| h_n(L) \|_{\mathcal{V}_n^*} = \sqrt{n!} \, \| L \|_{V_n^*}$$

for all $L \in V_n^*$. Thus we have justified the isometry indicated by the last column. The last statement follows from the reflexivity of the spaces $\{V_n, G_n, V_n^*\}$ and the fact that reflexivity is preserved under isomorphism. This completes the proof. •

An Example Let $D \subset R^d$ be an open bounded set with smooth boundary and $1 < q \leq 2 \leq p < \infty$ satisfying $(1/p) + (1/q) = 1$. For any $n \geq 1$, choose any number $m \geq (nd/p) + 1$ and take $V_n = W_0^{m,p}(D^n, M_n) \subset \hat{L}_2(D^n, M_n) \equiv G_n$ (M_n = symmetric n-linear forms as introduced in section 2.3) with the dual $V_n^* = W^{-m,q}(D^n, M_n^*)$ giving the Gelfand triple $V_n \hookrightarrow G_n \hookrightarrow V_n^*$ with continuous and dense embeddings. Clearly, the larger the value of m is the smaller are the spaces V_n and $\mathcal{V}_n$, and the larger are the corresponding duals V_n^* and $\mathcal{V}_n^*$. The space $\mathcal{V}_n^*$ consists of a very large class of n-homogeneous generalized functionals of Gaussian random fields.

This result generalizes the results of Ahmed [Ahmed (1968, 1973, 1969)], Hida [Hida (1978)], and Nualrat and Zakai [Nualart (1989)].

Now we are prepared to extend these results to cover nonhomogeneous generalized functionals. First we introduce the product spaces

$$\Sigma \equiv \prod_{n=0}^{\infty} V_n; \quad \Xi \equiv \prod_{n=0}^{\infty} V_n^*; \quad \mathcal{L}(\Sigma, \Xi) \equiv \prod_{n=0}^{\infty} \mathcal{L}(V_n, V_n^*).$$

Each of these spaces can be furnished with a countable family of suitable seminorms which turn them into Fréchet Spaces. Clearly, Ξ is the algebraic dual of Σ, and $L(\Sigma, \Xi)$ is the space of Linear operators not necessarily bounded though each component $\mathcal{L}(V_n, V_n^*)$ is the space of bounded linear operators. Using these spaces one can construct various locally convex topological vector spaces of generalized functionals of Gaussian random fields. For simplicity, here we consider only Hilbertian topologies. From now on, it is assumed that the sequence of spaces $\{V_n, V_n^*\}$ have Hilbertian structure and we can then introduce the following Hilbert spaces:

$$V \equiv \sum_{n \geq 0} \oplus \hat{V}_n \equiv \sum_{n \geq 0} \oplus \sqrt{n!} V_n \subset \Sigma$$

$$G \equiv \sum_{n \geq 0} \oplus \hat{G}_n \equiv \sum_{n \geq 0} \oplus \sqrt{n!} G_n \subset \mathcal{K}$$

$$V^* \equiv \sum_{n \geq 0} \oplus \hat{V}_n^* \equiv \sum_{n \geq 0} \oplus \sqrt{n!} V_n^* \subset \Xi$$

with the corresponding norms given by

$$\| R \|_X \equiv \left(\sum_{n \geq 0} \| R_n \|_{X_n}^2 \right)^{1/2}$$

where $X_n \equiv \{V_n, G_n, V_n^*\}$ and $X \equiv \{V, G, V^*\}$. Using these spaces and identifying h_n with its extension $\tilde{h}_n$, we construct the vector spaces of generalized Gaussian random fields as follows:

$$\mathcal{V} \equiv \{\phi : \phi = \sum_{n \geq 0} h_n(K_n), K \equiv \{K_0, K_1, \cdots\} \in V\} \qquad (7.32)$$

$$\mathcal{V}^* \equiv \{\psi : \psi = \sum_{n \geq 0} h_n(L_n), L \equiv \{L_0, L_1, \cdots\} \in V^*\}. \qquad (7.33)$$

By virtue of Lemma 7.5.1 and Lemma 7.5.2, the natural duality pairing between these spaces is given by

$$< \psi, \phi >_{\mathcal{V}^*, \mathcal{V}} \equiv \sum_{n \geq 0} < h_n(L_n), h_n(K_n) >_{\mathcal{V}_n^*, \mathcal{V}_n} = \sum_{n \geq 0} n! < L_n, K_n >_{V_n^*, V_n}$$

for $\psi \in \mathcal{V}^*$ and $\phi \in \mathcal{V}$. In view of Lemma 7.5.2, one can introduce norm topologies in $\mathcal{V}$ and $\mathcal{V}^*$ as follows.

$$\| \phi(K) \|_{\mathcal{V}} \equiv \left(\sum_{n \geq 0} \| h_n(K_n) \|_{\mathcal{V}_n}^2 \right)^{1/2}$$

$$\qquad (7.34)$$

$$\| \psi(L) \|_{\mathcal{V}^*} \equiv \left(\sum_{n \geq 0} \| h_n(L_n) \|_{\mathcal{V}_n^*}^2 \right)^{1/2}.$$

From the pairing defined above, one can easily verify that for each $e^* \in \mathcal{V}^*$ there exists a unique $e \in \mathcal{V}$ with $\| e \|_{\mathcal{V}} = 1$ such that $< e^*, e >_{\mathcal{V}^*, \mathcal{V}} = \| e \|_{\mathcal{V}^*}$.

Now we are ready to state the main result of this section.

Theorem 7.5.3 *The spaces $\mathcal{V}$ and $\mathcal{V}^*$ are isometrically isomorphic to V and V^* respectively. Further, if c_n is the embedding constant for the inclusion $V_n \hookrightarrow G_n$ and there exists a constant c such that $\infty > c \geq \sup\{c_n, n \geq 0\}$, then the following diagram holds:*

$$V \hookrightarrow G \hookrightarrow V^*$$

$$\updownarrow \quad \updownarrow \quad \updownarrow \qquad\qquad (7.35)$$

$$\mathcal{V} \hookrightarrow \mathcal{G} \hookrightarrow \mathcal{V}^*,$$

where the embeddings are continuous and dense with the embedding constant given by c.

Proof The isometries follow from the definition of norm topologies and the Lemma 7.5.1 and Lemma 7.5.2. For continuity of the inclusions, let $K \in V$. Then computing its G norm we have

$$\| K \|_G^2 = \sum_{n \geq 0} n! \, \| K_n \|_{G_n}^2 \leq \sum_{n \geq 0} n!(c_n)^2 \, \| K_n \|_{V_n}^2 \leq c^2 \, \| K \|_V^2 \,.$$

Hence $\| K \|_G \leq c \, \| K \|_V$ which proves continuity of the inclusion $V \hookrightarrow G$. For the continuity of the second inclusion, note that it follows from reflexivity of the Banach spaces that the imbedding constant for the inclusion $G_n \hookrightarrow V_n^*$ is also c_n. Hence $\| K \|_{V^*} \leq c \, \| K \|_G$ for every $K \in G$. Thus, we have shown that the embeddings $V \hookrightarrow G \hookrightarrow V^*$ are continuous with embedding constant c. Hence by virtue of the isomorphism, the inclusions $\mathcal{V} \hookrightarrow \mathcal{G} \hookrightarrow \mathcal{V}^*$ are also continuous with the same embedding constant. The density follows from those of the inclusions of the components $V_n \hookrightarrow G_n \hookrightarrow V_n^*$. This completes the proof. •

The Gelfand triple $\{\mathcal{V}, \mathcal{G}, \mathcal{V}^*\}$ has been constructed using $\mathcal{G}$ as the pivot space. If instead, we use $\mathcal{H}$ which is the isomorphic image of the Hilbert space given by

$$H \equiv \{K \in \mathcal{K} : \| K \|_H \equiv \Big(\sum_{n \geq 0} \| K_n \|_{L_2(D^n, M_n)}^2 \Big)^{1/2} < \infty\},$$

that is, $\mathcal{H} \equiv \Phi_\delta(H)$, we can construct yet another class of Gelfand triple. This is done as follows. Let $b : N \longrightarrow [1, \infty]$ be a monotone nondecreasing function satisfying $b(n) \geq 1$ for each $n \in N$. Introduce the vector space

$$Z_b \equiv \{K \in \mathcal{K} : K_n \in V_n \text{ and } \Big(\sum_{n \geq 0} b(n) \, \| K_n \|_{V_n}^2 \Big)^{1/2} < \infty\}.$$

Under the assumptions of Theorem 7.5.3, it is easy to verify that the injection $Z_b \hookrightarrow H$ is continuous and dense. Identifying H with its dual, as before, we obtain the Gelfand triple $Z_b \hookrightarrow H \hookrightarrow Z_b^*$ where Z_b^* is the topological dual of Z_b.

Now we can introduce the generalized functionals of Gaussian random fields through the isomorphism Φ_δ and its extension (which we denote by Φ_δ again) of section 2.5 of chapter 2 giving

$$\Phi_\delta(Z_b) = \mathcal{Z}_b, \quad \Phi_\delta(H) = \mathcal{H}, \quad \text{and} \quad \Phi_\delta(Z_b^*) = \mathcal{Z}_b^*.$$

Choice of the map $b(n), n \in N$, is determined by the requirements of applications. Note that the norm for the space Z_b^* is given by

$$\| L \|_{Z_b^*} = \left(\sum_{n \geq 0} (1/b(n)) \| L_n \|_{V_n^*}^2 \right)^{1/2} < \infty \qquad (7.36)$$

and the duality pairing between $\mathcal{Z}_b^*$ and $\mathcal{Z}_b$ is given by

$$< z^*, z >_{\mathcal{Z}_b^*, \mathcal{Z}_b} = \sum_{n \geq 0} (1/n!) < h_n(L_n), h_n(K_n) >_{V_n^*, V_n} . \qquad (7.37)$$

This leads us to the following result.

Theorem 7.5.4 *Suppose the embedding constants for $V_n \hookrightarrow G_n$ satisfy the conditions of Theorem 7.5.3 and the map $b : N \longrightarrow [1, \infty)$ is a nondecreasing function of its argument. Then the following diagram holds:*

$$Z_b \hookrightarrow H \hookrightarrow Z_b^*$$
$$\updownarrow \quad \updownarrow \quad \updownarrow \qquad (7.38)$$
$$\mathcal{Z}_b \hookrightarrow \mathcal{H} \hookrightarrow \mathcal{Z}_b^*,$$

where the embeddings are continuous and dense with embedding constant c.

Remark 7.5.5 Following identical procedure, the results presented in the preceding sections can be easily extended to cover generalized functionals of fractional Brownian motion and Lévy process.

Functionals of Brownian motion have many applications in physical and biological sciences. The results presented in this chapter have applications in stochastic differential and functional differential equations and systems theory [see [Ahmed (1968)], [Ahmed (1973)] [Ahmed (1969)] [Ahmed

(1970)] [Ahmed (1983)]], in Malliavin calculus [see Bell [Bell (1987)]], and in stochastic partial differential equations [see Nualart and Zakai [Nualart (1989)], Nualart and Rozovskii [Nualart (1997)], Ahmed [Ahmed (1983)], Ahmed, Fuhrman and Zabczyk [Ahmed (1997)]], [Oksendal (1996)], in quantum mechanics [see Hida, [Hida (1978)], [Hida (1980)]]. The problem of analysis and synthesis of input-output maps based on optimization in the Hilbert space $L_2(\Omega, \mathcal{B}, \mu_\lambda) \equiv G_\lambda$ for stochastic functional differential equations was studied in [Ahmed (1973)].

Interesting and new areas of applications are nonlinear filtering, control theory, stochastic Navier-Stokes equations and associated control problems. We discuss them in details in the next few sections after presenting some additional results.

7.6 Vector Measures with Values from Wiener-Itô Distributions

In this section we wish to consider vector measures like those considered in section 6.5 of Chapter 6. All the results of the preceding sections can be readily extended to Gaussian random fields. Particularly interesting is the space of vector measures similar to those of Theorems 6.5.6, and Theorem 6.5.7. Let D be any separable complete metric space and $\mathcal{B}_D$ the sigma algebra of Borel subsets of the set D. The most interesting space of vector measures is the space of finitely additive measures having bounded variation and taking values from the Banach space of generalized functionals of Brownian motion and random fields $\{\mathcal{Y}_{q,-s}^-, q \geq 1, s \geq 0\}$ as introduced in section 7.3. This is denoted by

$$\mathcal{M}_{fabv}(\mathcal{B}_D, \mathcal{Y}_{q,-s}^-).$$

This space can accommodate solutions of stochastic differential and integral equations which are measures taking values from the space of Wiener-Itô distributions. We just state the results similar to those of Theorem 6.5.6 and Theorem 6.5.7 as mentioned above. The proofs are identical.

Theorem 7.6.1 *Let $\{p, q\}$ be the conjugate pair with $1 \leq q \leq 2 \leq p < \infty$, $s \geq 0$, and $\{\mathcal{Y}_{p,s}^+, \mathcal{Y}_{q,-s}^-\}$ be the dual pair of Banach spaces as introduced* *above. Then, the following representation (isometric isomorphism) holds*

$$(B_\infty(D, \mathcal{Y}_{p,s}^+))^* \cong \mathcal{M}_{fabv}(\mathcal{B}_D, \mathcal{Y}_{q,-s}^-).$$

Proof Proof is similar to that of Theorem 6.5.6.

Closely related to Theorem 7.6.1, is the following result for stochastic processes defined on $I \equiv [0, T]$ with values in $\mathcal{M}_{fabv}(\mathcal{B}_D, \mathcal{Y}_{q,-s}^-)$. We have the following result.

Theorem 7.6.2 *The topological dual of the Banach space*

$$L_1(I, B_\infty(D, \mathcal{Y}_{p,s}^+))$$

is the space of weakly measurable finitely additive bounded vector measures with values in $\mathcal{Y}_{q,-s}^-$ which is denoted by $L_\infty^w(I, \mathcal{M}_{fabv}(\mathcal{B}_D, \mathcal{Y}_{q,-s}^-))$. In other words,

$$\left(L_1(I, B_\infty(D, \mathcal{Y}_{p,s}^+)) \right)^* \cong L_\infty^w(I, \mathcal{M}_{fabv}(\mathcal{B}_D, \mathcal{Y}_{q,-s}^-)).$$

Proof Proof is identical to that of Theorem 6.5.7.

Similar results hold for the spaces $\{\mathcal{Z}_b, \mathcal{H}, \mathcal{Z}_b^*\}$ as presented below.

Theorem 7.6.3 *Consider the dual pair of Banach spaces $\mathcal{Z}_b, \mathcal{Z}_b^*$ as introduced above. Then the following representation holds*

$$(B_\infty(D, \mathcal{Z}_b))^* \cong \mathcal{M}_{fabv}(\mathcal{B}_D, \mathcal{Z}_b^*).$$

Proof Proof is similar to that of Theorem 6.5.6.

Again, for stochastic processes, closely related to Theorem 7.6.3 is the following result.

Theorem 7.6.4 *The topological dual of the Banach space $L_1(I, B_\infty(D, \mathcal{Z}_b))$ is the space of weakly measurable finitely additive bounded vector measures with values in $\mathcal{Z}_b^*$ which is denoted by $L_\infty^w(I, \mathcal{M}_{fabv}(\mathcal{B}_D, \mathcal{Z}_b^*))$. In other words,*

$$\left(L_1(I, B_\infty(D, \mathcal{Z}_b)) \right)^* \cong L_\infty^w(I, \mathcal{M}_{fabv}(\mathcal{B}_D, \mathcal{Z}_b^*)).$$

Proof Proof is identical to that of Theorem 6.5.7.

Remark 7.6.5 If D is compact, then finitely additive measures have simple countably additive extensions. Hence, in this situation we may replace $\mathcal{M}_{fabv}(\mathcal{B}_D, \mathcal{Y}_{q,-s}^-)$ by $\mathcal{M}_{cabv}(\mathcal{B}_D, \mathcal{Y}_{q,-s}^-)$ and $\mathcal{M}_{fabv}(\mathcal{B}_D, \mathcal{Z}_b^*)$ by $\mathcal{M}_{cabv}(\mathcal{B}_D, \mathcal{Z}_b^*)$.

We expect these vector measures to find applications in filtering and in the study of measure solutions [Ahmed (2005)] [Ahmed (2007)] [Ahmed (1996)] [Ahmed (1999)] of equations of population dynamics, plasma dynamics, Navier-Stokes equations [Ahmed (1998)], and McKean-Vlasov and Maxwell-Vlasov equations, and in general Physics. In the following two sections we consider application to Nonlinear Filtering and stochastic Navier-Stokes equations in some details.

7.7 Application to Nonlinear Filtering

Here we consider Nonlinear Filtering equations in infinite dimensional space. We discuss its approximation, representation and optimal choice of observer.

Generalized Filtering GF:

We consider the filtering problem in the correlated case where the observation operator $\mathcal{B}$ is a differential operator (thus involving unbounded observation operator). In particular, the filtering example considered in section 6.5, can be substantially generalized by admitting solutions $\mu \in L_\infty^w(I, \mathcal{M}_{fabv}(\mathcal{B}_D, \mathcal{Y}_{q,-s}^-))$ or $\mu \in L_\infty^w(I, \mathcal{M}_{fabv}(\mathcal{B}_D, \mathcal{Z}_b^*))$. If the measurement noise affects the dynamics, the filtering equations are given by

$$d\xi = A\xi dt + B(\xi)dt + R_1^{1/2}dW_1 + CdW_2 \qquad (7.39)$$

$$dy = h(\xi)dt + R_2^{1/2}dW_2 \qquad (7.40)$$

where the first equation is the state equation and the second is the measurement equation. This is the correlated case. Note the distinction between equations (1.39) and (7.39). In this case the filtering equation has the same form as that of Chapter 6, given by (6.39). Written in the weak form, this is given by,

$$d\mu_t(\varphi) = \mu_t(\mathcal{A}\varphi)dt + \mu_t(\mathcal{B}\varphi) \cdot dV, t \geq 0, \mu_0(\varphi) = \nu(\varphi) \qquad (7.41)$$

with $\varphi \in D(\mathcal{A}) \cap B_\infty(D), D = H$, and V an R^d valued innovation process. But now the operator $\mathcal{B}$ is an unbounded operator and it is given by

$$\mathcal{B}\varphi \equiv (R_2^{-1/2}h + C^*D_F)\varphi \equiv (\Gamma + C^*D_F)\varphi$$

where $\Gamma \equiv R_2^{-1/2}h$ is the observation (or measurement) operator which is multiplicative and D_F a first order differential operator in the sense of Fréchet. For this problem, the space of measure valued processes introduced above are more suitable than those used in section 6.5. It will be interesting to prove existence of measure valued solutions for the equation (7.41) in either of the spaces $L_\infty^w(I, \mathcal{M}_{fabv}(\mathcal{B}_D, \mathcal{Y}_{q,-s}^-))$ and $L_\infty^w(I, \mathcal{M}_{fabv}(\mathcal{B}_D, \mathcal{Z}_b^*))$. We leave this as an interesting problem for future research.

Yosida Approximation YA:

Note that the unbounded operator $\mathcal{B}$ may be approximated by a sequence of bounded operators given by

$$\mathcal{B}_n \equiv n\mathcal{B}R(n, \mathcal{A})$$

where $R(\lambda, \mathcal{A})$ denotes the resolvent operator corresponding to the principal operator $\mathcal{A}$ and $\lambda \in \rho(\mathcal{A})$, the resolvent set of $\mathcal{A}$. Clearly, $\{\mathcal{B}_n\}$ is a sequence of bounded linear operators in $B_\infty(H)$ and it converges in the strong operator topology to $\mathcal{B}$ on its domain. In view of the bounded case, as in Example 6.7.8 of Chapter 6, with $\mathcal{B}_n$ replacing $\mathcal{B}$ of equation (7.41), we have for each $\varphi \in D(\mathcal{A}) \cap B_\infty(H)$, $\mu_t^n(\varphi) \in L_2(\mu^w, R)$ for each finite $n \in N_+$. And its projection to m-th homogeneous chaos is given by

$$\pi_m(\mu_t^n(\varphi)) = f_m(K_m^n(t, \varphi))$$

$$\equiv \int_{I_t^n} K_m^n(t, \varphi; \tau_1, \cdots, \tau_m)dV(\tau_1) \otimes \cdots \otimes dV(\tau_m). \qquad (7.42)$$

Thus, for each $t \in I$, and $n \in N_+$, $\varphi \longrightarrow K^n(t, \varphi)$ is a bounded linear map from $B_\infty(H)$ to G_β, and so for each $t \in I$, and $m \in N$, $\varphi \longrightarrow K_m^n(t, \varphi)$ is a continuous linear map from $B_\infty(H)$ to $\hat{L}_2(I^m, M_m)$. Further, by virtue of the isometry, we have

$$\| \mu_t^n(\varphi) \|_{\mathcal{G}_\beta}^2 = E^{\mu^w}\{|\mu_t^n(\varphi)|^2\} = \sum_{m=0}^{\infty} m! \| K_m^n(t, \varphi) \|_{L_2(I^m, M_m)}^2 < \infty,$$

$$(7.43)$$

for each $n \in N_+$, $t \in I \equiv [0,T]$, $T < \infty$, and $\varphi \in D(\mathcal{A}) \cap B_\infty(H)$. For $n \in N_+$, large enough, μ_t^n is a good approximation of the true solution of the equation (7.41). Since the observation operator $\mathcal{B}$ is unbounded, it is possible that the limit may escape the space of vector measures $\mathcal{M}_{fabv}(\mathcal{B}_D, \mathcal{G}_\beta)$. This is where the space of generalized functionals of Brownian motion is most appropriate. For example, the space $\mathcal{M}_{fabv}(\mathcal{B}_D, \mathcal{G}_\beta^*)$ which may contain the limit. In this case the expression (7.43), measured in terms of the norm topology of $\mathcal{G}_\beta^*$, is given by

$$\| \mu_t^n(\varphi) \|_{\mathcal{G}_\beta^*}^2 = \sum_{m \geq 0} (1/m!)^2 \mathbb{E}\{|\pi_m(\mu_t^n(\varphi))|^2\}$$

$$= \sum_{m=0}^{\infty} (1/m!) \| K_m^n(t, \varphi) \|_{L_2(I^m, M_m)}^2 . \qquad (7.44)$$

If this series is uniformly convergent with respect to $n \in N_+$, then we have $\mu_t(\cdot) \in \mathcal{M}_{fabv}(\mathcal{B}_E, \mathcal{G}_\beta^*)$ or equivalently $\mu_t(\varphi) \in \mathcal{G}_\beta^*$. Clearly, this shows that the spaces of vector valued measures $\mathcal{M}_{fabv}(\mathcal{B}_E, \mathcal{G}_\beta^*)$, $\mathcal{M}_{fabv}(\mathcal{B}_E, \mathcal{Z}_b^*)$ and $\mathcal{M}_{fabv}(\mathcal{B}_E, \mathcal{Y}_{p,s}^*)$ offer a much broader scope for the existence of solutions of the Zakai equation (7.41) including the correlated case where $\mathcal{B}$ is unbounded.

Optimal Choice of Observation Operator Γ:

We return to the problem concerning optimal choice of the observation operator Γ for the infinite dimensional filtering problem. Let $BM(H, R^d)$ denote the space of bounded Borel measurable functions from the Hilbert space H to the Euclidean space R^d. As seen in Chapter 4, the performance measure is given by the mean square error between the actual process $\phi(x(t))$ and its best estimate $E\{\phi(x(t))|\mathcal{F}_t^y\}$ corresponding to a given observation operator Γ. We are interested to find an observation (measurement) operator that minimizes the terminal estimation error. Let μ^Γ denote the measure solution of the filter equation (7.41) corresponding to the choice of an observation operator Γ as defined above by the expression following equation (7.41). Considering any finite $T > 0$, the objective functional denoted by ϱ is given by the expression

$$\varrho(\Gamma) = \mathbb{E}\left\{(1/\mu_T^\Gamma(H))^2\left(\mu_T^\Gamma(H)\int_H \phi^2(x)\mu_T^\Gamma(dx) - \left(\int_H \phi(x)\mu_T^\Gamma(dx)\right)^2\right)\right\},$$
$$= \mathbb{E}J(\mu_T^\Gamma) = \int_\Omega J(\mu_T^\Gamma)(v)W(dv) \tag{7.45}$$

where (Ω, W) is the classical wiener measure space induced by the innovation process V as seen in equation (7.41) and W is the Wiener measure. Since $t \longrightarrow \mu_t^\Gamma$ is weakly continuous $J(\mu_T^\Gamma)$ is well defined. The form of the expression is identical to that of equation (4.21) for the finite dimensional case. The expectation is with respect to the Wiener measure denoted here by $W(\cdot)$. Clearly, $J(\mu_T^\Gamma)(\cdot)$ is nonnegative and our objective now is to choose an operator Γ from an admissible class $\mathcal{H}_{ad} \subset BM(H, R^d)$ that minimizes the functional $\varrho(\Gamma)$. We follow the same procedure as in the finite dimensional case treated in section 4.6 of Chapter 4.

To proceed further, we must choose a suitable class for the admissible set $\mathcal{H}_{ad}$.

The admissible set $\mathcal{H}_{ad}$ is a family of continuous functions from H to R^d furnished with the compact open topology see [[Willard (1970)], p278-287]. More precisely we use the following abstract Ascoli theorem.

Theorem 7.7.1 (Ascoli) *The set $\mathcal{H}_{ad}$ is compact in the compact-open topology if and only if*

(a): $\mathcal{H}_{ad}$ is point wise closed

(b): for each $x \in H$, the closure of the set $\{\Gamma(x), \Gamma \in \mathcal{H}_{ad}\}$ is compact,

(c): $\mathcal{H}_{ad}$ is equicontinuous on each compact subset of H.

Using this result we can prove existence of optimal observer Γ for the uncorrelated filter equation which, by definition, is given by equation (7.41) with $C = 0$. This is stated in the following theorem.

Theorem 7.7.2 *Consider the Filter equation (7.41) with the observation operator Γ as defined above and suppose the admissible set of observers $\mathcal{H}_{ad}$ satisfies the conditions (a)-(c) of Ascoli Theorem. Then there exists an optimal observer $\Gamma^o \in \mathcal{H}_{ad}$ at which ϱ attains its minimum.*

Proof Let $C_0(H)$ denote the space of bounded continuous functions on H

having compact supports. Using the Markov semigroup $S(t), t \geq 0$, generated by the operator $\mathcal{A}$ (more precisely its extension), and the variation of constants formula, we can write equation (7.41) as the integral equation

$$\mu_t(\varphi) = \nu(S(t)\varphi) + \int_0^t \mu_r(S(t-r)\Gamma\varphi) \cdot dV, t \in I, \varphi \in C_0(H). \quad (7.46)$$

Let $\| \mu \|_v$ denote the variation norm of any bounded Borel measure μ on H. Since $C_0(H)$ is dense in $C_b(H)$ and the set $\mathcal{H}_{ad}$ is bounded, and $S(t), t \geq 0$, is a contraction semigroup, using Gronwall inequality it is easy to verify that

$$\sup\left\{ \mathbb{E}\{\| \mu_t^\Gamma \|_v^2\}, \Gamma \in \mathcal{H}_{ad}, t \in I \right\} < \infty.$$

Now we verify that $\Gamma \longrightarrow \mu^\Gamma$ is continuous with respect to the compact open topology on $\mathcal{H}_{ad}$ and weak convergence topology on the space of bounded regular Borel measures on H. Let Γ_α be a net in $\mathcal{H}_{ad}$ that converges to Γ_o and let μ^α and μ^o be the corresponding solutions of equation (7.46). Then it is easy to see that

$$\mu_t^o(\varphi) - \mu_t^\alpha(\varphi) = \int_0^t (\mu_r^o - \mu_r^\alpha)(S(t-r)\Gamma_o\varphi) \cdot dV(r)$$

$$+ \int_0^t \mu_r^\alpha(S(t-r)(\Gamma_o - \Gamma_\alpha)(\varphi)) \cdot dV(r), t \in I. \quad (7.47)$$

Since $\{\mu^\alpha\}$ is bounded as shown above and Γ_α net converges to Γ_o on compacts, we have

$$\lim_\alpha \int_0^t \mu_r^\alpha(S(t-r)(\Gamma_o - \Gamma_\alpha)(\varphi)) \cdot dV(r) = 0 \quad W - a.s \quad (7.48)$$

for every φ with compact support and $t \in I$. Thus taking the limit in (7.47) along a sub net, if necessary, we arrive at the following identity

$$\vartheta_t(\varphi) = \int_0^t \vartheta_r(S(t-r)\Gamma_o\varphi) \cdot dV(r), t \in I, \quad (7.49)$$

which holds W-almost surely for all φ having compact supports. Since this is a linear homogeneous equation on the space of bounded regular Borel measures, with $\vartheta_0 = 0$, we have $\vartheta_t \equiv 0, t \in I$, W-almost surely. Thus we have shown that $\mu^\alpha \xrightarrow{w} \mu^o$ W-almost surely. This proves that $\Gamma \longrightarrow \mu^\Gamma$ is continuous with respect to the compact open topology on $\mathcal{H}_{ad}$ and weak convergence topology on the space of bounded regular Borel measures on H. Now considering the objective functional (7.45) given by

$$\varrho(\Gamma) \equiv \int_\Omega J(\mu_T^\Gamma)(v) \; W(dv)$$

and noting that $J(\mu_T^\Gamma)(v)$ is W-almost surely bounded from above by the quantity $\| \phi \|_\infty^2$, we use Lebesgue bounded convergence theorem, as seen in the finite dimensional case of Chapter 4, to conclude that $\Gamma \longrightarrow \varrho(\Gamma)$ is continuous on $\mathcal{H}_{ad}$. Since the admissible set $\mathcal{H}_{ad}$ is compact, ϱ attains its minimum on it proving the existence of an optimal observer. This completes the proof. •

Remark 7.7.3 In case we are interested in cumulative filtering error over a finite time interval I, the proper objective functional to consider is

$$\bar{\varrho}(\Gamma) = \int_\Omega \int_I J(\mu_t^\Gamma)(v)\lambda(dt)W(dv)$$

where λ is a bounded positive measure designed to assign weights according to importance of time periods. We leave it for the reader to verify that this functional also admits an optimal observer.

7.8   Application to Stochastic Navier-Stokes Equation

Here we consider measure valued solutions for stochastic Navier-Stokes equations and their optimal control. For convenience of the reader we first present the deterministic model. Navier-Stokes equation is derived from the basic principle of conservation of momentum. For incompressible fluid with nonslip boundary condition, the Navier-Stokes equation is given by

$$\partial v/\partial t - \nu \triangle v + (v \cdot \nabla)v + \nabla p = g, t > 0, \xi \in D \subset R^n;$$
$$div(v) = 0, v(0, \xi) = v_0(\xi), \xi \in D, v|_{\partial D} = 0, \qquad (7.50)$$

where D is an open connected bounded domain in R^n ($n = 2, 3$) with smooth boundary, v is the fluid velocity vector, ν is the kinematic viscosity, $\triangle$ is the Laplacian , p is the pressure and finally g is the volume force. This equation can be reformulated as an abstract evolution equation by projection to the space of divergence free vector fields. Define the following vector spaces

$$V \equiv \overline{closure}^{H_1(D,R^n)} \{v \in C_0^\infty(D,R^n) : div(v) = 0\}$$
$$H \equiv \overline{closure}^{L_2(D,R^n)} \{v \in C_0^\infty(D,R^n) : div(v) = 0\}.$$

Clearly, $V \subset H_0^1$ and one can easily verify that the embedding $V \hookrightarrow H$ is continuous and dense. Let i_c denote the corresponding embedding constant. Identifying H with its dual, we arrive at the Gelfand triple $V \hookrightarrow H \hookrightarrow V^*$ where V^* is the topological dual of V. The injections are also compact. Let P denote the orthogonal projection of $L_2(D,R^n)$ onto H, the Hilbert space of divergence free vector fields. Define

$$A\psi = -P(\triangle\psi),\ b(\varphi,\psi) = P(\varphi \cdot \triangledown\psi), B(\varphi) = b(\varphi,\varphi), f \equiv P(g)$$

for $\varphi, \psi \in V \cap H^2(D)$. The operator A, known as the Stokes operator, is an unbounded positive self adjoint operator in H with strictly positive eigen values. Hence, A has continuous inverse A^{-1} which is a compact operator in H. The reader can easily verify that $P(\triangledown p) = 0$. The most important component in the dynamics is the convective term which is a nonlinear operator denoted by B. The properties of this operator is studied through the trilinear form

$$L(x,y,z) \equiv \int_D < (x(\xi) \cdot \triangledown)y(\xi), z(\xi) >_{R^n} d\xi.$$

This is a well defined function on $V \times V \times V$ and the reader can easily verify that on the divergence free vector fields, $L(x,y,y) = 0$ for all $x,y \in V$. Further, it follows from this that the operator $B(\cdot) = b(\cdot,\cdot)$ maps from V to V^*. For further details on the properties of this trilinear form see Temam [Temam (1988)] and [Foias (2001)]. Now applying the projection operator P on both sides of equation (7.50), we obtain the following abstract differential equation on the Hilbert space H

$$(d/dt)v + \nu Av + B(v) = f, v(0) = v_0. \tag{7.51}$$

For simplicity, we assume that the volume force f is given by the sum of there components $f = f_0 + \dot{W} + Cu$ where f_0 is the deterministic component of natural forces, $\dot{W}$ is the random fluctuation part given by

$$\dot{W} \equiv \sum_{i=1}^d \alpha_i v_i \dot{w}_i \tag{7.52}$$

where α_i are locally square integrable functions, $\{v_i\} \in H$ is a complete orthonormal system (normalized eigen functions of the positive self adjoint operator A), and $w = (w_1, w_2, \cdots, w_d)$ is an R^d valued standard Brownian motion on a complete filtered probability space denoted by $(\Omega, \Sigma, \mathcal{F}_t^w, P)$, with $\mathcal{F}_t^w \subset \Sigma, t \geq 0$, being an increasing family of right continuous subsigma algebras generated by the Brownian motion w. The last term of f is the control force. Substituting these in equation (7.51) we obtain the stochastic differential equation

$$dv + (\nu Av + B(v))dt = F_0(t, u)dt + F(t)dw, t \geq 0,$$

$$v(0) = v_0, \tag{7.53}$$

where w is the d-dimensional standard Brownian motion and $F \in L_2(I, \mathcal{L}_2(R^d, H))$ determined by $\{\alpha_i, 1 \leq i \leq d\}$ and $\{v_i\}$ through the expression (7.52). Note that $F_0(t, u) = f_0(t, \cdot) + Cu$, where $C \in \mathcal{L}(U, H)$ with U being another Hilbert space and F is a bounded operator valued function with values in $\mathcal{L}(R^d, H)$ generating the random volume force given by the expression (7.52). It is a well known fact that, under standard assumptions, the deterministic $3 - d$ Navier-Stokes equation may not have a strong solution globally though weak solutions do exist [Leray (1934)] [Leray (1933)] for all finite intervals $[0, T]$. Here, we are interested in measure valued solutions. Measure solutions for infinite dimensional systems, both deterministic and stochastic, have been studied by Fattorini [Fattorini (1997)] and the author in several papers [Ahmed (2005)],[Ahmed (1996)] [Ahmed (1999)][Ahmed (1999)] [Ahmed (2006)]. See also the references therein. For deterministic Navier-Stokes equations, the author has studied measure solutions and their optimal control in [Ahmed (1998)] using semigroup approach. Fursikov has studied stabilization problems of strong solutions of deterministic Navier-Stokes systems using boundary feedback controls [Fursikov (2004)]; see also the references therein. Using energy spaces, here we consider the stochastic version. Though it is a special case of the general results of [Ahmed (2005)], it has some interesting and physically important characteristics not found in the abstract case. To proceed further, we must introduce the following operators,

$$\mathcal{A}(t)\varphi(\xi) \equiv (1/2)Tr[D^2\varphi(\xi)(F(t)F(t)^*)],$$

$$\mathcal{B}(\varphi)(\xi) \equiv < -\nu A\xi - B(\xi), D\varphi(\xi) >_{V^*,V},$$

$$\mathcal{C}_1(u)(t)\varphi(\xi) \equiv < F_0(t,u), D\varphi(\xi) >_H, \quad \mathcal{C}_2(t)\varphi(\xi) \equiv F^*(t)D\varphi(\xi),$$

where φ is any test function on H, twice Fréchet differentiable, so that the functions defined above belong to $C_b(H)$. Using these operators and the NSE, we can now write the evolution equation in its weak form on the space of measures as follows

$$d\mu_t(\varphi) = \mu_t(\mathcal{A}\varphi)dt + \mu_t(\mathcal{B}\varphi)dt$$

$$+ \mu_t(\mathcal{C}_1(u)\varphi)dt + < \mu_t(\mathcal{C}_2\varphi), dw >, \mu_0(\varphi) = \nu_0(\varphi), \quad (7.54)$$

for $t \in I = [0,T]$. The measure ν_0 represents the probability law of the initial velocity distribution, $\mathcal{L}(v_0) = \nu_0$. If $v_0 \in H$ is known, ν_0 is a Dirac measure on H and $\nu_0(\varphi) = \varphi(v_0)$. To solve the NSE (7.51) it is necessary to specify the initial velocity v_0 throughout the domain D and this is not always practically feasible, for example the wind velocity in a large geographical region. However, an approximate distribution of initial velocity is not difficult to specify.

For definition of measure solutions, and their existence and regularity properties, see Chapter 1, section 1.4. The questions of existence and regularity properties of measure valued solutions for stochastic infinite dimensional systems in general have been studied by the author in several papers [Ahmed (2005)][Ahmed (1999)] [Ahmed (2006)] (see also the references therein). For measure solutions of deterministic NSE see [Ahmed (1998)] and for general nonlinear deterministic systems see [Ahmed (1999)]. As seen in Definition 1.4.5, a random process $\{\mu_t, t \in I\}$, satisfying

$$\mu \in M_{\infty,2}^w(I \times \Omega, , \Pi_{rba}(H^+)) \subset M_{\infty,2}^w(I \times \Omega, \mathcal{M}_{rba}(H^+)),$$

is a measure solution of equation (7.54) if, for every $\varphi \in D(\mathcal{A}) \cap D(\mathcal{B})$, the following identity holds with probability one

$$\mu_t(\varphi) = \nu_0(\varphi) + \int_0^t \mu_s(\mathcal{A}\varphi)ds + \int_0^t \mu_s(\mathcal{B}\varphi)ds$$

$$+ \int_0^t \mu_s(\mathcal{C}_1(u)\varphi)ds + \int_0^t < \mu_s(\mathcal{C}_2\varphi), dw(s) >, (7.55)$$

for all $t \in I$. This is the abstract case where the process takes values from the space of regular, bounded and finitely additive measures. In the case of NSE, we have countably additive measures given that the initial measure is so. It is important to mention that in the infinite dimensional spaces even a deterministic system may not have a pathwise solution while it may have a measure valued solution [Godunov (1974)], [Fattorini (1997)]. In any case, pathwise solutions are also measure valued solutions (Dirac measure along the path) and therefore can be embedded into the class of measure valued processes. For measure solutions of deterministic Navier-Stokes equations, semigroup approach was used in [Ahmed (1998)] following the general technique developed in [Ahmed (1999)]. Using this approach combined with stopping time, one can prove the existence of measure solutions for the stochastic Navier-Stokes equation. Another technique is based on the well known fact that for every initial state $v_0 \in H$, the NSE has a weak solution (a path process). This is due to the celebrated result of Jean Leray of 1933-34 [Leray (1934)] [Leray (1933)]. Hence, given the distribution of initial states, one can use the Krein-Milman theorem to approximate it by Dirac measures and take the convex combinations of the Dirac measures corresponding to each of the path processes to construct the measure valued solution. This is the approach taken by Foias, Manley, Rosa and Temam [Foias (2001)].

Here we concentrate on the regularity properties and feedback control of the measure valued process. For existence and uniqueness of measure solutions we refer to the papers [36, 60, 84] [Ahmed (2005)] [Ahmed (1999)], [Ahmed (2000)].

For correct physical interpretation, it is better to consider the natural energy space $H \times V$ and the associated norms. We show that the measure solutions for the NSE have better and interesting properties. Let $\mathcal{M}(H)$ and $\mathcal{M}(V)$ denote the space of finite signed Borel measures defined on the respective Borel fields. These are Banach spaces with respect to total variation norm. By virtue of Jordan decomposition, any signed measure μ can be written as $\mu = \mu^+ - \mu^-$ where μ^+ and μ^- are positive measures. The total variation norm is then given by $\| \mu \|_v = \mu^+(H) + \mu^-(H)$. We use $|\mu| = |\mu^+| + |\mu^-|$ to denote the nonnegative measure induced by the variation of μ. If μ is nonnegative, $|\mu| = \mu$ itself. We introduce the following linear subspaces of signed measures

$$M_2(H) \equiv \{\mu \in \mathcal{M}(H) : \int_H |\xi|_H^2 \, |\mu|(d\xi) < \infty\},$$

$$M_2(V) \equiv \{\mu \in \mathcal{M}(V) : \int_V \| \xi \|_V^2 \, |\mu|(d\xi) < \infty\},$$

and define the associated norms by

$$|\mu|_{M_2(H)} = \left(\int_H |\xi|_H^2 \, |\mu|(d\xi)\right) \text{ and } \| \mu \|_{M_2(V)} = \left(\int_V \| \xi \|_V^2 \, |\mu|(d\xi)\right)$$

respectively. These are closed subspaces of the Banach space of signed measures and hence are Banach spaces. Let $L_\infty^{wa}(I, L_1(\Omega, M_2(H)))$ $\left(L_1^{wa}(I, L_1(\Omega, M_2(V)))\right)$ denote the space of stochastic processes $\{\mu_t, t \geq 0\}$ taking values from the space $M_2(H)\left(M_2(V)\right)$ and weakly adapted to $\mathcal{F}_t^w, t \geq 0$, that is, for each test function φ, $t \longrightarrow \mu_t(\varphi)$ is $\mathcal{F}_t^w$ measurable (adapted). We have used $\mathcal{F}_t^w, t \geq 0$, to denote the filtration generated by the Brownian motion w of equation (7.53).

We prove the following existence result and the energy estimates giving useful bounds and also characterizing the measure solutions of NSE. Let U be a separable Hilbert space and $L_2^a(I, U)$ denote the space of $\mathcal{F}_t^w$-adapted processes with values in U. Endowed with the standard norm topology, $\| u \| \equiv \left(\mathbb{E} \int_I |u(t)|_U^2 dt\right)^{1/2}$, this is a Hilbert space and it is chosen as the control space.

Theorem 7.8.1 *Consider the evolution equation (7.54) in the weak form (7.55) and suppose the initial measure ν_0 is a probability measure (possibly random and weakly $\mathcal{F}_0$ adapted) satisfying $\mathbb{E}|\nu_0|_{M_2(H)} < \infty$ and that $F \in L_2(I, \mathcal{L}_2(R^d, H))$. Then, for every admissible control $u \in \mathcal{U}_{ad} \subset L_2^a(I, U)$ satisfying $\mathbb{E} \int_I |F_0(t, u)|_H^2 dt < \infty$, equation (7.53) equivalent to the equation (7.55) has a probability measure valued solution μ satisfying the following energy estimate,*

$$\mathbb{E}|\mu_t|_{M_2(H)} + \nu\mathbb{E} \int_0^t \| \mu_s \|_{M_2(V)} \, ds \leq \mathbb{E}|\nu_0|_{M_2(H)}$$

$$+ \int_0^t Tr(FF^*)ds + ((i_c)^2/\nu)\mathbb{E} \int_0^t |F_0(s, u)|_H^2 ds,$$

$$\tag{7.56}$$

where i_c is a positive constant associated with the embedding $V \hookrightarrow H$. Hence, $\mu \in L^{wa}_{\infty}(I, L_1(\Omega, M_2(H))) \cap L^{wa}_1(I, L_1(\Omega, M_2(V)))$ and further, $\mu \in C^w(I, M_2(H))$ almost surely where $C^w(I, M_2(H))$ denotes the space of weakly continuous functions on I with values in $M_2(H)$.

Proof Fix a control $u \in \mathcal{U}_{ad}$. Now, taking $\varphi(\xi) = |\xi|^2_H, \xi \in V$ and substituting in the expression (7.55) and using the fundamental property of the trilinear form giving $L(x, y, y) = 0$ for all $x, y \in V$, we arrive at the following identity

$$|\mu_t|_{M_2(H)} + 2\nu \int_0^t \| \mu_s \|_{M_2(V)} \, ds = |\nu_0|_{M_2(H)} + \int_0^t \mu_s(Tr(FF^*))ds$$
$$+ \int_0^t \mu_s((F_0(s, u), D\varphi))ds + \int_0^t < \mu_s(F^*D\varphi), dw(s) >,$$

$$(7.57)$$

which holds with probability one. Since by our assumption $F \in L_2(I, \mathcal{L}_2(R^d, H))$, independent of ξ, and the measure solution μ_t is a probability measure valued random process, we have

$$\int_0^t \mu_s(Tr(FF^*))ds = \int_0^t \| F \|^2_{HS} \, ds \qquad P - a.s. \qquad (7.58)$$

Using Cauchy inequality and our assumption that F_0 is independent of the state, the integrand of the third term on the right hand side of the identity (7.57) satisfies

$$|\mu_s((F_0(s, u), D\varphi))| = 2| \int_H (F_0(s, u), \xi)_H \mu_s(d\xi)|$$
$$\leq 2 \int_H |F_0(s, u)|_H |\xi|_H \mu_s(d\xi) \leq (i_c/\varepsilon)|F_0(s, u)|^2_H + (i_c\varepsilon) \| \mu_s \|_{M_2(V)},$$

P-almost surely for any $\varepsilon > 0$, where $i_c(> 0)$ denotes the embedding constant $V \hookrightarrow H$. Thus the third integral is bounded above as shown below,

$$| \int_0^t \mu_s((F_0(s, u), D\varphi))ds| \leq (i_c/\varepsilon) \int_0^t |F_0(s, u)|^2_H ds$$
$$+ (i_c\varepsilon) \int_0^t \| \mu_s \|_{M_2(V)} \, ds \qquad (7.59)$$

P-almost surely. Now considering the stochastic integral (last term of (7.57)), it is easy to verify that the integrand satisfies the following inequality

$$|\mu_s(F^*D\varphi)|^2_{R^d} \leq 4 \int_H |F^*\xi|^2_{R^d} \, \mu_s(d\xi)$$

$$\leq 4 \parallel F(s) \parallel^2_{H.S} \, |\mu_s|_{M_2(H)}, s \in I, \quad P-a.s, \quad (7.60)$$

where H.S denotes the Hilbert-Schmidt norm. It follows from this estimate that the stochastic integral is well defined provided

$$\sup\{\mathbb{E}|\mu_t|_{M_2(H)}, t \in I\}$$

is finite. Thus, a-priori, we cannot set the expected value of the stochastic integral to zero. So we proceed as follows. Using (7.60) it is easy to verify that

$$\mathbb{E}\left\{\int_0^t <\mu_s(F^*D\varphi), dw(s)>\right\} \leq 2\sqrt{\int_0^t \parallel F(s) \parallel^2_{H.S} \mathbb{E}|\mu_s|_{M_2(H)}ds}.$$

Applying the expectation operation on both sides of equation (7.57) and choosing $\varepsilon = (\nu/i_c)$ and using the above estimate we arrive at the following inequality

$$\mathbb{E}|\mu_t|_{M_2(H)} + \nu\mathbb{E}\int_0^t \parallel \mu_s \parallel_{M_2(V)} ds \leq \mathbb{E}|\nu_0|_{M_2(H)} + \int_0^t \parallel F \parallel^2_{H.S} ds$$

$$+ (i_c^2/\nu)\mathbb{E}\int_0^t |(F_0(s,u)|^2_H ds + 2\sqrt{\int_0^t \parallel F \parallel^2_{H.S} \mathbb{E}|\mu_s|_{M_2(H)}ds}.$$

$$(7.61)$$

It follows from our assumptions on the data that

$$C_1 \equiv \mathbb{E}|\nu_0|_{M_2(H)} + \int_0^T \parallel F \parallel^2_{H.S} ds + (i_c^2/\nu)\mathbb{E}\int_0^T |(F_0(s,u)|^2_H ds < \infty.$$

Thus by setting $h(t) \equiv \parallel F(t) \parallel^2_{H.S}$, $z(t) \equiv \mathbb{E}|\mu_t|_{M_2(H)}$, and noting that all the terms in the expression (7.61) are nonnegative, it follows from this inequality that

$$z(t) \leq C_1 + 2\sqrt{\int_0^t h(s)z(s)ds}, t \in I \equiv [0,T].$$

Now we use the well known Winter's inequality [[Ahmed (1981)], Lemma 5.2.1, p338]. Define $v(t) \equiv \int_0^t h(s)z(s)ds$. Clearly, this is a nondecreasing continuous function on I. Multiplying either side of the above inequality by h and integrating over the interval $[0, t]$, we obtain the following inequality

$$v(t) \leq C_1 \parallel h \parallel_{L_1(I)} + 2 \int_0^t h(s)\sqrt{v(s)}ds, t \in I \equiv [0, T].$$

Define $b \equiv \sup\{v(t), t \in I\} = v(T)$ and $\tilde{C}_1 = C_1 \parallel h \parallel_{L_1(I)}$. Now, if $b \leq \tilde{C}_1$, z is uniformly bounded on I and the conclusion follows. On the other hand, if $b > \tilde{C}_1$, we have the inequality,

$$\int_{\tilde{C}_1}^b dv/\sqrt{v} \leq 2 \int_0^T h(s)ds.$$

Hence,

$$\sqrt{b} \equiv \sqrt{v(T)} \leq \sqrt{\tilde{C}_1} + \int_0^T h(s)ds < \infty$$

and therefore

$$\sup\{z(t), t \in I\} \leq C_1 + 2\sqrt{b} < \infty.$$

This proves that

$$\sup\{\mathbb{E}|\mu_t|_{M_2(H)}, t \in I\} < \infty,$$

and hence the expected value of the stochastic integral in the identity (7.57) vanishes and therefore the inequality (7.61) also holds without the last term. Thus we have proved the estimate as stated in the theorem, and hence $\mu \in L_\infty^{wa}(I, L_1(\Omega, M_2(H))) \cap L_1^{wa}(I, L_1(\Omega, M_2(V)))$. Weak continuity of the map $t \longrightarrow \mu_t$, follows directly from the expression (7.55) and the estimates. Existence follows from the same basic technique as used in the proof of existence of measure solutions for general stochastic differential equations such as [Theorem 3.3 [Ahmed (2005)]]. See also [Ahmed (1999)] [Ahmed (2000)]. This completes the proof. •

Following result is an immediate consequence of the above result.

Corollary 7.8.2 *Consider the evolution equation (7.54) and suppose the assumptions of Theorem 7.8.1 hold. Then, for each fixed $T \in [0, \infty), I \equiv [0, T]$, and $u \in \mathcal{U}_{ad}$ and $\varphi \in BC(H)$, $\mu_T^u(\varphi) \in L_2(\Omega)$ and it possesses the Wiener-Itô expansion given by*

$$\mu_T^u(\varphi) = \sum_{n \geq 0} f_n(K_n^u(\varphi)) = \sum_{n \geq 0} \int_{I^n} K_n^u(\varphi, \tau_1, \cdots, \tau_n)dw(\tau_1) \otimes \cdots \otimes dw(\tau_n)$$

$$(7.62)$$

with $K^u(\varphi) \in G_\beta$ that is

$$\| K^u(\varphi) \|^2_{G_\beta} = \sum_{n \geq 0} n! \| K^u_n(\varphi) \|^2_{L_2(I^n, M_n)} < \infty, \qquad (7.63)$$

where $M_n \equiv \mathcal{L}_2(R^{d \otimes n}, R)$.

Proof For each fixed $u \in \mathcal{U}_{ad}$ and $\varphi \in BC(H)$, it is clear from the above theorem that $t \longrightarrow \mu^u_t(\varphi)$ is an $\mathcal{F}^w_t$ adapted continuous scalar valued random process satisfying

$$\sup_{t \in I} \mathbb{E}(\mu^u_t(\varphi))^2 < \infty.$$

Hence $\mu^u_T(\varphi) \in L_2(\Omega)$ and therefore it follows from Theorem 2.3.4 that there exists a $K^u(\varphi) \in G_\beta$ and hence a sequence $\{K^u_n(\varphi) \in \hat{L}_2(I^n, M_n)\}, n \in N$, such that

$$\mathbb{E}|\mu^u_T(\varphi)|^2 = \sum_{n \geq 0} n! \| K^u_n(\varphi) \|^2_{L_2(I^n, M_n)} < \infty.$$

This completes the proof. •

Remark 7.8.3 By virtue of Theorem 7.8.1, for each $u \in \mathcal{U}_{ad}$ and $T < \infty$, $\mu^u_T \in M_2(H)$ with probability one. Thus, the above Corollary holds not only for $\varphi \in BC(H)$ but also for any $\varphi \in C(H)$ having at most quadratic growth. Further, it follows from this result that the kernels are vector valued measures, $K^u_n : \mathcal{B}_H \longrightarrow \hat{L}_2(I^n, M_n)$.

This remark prompts the following definition. Let $\kappa : H \longrightarrow [0, \infty]$ be given by $\kappa(\xi) \equiv (1 + |\xi|^2_H)$. Let $C_\kappa(H)$ denote the space of (not necessarily bounded) continuous scalar valued functions defined on H characterized as follows:

$$C_\kappa(H) \equiv \left\{ \varphi \in C(H) : \sup\{(|\varphi(x)|/\kappa(x)), x \in H\} < \infty \right\}. \qquad (7.64)$$

Endowed with the sup-norm topology $\| \varphi \|_{C_\kappa(H)} \equiv \sup\{(|\varphi(x)|/\kappa(x)), x \in H\}$, $C_\kappa(H)$ is a Banach space. It is interesting to note that its dual $(C_\kappa(H))^* = M_2(H)$, the space of finite signed Borel measures on $\mathcal{B}_H$ having finite second moments. Since the dual of any Banach space is a Banach space, $M_2(H)$ is a Banach space as mentioned earlier. In view of this, for each $n \in N$, $K^u_n \in M_2(H, \hat{L}_2(I^n, M_n))$. In other words, the kernels are vector valued measures on $\mathcal{B}_H$ with values in the Hilbert space $\hat{L}_2(I^n, M_n)$.

Some Optimal Feedback Control Problems:

Consider the controlled NSE given by (7.53). Suppose we want to design a feedback controller with measured data as its input contained in a suitable function space. Suppose it is possible to measure the flow velocity in a given region $D_0 \subset D$ and use this information as the input to a linear controller giving u. The problem is to find the best linear controller (from a given class) that minimizes the excursion of flow velocity beyond a pre-specified regime. Consider the observation space (space of measured data) as

$$Y \equiv \{y \in L_2(D_0, R^n) : div(y) = 0\}$$

furnished with the standard Hilbertian norm. Clearly, by extending each element of Y outside D_0 by setting it equal to zero, we have $Y \subset H$. Let χ_{D_0} denote the characteristic function of the set D_0. Define the operator L_0 from H to Y by

$$(L_0 \zeta)(x) \equiv \chi_{D_0}(x)\zeta(x), x \in D, \zeta \in H.$$

Clearly, this is a bounded linear operator from H to Y. The observed data is given by $y(t,x) \equiv (L_0 v)(t,x)$ for a period of time say $I = [0,T]$ where v is the velocity field. This data is then passed into a controller giving $u = \Gamma y$ where $\Gamma \in \mathcal{L}(Y, U)$ is a linear (time invariant) operator mapping Y to U. Thus the volume force generated by the feedback control law Γ alone is $C\Gamma(y) \equiv C\Gamma(L_0 v)$. Hence the deterministic component of the total volume force is given by

$$F_0(t, u) = F_0(t, \Gamma y) = f_0 + C\Gamma(y) = f_0 + C\Gamma(L_0 v). \qquad (7.65)$$

For admissible feedback-control laws denoted by $\mathcal{FC}_{ad}$, we choose, without any loss of generality, the closed unit ball $B_1(\mathcal{L}(Y, U)) \subset \mathcal{L}(Y, U)$. Clearly, the composition map $C\Gamma L_0 \in \mathcal{L}(H)$. In this case the estimate (7.56) is slightly modified and the upper bound now follows from Gronwall Lemma applied to the following inequality

$$\mathbb{E}|\mu_t|_{M_2(H)} + \nu\mathbb{E}\int_0^t \| \mu_s \|_{M_2(V)} \, ds \leq \mathbb{E}|\nu_0|_{M_2(H)} + \int_0^t Tr(FF^*)ds$$

$$+ ((i_c)^2/\nu)\int_0^t |f_0(s)|_H^2 ds + 2\gamma\mathbb{E}\int_0^t |\mu_s|_{M_2(H)} ds$$

$$(7.66)$$

where $\gamma \equiv \| C\Gamma L_0 \|_{\mathcal{L}(H)}$.

Now we introduce the objective functional. Choose any desired $r > 0$, and consider the closed ball $B_r(H)$ of radius r in H around the origin and define $\psi_r(\xi) \equiv \beta d^2(\xi, B_r(H)), \xi \in H$ and any $\beta \geq 1$, where $d(\xi, B_r(H))$ denotes the distance of the point ξ from the ball $B_r(H)$. For each $\Gamma \in \mathcal{FC}_{ad}$, define the objective functional as

$$J(\Gamma) = \mathbb{E} \int_I \mu_t^\Gamma(\psi_r) \lambda(dt), \qquad (7.67)$$

where μ^Γ is the solution of the evolution equation (7.54) corresponding to $u = \Gamma y \equiv \Gamma L_0 \xi$ for $\Gamma \in \mathcal{FC}_{ad}$, and λ is any nonnegative countably additive measure having bounded total variation on I. The problem is to find a $\Gamma \in \mathcal{FC}_{ad}$ that minimizes the above functional. This is an average measure of excursion of fluid (kinetic) energy outside the desired level $B_r(H)$. Our objective is to find a feedback control law $\Gamma \in \mathcal{FC}_{ad}$ that minimizes this energy (reduce turbulence).

For any positive measure $\mu \in M_2(H)$, define the covariance operator Q^μ as

$$(Q^\mu e, e) \equiv \int_H (\xi, e)^2 \mu(d\xi), e \in H.$$

For (nonnegative) measure valued stochastic processes $\{\mu_{t,\omega}\}$, the covariance operator is denoted by $Q_{t,\omega}^\mu$ and it is given by

$$(Q_{t,\omega}^\mu e, e) \equiv \int_H (\xi, e)^2 \mu_{t,\omega}(d\xi), e \in H.$$

Consider the controlled evolution equation (7.54) with the operator $\mathcal{C}_1(u)$ given by

$$\mathcal{C}_1(u)(\varphi) \equiv (f_0, D\varphi)_H + (C\Gamma L_0(\cdot), D\varphi(\cdot))_H$$

for $\Gamma \in \mathcal{FC}_{ad}$. This turns (7.54) into a feedback system. The reachable set for this system is given by the set of all (probability) measure valued stochastic processes determined by equation (7.54) as Γ describes the admissible set $\mathcal{FC}_{ad}$. This is denoted by

$$\mathcal{R} \equiv \Big\{ \mu \in L_\infty^{wa}(I, L_1(\Omega, M_2(H))) \cap L_1^{wa}(I, L_1(\Omega, M_2(V))) :$$

$$\mu = \mu^\Gamma, \text{ sol. of (7.54) for some } \Gamma \in \mathcal{FC}_{ad} \Big\}. \quad (7.68)$$

Theorem 7.8.4 *Consider the system (7.54) with the cost functional (7.67) and control laws given by $\mathcal{FC}_{ad}$ which is endowed with the relative weak operator topology of $\mathcal{L}(Y, U)$. Suppose the assumptions of Theorem 7.8.1 hold, the operator C(of the expression (7.65)) $\in \mathcal{L}(U, H)$, λ is a countably additive positive measure having bounded variation, $\{e_i\}$ any complete orthonormal basis of H and the reachable set $\mathcal{R}$ satisfies the following properties:*

(P1): For every $\varepsilon > 0$, there exists a finite positive number C_ε such that for all $\mu \in \mathcal{R}$,

$$\mu_{t,\omega}\{\xi \in H : |\xi|_H > C_\varepsilon\} < \varepsilon, \quad \text{for } \lambda \times P \text{ almost all } (t, \omega) \in I \times \Omega.$$

(P2): $\lim_{n \to \infty} \sum_{i \geq n}^\infty (Q_{t,\omega}^\mu e_i, e_i) = 0$ uniformly with respect to $\mu \in \mathcal{R}$, for $\lambda \times P$ almost all $(t, \omega) \in I \times \Omega$.

Then, there exists an optimal feedback control law $\Gamma^o \in \mathcal{FC}_{ad}$ at which the functional (7.67) attains its minimum.

Proof   First, we prove that the reachable set $\mathcal{R}$ is weakly compact in the sense that every generalized sequence $\{\mu^n\}$ from $\mathcal{R}$ has a weakly convergent subsequence converging to some μ^o. Let ℓ denote the Lebesgue measure on R and P the probability measure on Ω. It suffices to show that for every $\varphi \in C_\kappa \cap C_F^2$,

$$\lim_{n \to \infty} \eta(\mu^n) \equiv \lim_{n \to \infty} \int_\Omega \int_I \mu_{t,\omega}^n(\varphi) \, \ell(dt) \, P(d\omega)$$
$$= \int_\Omega \int_I \mu_{t,\omega}^o(\varphi) \, \ell(dt) \, P(d\omega) = \eta(\mu^o). \quad (7.69)$$

For any $\varphi \in C_\kappa \cap C_F^2$, define the linear functional

$$\eta(\mu) \equiv \mathbb{E} \int_I \mu_t(\varphi)\ell(dt) = \int_\Omega \int_I \mu_{t,\omega}(\varphi) \, \ell(dt) \, P(d\omega)$$

on $L_\infty^{wa}(I, L_1(\Omega, M_2(H)))$ evaluated at $\mu \in \mathcal{R}$. Since $t \to \mu_t$ is weakly continuous P-almost surely, and $\varphi \in C_\kappa$, the function $t \longrightarrow \mu_t(\varphi)$ is $\mathcal{F}_t$ adapted and so $(t, \omega) \longrightarrow \mu_{t,\omega}(\varphi)$ is a measurable scalar valued function (with respect to the predictable sigma field). Hence, the integrand of the above functional is a measurable scalar valued function. Since the reachable set $\mathcal{R}$ satisfies the properties (P1) and (P2), it follows from a well known theorem on weak compactness of bounded measures on infinite dimensional

Hilbert spaces [Gihman and Skorohod, [Gihman (1971)], Chapter 6, Theorem 2, p377] that $\{\mu_{t,\omega}, \mu \in \mathcal{R}\}$ is weakly compact for $\ell \times P$ almost all $(t,\omega) \in I \times \Omega$. Now, let $\{\mu^n\} \subset \mathcal{R}$ be any sequence. Then due to weak compactness $\ell \times P$ almost surely, there exists a generalized subsequence (subnet), relabeled as the original sequence, and an element μ^o such that $\mu^n_{t,\omega}(\varphi) \longrightarrow \mu^o_{t,\omega}(\varphi)$ $\ell \times P$ almost every where. Since almost every where convergence preserves measurability, the limit $\mu^o(\varphi)$ is a measurable function on $I \times \Omega$ (with respect to the predictable sigma field) and hence $\mu^o \in L^{wa}_\infty(I, L_1(\Omega, M_2(H)))$. Using the identity (7.57) and the estimate (7.66) and recalling that $\varphi \in C_\kappa$, it is not difficult to verify that the sequence $\{\mu^n_{t,\omega}(\varphi)\}$ is dominated by a function in $L_1(I \times \Omega, \ell \times P)$. Thus, it follows from Lebesgue dominated convergence theorem that

$$
\begin{aligned}
\lim_{n\to\infty} \eta(\mu^n) &= \lim_{n\to\infty} \int_\Omega \int_I \mu^n_{t,\omega}(\varphi)\ \ell(dt)\ P(d\omega) \\
&= \int_\Omega \int_I \mu^o_{t,\omega}(\varphi)\ \ell(dt)\ P(d\omega) = \eta(\mu^o).
\end{aligned}
$$
(7.70)

This shows that $\mathcal{R}$ is weakly compact. We use this result to prove that the functional given by (7.67) is continuous on $\mathcal{FC}_{ad}$. Let $\{\Gamma^n\} \subset \mathcal{FC}_{ad}$ be any generalized sequence (net). Recall that, for any pair of Banach spaces $\{X, Z\}$, the closed unit ball in $\mathcal{L}(X, Z)$ is compact in the weak.operator topology if and only if Z is reflexive [Dunford (1958)]. Since $Y(\subset H)$ and U are Hilbert spaces, it is clear that $\mathcal{FC}_{ad} \equiv B_1(\mathcal{L}(Y, U))$ is compact in the weak operator topology τ_{wo}. Thus, there exists a generalized subsequence (subnet), relabeled as the original sequence, and an element $\Gamma^o \in \mathcal{FC}_{ad}$ such that $\Gamma^n \xrightarrow{\tau_{wq}} \Gamma^o$. Let $\{\mu^n, \mu^o\}$ denote the corresponding weak solutions of equation (7.54) which is equivalent to the integral equation (7.55). Clearly, by definition of the reachable set, $\{\mu^n, \mu^o\} \subset \mathcal{R}$. We show that $\mu^n \xrightarrow{w} \mu^o$. Let $\mathcal{B}_0$ denote the operator

$$
\mathcal{B}_0(\varphi)(\xi) \equiv\ < f_0 + C\Gamma^o L_0(\xi), D\varphi(\xi) >_H
$$

and $\{C^n_3\}$ the sequence of operators given by

$$
(C^n_3 \varphi)(\xi) \equiv\ < C(\Gamma^o - \Gamma^n)L_0\xi, D\varphi(\xi) >_H
$$
(7.71)

defined on $C_\kappa \cap C_F^2$. Define $\nu^n \equiv \mu^o - \mu^n$, $\mathcal{C} \equiv \mathcal{A} + \mathcal{B} + \mathcal{B}_0$. Clearly, these are well defined for each $\varphi \in C_\kappa \cap C_F^2$. Then using the expression (7.55) corresponding to $\{\Gamma^n, \mu^n\}$ and $\{\Gamma^o, \mu^o\}$ respectively, and taking the difference, we obtain the following expression

$$\nu_t^n(\varphi) = \int_0^t \nu_s^n(\mathcal{C}\varphi)ds + \int_0^t \nu_s^n(\mathcal{C}_2\varphi) \cdot dw + \int_0^t \mu_s^n(C_3^n\varphi)ds, \quad (7.72)$$

$t \in I$. Clearly, this identity holds for any $\varphi \in C_\kappa \cap C_F^2$ having compact support. Since $\mathcal{FC}_{ad}$ is compact in the weak operator topology, it follows from the expression (7.71) that $\lim_{n\to\infty} C_3^n\varphi \to 0$ uniformly on H for every $\varphi \in C_\kappa \cap C_F^2$ having compact support. By virtue of weak compactness of the reachable set $\mathcal{R}$, along a subsequence if necessary, μ^n converges weakly to some $m^o \in L_\infty^{wa}(I, L_1(\Omega, M_2(H)))$. Therefore it follows from these strong and weak convergence properties that $\mu_s^n(C_3^n\varphi) \longrightarrow 0$ for all $s \in [0, t]$ and hence the third term of equation (7.72) converges to zero for each $t \in I$ P-a.s. Thus ν^n also converges weakly to some element, say ν which is $\mu^o - m^o$. Letting $n \to \infty$, it follows from the above equation that ν satisfies the following linear homogeneous functional equation,

$$\nu_t(\varphi) = \int_0^t \nu_s(\mathcal{C}\varphi)ds + \int_0^t \nu_s(\mathcal{C}_2\varphi) \cdot dw, t \in I, \quad (7.73)$$

for every $\varphi \in C_\kappa \cap C_F^2$ having compact support. Hence $\nu = 0$ implying that μ^n converges weakly to μ^o. This proves the weak continuity of the map $\Gamma \longrightarrow \mu^\Gamma$. Clearly the distance function $\psi_r \in C_\kappa$ and hence $\mu_t^\Gamma(\psi_r)$ is well defined. Since λ is a countably additive positive measure having bounded total variation, and $t \longrightarrow \mu_t^\Gamma(\psi_r)$ is P almost surely continuous, the objective functional given by (7.67) is continuous in the weak operator topology on $\mathcal{FC}_{ad}$ which is compact in the weak operator topology of $\mathcal{L}(Y, U)$. Hence there exists a feedback control law $\Gamma^o \in \mathcal{FC}_{ad}$ at which J attains its minimum. This completes the proof. $\bullet$

Remark 7.8.5 For admissible feedback controls, we have considered constant operators from $B_1(\mathcal{L}(Y, U))$ for the entire period I. By partitioning I into a finite number of disjoint measurable sets covering $I = \cup I_i$ and choosing any $r > 0$, the above result can be easily extended to admit piecewise constant operator valued functions as controls such as

$$\tilde{\mathcal{FC}}_{ad} \equiv \{\Gamma : I \longrightarrow B_r(\mathcal{L}(Y,U)) :$$
$$\Gamma(t) = \Gamma_i, t \in I_i, \Gamma_i \in B_r(\mathcal{L}(Y,U))\} \subset \prod_i B_r(\mathcal{L}(Y,U)).$$

Since each factor space is compact (in the weak operator topology), it follows from Tychonoff product theorem [[Willard (1970)], Theorem 17.8] that this set is also compact. We leave it to the reader to consider the limiting case allowing arbitrary refinement of the partition.

Another problem of significant interest is the terminal control problem as described below. Let $f : R^k \longrightarrow [0, \infty]$ be a continuous function bounded on bounded sets and $\{\varphi_i, 1 \leq i \leq k\} \in C_\kappa \cap C_F^2$ a family of nonnegative functions. Define the cost functional,

$$J(\Gamma) = \mathbb{E}\left\{f(\mu_T^\Gamma(\varphi_1), \mu_T^\Gamma(\varphi_2) \cdots \mu_T^\Gamma(\varphi_k))\right\}. \tag{7.74}$$

The problem is to find a feedback control law $\Gamma \in \mathcal{FC}_{ad}$ that minimizes this functional.

Corollary 7.8.6 *Consider the system (7.54) with the terminal cost given by (7.74) and suppose the assumptions of Theorem 7.8.4 hold. Let $f : R^k \longrightarrow [0, \infty]$ be a continuous nonnegative function satisfying*

$$f(x_1, x_2, \cdots, x_k) \leq c\{1 + \sum_{i=1}^{k} |x_i|\} \tag{7.75}$$

for certain constant $c > 0$. Then, there exists a feedback control law Γ^o at which J attains its minimum solving the terminal control problem.

Proof By Theorem 7.8.4, $\Gamma \longrightarrow \mu^\Gamma$ is weakly continuous, and for each $\varphi_i \in C_\kappa \cap C_F^2$, $t \longrightarrow \mu_t^\Gamma(\varphi_i)$ is P-almost surely continuous and hence $\mu_T^\Gamma(\varphi_i)$ is well defined. Let $\{\Gamma_\alpha\} \subset \mathcal{FC}_{ad}$ be a net. Since $\mathcal{FC}_{ad}$ is compact in the weak operator topology, there exists a subnet, relabeled as the original net, and a $\Gamma_o \in \mathcal{FC}_{ad}$ so that $\Gamma_\alpha \xrightarrow{\tau_{wo}} \Gamma_o$. Letting $\{\mu^\alpha, \mu^o\}$ denote the corresponding solutions of the system (7.54/7.55) with feedback law given by (7.65), it follows from the weak continuity, as proved in Theorem 7.8.4, that along a subnet, if necessary, $\mu^\alpha \xrightarrow{w} \mu^o$. Since f is continuous, this implies that

$$g_\alpha \equiv f(\mu_T^\alpha(\varphi_1), \mu_T^\alpha(\varphi_2) \cdots \mu_T^\alpha(\varphi_k)) \longrightarrow f(\mu_T^o(\varphi_1), \mu_T^o(\varphi_2) \cdots \mu_T^o(\varphi_k)) \equiv g_o$$

P-almost surely. Clearly, it follows from the property of f given by (7.75) that

$$|g_\alpha(\omega)| \le c(1 + \sum_{i=1}^k |\mu_T^\alpha(\varphi_i)(\omega)|, \ \forall \ \omega \in \Omega.$$

Since $\varphi_i \in C_\kappa$, there exists a finite positive number k_i such that

$$|\mu_T^\alpha(\varphi_i)(\omega)| \le k_i\{1 + |\mu_T^\alpha|_{M_2(H)}(\omega)\} \ \forall \ \omega \in \Omega.$$

Hence there exists a finite positive number $\tilde{C}$ such that

$$|g_\alpha(\omega)| \le \tilde{C}\{1 + |\mu_T^\alpha|_{M_2(H)}(\omega)\} \ \forall \ \omega \in \Omega.$$

By virtue of the estimate (7.57), the expression (7.65), boundedness of the set $\mathcal{FC}_{ad}$, and the above estimate, it follows from the same arguments as in Theorem 7.8.4 that $\{g_\alpha\}$ is dominated by an element of $L_1(\Omega)$. Hence, again it follows from Lebesgue dominated convergence theorem that

$$\lim_\alpha \mathbb{E}g_\alpha \equiv \lim_\alpha \int_\Omega g_\alpha(\omega)P(d\omega) = \int_\Omega g_o(\omega)P(d\omega) = \mathbb{E}g_o. \qquad (7.76)$$

This shows that the functional J given by (7.74) is weakly continuous. Since $\mathcal{FC}_{ad}$ is compact in the weak operator topology, J attains its minimum on $\mathcal{FC}_{ad}$ proving existence of an optimal linear feedback control law. •

Another terminal control problem of significant interest is described as follows. Suppose at the end of the time horizon $[0, T]$, it is required to reach a target set C, a closed subset of H. We can formulate this as an optimal control problem: Find $\Gamma \in \mathcal{FC}_{ad}$ that maximizes the functional

$$J(\Gamma) \equiv \mathbb{E}\{\mu_T^\Gamma(C)\}. \qquad (7.77)$$

If $C \subset V$, the above problem is equivalent to demanding smoothness of the flow at the end of the active period. If $C \subset H\backslash V$, the problem is equivalent to requiring the kinetic energy to satisfy certain energy constraints. If C is a finite dimensional closed linear subspace of H, the problem is equivalent to demanding certain limits on the Hausdorff (or Fractal) dimension of the flow.

For the benefit of the reader we digress a little and introduce what is known as the Hausdorff dimension. First, let us consider R^n and let K be a bounded subset of it and $r > 0$ and let $N_K(r)$ denote the minimum number of balls of radius r required to cover K. Then define the Hausdorff dimension d_H of K to be

$$d_H(K) := -\lim_{r \to 0} \frac{\log N_K(r)}{\log r}.$$

For example, if K is the unit cube in R^n, it is clear that the number of r cubes (cubes of side r) required to cover it is $N(r) = (1/r^n)$. Thus it follows from the above expression that $d_H(K) = n$.

In fact the Hausdorff dimension is well defined in any metric space. Let (X, ρ) be a metric space with the distance function ρ. Define the diameter of any bounded set $C \subset X$ by $D(C) \equiv \sup\{\rho(x,y), x, y \in C\}$. For any $\delta > 0$, let Π_δ denote the class of all countable δ covers of C, that is, a collection of sets $\{C_i(\delta), i \in N\}$ each of diameter less than δ such that

$$C \subseteq \bigcup_{i \in N} C_i(\delta).$$

For each number $d > 0$, define the function

$$H_\delta^d(C) = \inf_{\Pi_\delta} \sum_{i \in N} [D(C_i(\delta))]^d$$

where the infimum is taken over all countable δ-covers Π_δ of the set C. Note that $\delta \longrightarrow H_\delta^d(C)$ is a monotonically decreasing function and hence the limit

$$\lim_{\delta \downarrow 0} H_\delta^d(C) \equiv H^d(C)$$

exists with values in $R_+ \equiv [0, \infty]$. This is called the d-dimensional Hausdorff measure of the set C and it is well defined on the Borel sets of (X, ρ). If X is R^n with any of the equivalent metrics, and $d < n$ it is easy to see that $H^d(C) = +\infty$. On the other hand, if $d > n$ then $H^d(C) = 0$. In the general case, it was shown by Hausdorff that there exists a critical number d^*, not necessarily an integer, for which $H^d(C) = +\infty$ for all $d < d^*$ and that $H^d(C) = 0$ for all $d > d^*$. This very critical number is called the Hausdorff dimension of the set C and it is formally defined as

$$d_H(C) = \inf\{d \geq 0 : H^d(C) = 0\} \equiv d^*(C).$$

In the study of dynamic systems one is interested in the dimension of compact attractors. For stochastic systems, one is interested in the Hausdorff dimension of the support of the measure induced by the system. More precisely, in the case of infinite dimensional state space, one is interested to determine the Hausdorff dimension of the ε-support, that is, a compact set K_ε such that $\mu(K_\varepsilon') < \varepsilon$. For example, any tight measure on a metric space has a compact ε-support and it is interesting to find the Hausdorff dimension of the set. It is intuitively clear that the larger the dimension is the greater the complexity (measured in terms of degrees of freedom). We return to this problem later. The reader may be familiar with the complexity of determining these geometric dimensions [Temam (1988)] of attractors for dynamical systems.

So, for practical applications, it may be sufficient to choose a finite dimensional subspace C of the state space H and design a controller that maximizes the expected mass of the measure on C. The following general result can be interpreted in this sense.

Theorem 7.8.7 *Consider the feedback control problem with the objective functional (7.77). Let C be a closed subset of H and suppose the assumptions of Theorem 7.8.4 hold. Then there exists a feedback control $\Gamma^o \in \mathcal{F}\mathcal{C}_{ad}$ such that*

$$J(\Gamma^o) \geq J(\Gamma) \forall \; \Gamma \in \mathcal{F}\mathcal{C}_{ad}.$$

Proof It suffices to prove that the functional J given by (7.77) is upper semicontinuous in the weak operator topology on $\mathcal{F}\mathcal{C}_{ad}$. Let $\{\Gamma^\alpha\}$ be a net in $\mathcal{F}\mathcal{C}_{ad}$ and $\Gamma^o \in \mathcal{F}\mathcal{C}_{ad}$ such that $\Gamma^\alpha \xrightarrow{\tau_{wo}} \Gamma^o$. Let $\{\mu^\alpha\}, \mu^o$ be the corresponding set of solutions of equation (7.55). Following the same arguments as given in the proof of Theorem 7.8.4, we conclude that there exists a subnet, relabeled as the original net, such that $\mu^\alpha \xrightarrow{w} \mu^o$. Since these measures are weakly continuous on I with probability one, we have

$$\mu_T^\alpha \xrightarrow{w} \mu_T^o \;\; P - a.s.$$

Then it follows from a well known theorem given in Parthasarathy [[Parthasarthy (1967)], Theorem 6.1,p40] that

$$\overline{\lim_\alpha} \, \mu_T^\alpha(C) \le \mu_T^o(C) \quad P - a.s.$$

Clearly this implies that

$$\overline{\lim_\alpha} \, \mathbb{E}\{\mu_T^\alpha(C)\} \le \mathbb{E}\{\overline{\lim_\alpha} \, \mu_T^\alpha(C)\} \le \mathbb{E}\{\mu_t^o(C)\}.$$

Thus by definition of the objective functional (7.77), we have

$$\overline{\lim_\alpha} \, J(\Gamma^\alpha) \le J(\Gamma^o)$$

and hence J is upper semicontinuous on $\mathcal{FC}_{ad}$ in the weak operator topology. Since $\mathcal{FC}_{ad} = B_1(\mathcal{L}(Y,U))$ is compact in the weak operator topology, J attains its supremum on it. This completes the proof. •

However the above result does not solve the problem that demands a control law which minimizes a cost functional subject to a constraint on the Hausdorff dimension of the support of the measure valued process. Towards this goal we consider an objective functional that includes the Hausdorff dimension.

Let $\mathcal{K}(H)$ denote the class of nonempty compact subsets of the Hilbert space H and suppose that this is endowed with a metric topology ρ_H so that $(\mathcal{K}(H), \rho_H)$ is a complete metric space. The most popular topology on $\mathcal{K}(H)$ is the metric topology given by the Hausdorff metric ρ_H which is defined by

$$\rho_H(K, L) \equiv max\{\sup\{d(x, K), x \in L\}, \sup\{d(y, L), y \in K\}\}$$

for $K, L \in \mathcal{K}(H)$. Since H is assumed to be separable, $(\mathcal{K}(H), \rho_H)$ is a Polish space (complete separable metric space).

Now we can introduce the appropriate cost functional. Let μ^Γ denote the solution of equation (7.55) corresponding to the feedback control $\Gamma \in \mathcal{FC}_{ad}$. The cost functional corresponding to Γ is then given by

$$J(\Gamma) \equiv \inf\left\{d_H(C) + \lambda\left(\frac{1}{T}\int_0^T\int_\Omega \mu_{t,\omega}^\Gamma(H \setminus C)dtP(d\omega)\right) : C \in \overline{\mathcal{K}}(H)\right\}$$

$$(7.78)$$

where $\overline{\mathcal{K}}(H) \equiv \mathcal{K}(H) \cup \emptyset$ and λ is a large positive number weighing the leakage of mass from compacts. The problem is to find a control law $\Gamma^o \in \mathcal{FC}_{ad}$

for the system (7.54) (equivalently (7.55)) that minimizes the functional J. In other words, we would like to prove the existence of an optimal control law. Unfortunately there are technical problems. It is shown by Mattila and Mauldin [Mattila (1997)] that the Hausdorff dimension function $d_H(\cdot)$ is not lower semicontinuous with respect to the Hausdorff metric ρ_H. It is a function belonging to the Baire's class 2 [[Mattila (1997)], Theorem 2.1, p83]. It is also known that the Baire class 1 is dense in Baire class 2. Baire class 1 consists of semi continuous functions. However, it is not known if the function $d_H(\cdot)$ can be approximated by lower semicontinuous functions. To avoid this technical difficulties we choose a set function which is lower semicontinuous with respect to the Hausdorff metric and also carries some relevant properties of the Hausdorff dimension function.

Let $\mathcal{N}$ denote the class of sets consisting of singletons and finite sets including the empty set. Consider the set function $\nu : \overline{\mathcal{K}}(H) \longrightarrow [0, \infty]$ satisfying the following properties:

(P1): $\nu(F) = 0$ for every $F \in \mathcal{N}$, (P2): $\nu(K_1) \leq \nu(K_2)$ for $K_1 \subset K_2$ and $K_1, K_2 \in \overline{\mathcal{K}}(H)$, (P3): $\lim_{d_H(L) \to \infty} \nu(L) = \infty, L \in \overline{\mathcal{K}}(H)$.

We replace the functional (7.78) by the following one

$$J(\Gamma) \equiv \inf \left\{ \nu(C) + \frac{\lambda}{T} \int_0^T \int_\Omega \mu_{t,\omega}^\Gamma (H \setminus C) dt dP, C \in \overline{\mathcal{K}}(H) \right\} \quad (7.79)$$

which serves the original purpose. Here we prove an existence theorem.

Theorem 7.8.8 *Consider the system (7.55) with the admissible controls $\mathcal{FC}_{ad}$, and the objective functional given by (7.79). Suppose the assumptions of Theorem 7.8.4 hold. Further, suppose the set function $\nu(\cdot)$ satisfies the properties (P1)-(P3) and that it is lower semi continuous on $\mathcal{K}(H)$ with respect to the Hausdorff metric ρ_H. Then there exists an optimal control law minimizing the cost function (7.79).*

Proof Define the functional $\ell : \mathcal{FC}_{ad} \times \overline{\mathcal{K}}(H) \longrightarrow [0, \infty]$ as follows

$$\ell(\Gamma, C) \equiv \nu(C) + \frac{\lambda}{T} \int_0^T \int_\Omega \mu_{t,\omega}^\Gamma (H \setminus C) \, dt \, dP,$$

and note that $J(\Gamma) = \inf \{ \ell(\Gamma, C), C \in \mathcal{K}(H) \}$. Let $\mathcal{N}_H \subset \overline{\mathcal{K}}(H)$ denote the class of finite subsets of H including the singletons and empty set. It is clear that ℓ satisfies the following properties: (1): $\ell(\Gamma, C) \geq 0$,   (2): $\ell(\Gamma, F) = \lambda$ for every $F \in \mathcal{N}_H$, and (3): $\overline{\lim}_{C \in k(H)} \ell(\Gamma, C) = \infty$. The first property is obvious. The second property follows from the property (P1) and the fact that the elements of $\mathcal{R}$ are non atomic. It is clear that $\ell(\Gamma, C) \leq \nu(C) + \lambda$ for all $\Gamma \in \mathcal{FC}_{ad}$ and all $C \in \overline{\mathcal{K}}(H)$. It follows from this and the property (P3), that $C \longrightarrow \ell(\Gamma, C)$ is coercive on $\mathcal{K}(H)$ in the sense that $\lim_{d_H(C) \to \infty} \ell(\Gamma, C) = +\infty$. By assumption ν is lower semicontinuous. We prove that for any fixed $\Gamma \in \mathcal{FC}_{ad}$, the second term of ℓ is also lower semicontinuous. Let $C_n \xrightarrow{\rho_H} C_0$. Clearly, for any $\varepsilon > 0$, there exists an $n_\varepsilon \in N$ such that $C_n \subset C_0^\varepsilon$ for all $n \geq n_\varepsilon$ where $C_0^\varepsilon \equiv \{\xi \in H : d(\xi, C_0) \leq \varepsilon\}$. Hence for any fixed Γ

$$\frac{\lambda}{T} \int_0^T \int_\Omega \mu_{t,\omega}^\Gamma(H \setminus C_0^\varepsilon) dt dP \leq \frac{\lambda}{T} \int_0^T \int_\Omega \mu_{t,\omega}^\Gamma(H \setminus C^n) dt dP$$

for all $n \geq n_\varepsilon$ and therefore

$$\frac{\lambda}{T} \int_0^T \int_\Omega \mu_{t,\omega}^\Gamma(H \setminus C_0^\varepsilon) dt dP \leq \lim_n \frac{\lambda}{T} \int_0^T \int_\Omega \mu_{t,\omega}^\Gamma(H \setminus C^n) dt dP.$$

Since this is true for every $\varepsilon > 0$, we have proved that the second term of ℓ is lower semicontinuous in C on $\mathcal{K}(H)$. Hence, for each $\Gamma \in \mathcal{FC}_{ad}$, $C \longrightarrow \ell(\Gamma, C)$ is lower semicontinuous with respect to the Hausdorff metric ρ_H. Thus it follows from the coercivity of the map $C \longrightarrow \ell(\Gamma, C)$ and its lower semicontinuity that it attains its infimum on $\overline{\mathcal{K}}(H)$. This shows that $J(\Gamma)$ is well defined as a functional from $\mathcal{FC}_{ad}$ to $[0, \infty]$. Let $\{\Gamma_\alpha\} \subset \mathcal{FC}_{ad}$, $\alpha \in \Lambda$ (directed set), be a minimizing net for $J(\Gamma)$. It follows from the preceding analysis that $\exists$ a net $\{C_\alpha\}_{\alpha \in \Lambda} \in \overline{\mathcal{K}}(H)$ such that $J(\Gamma_\alpha) = \ell(\Gamma_\alpha, C_\alpha)$ and that

$$J(\Gamma_\alpha) \to \inf\{J(\Gamma), \Gamma \in \mathcal{FC}_{ad}\} \equiv m.$$

In fact it follows from the coercivity property of ℓ (in its second argument) that there exists a set $X \in \mathcal{K}(H)$ such that $\{C_\alpha\}_{\alpha \in \Lambda} \subset X$. Clearly, $\mathcal{K}(X)$ with the relative metric topology ρ_H (inherited from $\mathcal{K}(H)$ containing $\mathcal{K}(X)$) is a compact metric space $(\mathcal{K}(X), \varrho_H)$ [[Papageorgiou (1997)], Theorem 1.34, p15; Theorem 2.11, p687]. It follows from this and the fact that $B_1(\mathcal{L}(Y, U)) \equiv \mathcal{FC}_{ad}$ is compact in the weak operator topology, that there exists a subnet of the net $\{\Gamma_\alpha, C_\alpha\}$, relabeled as the original net $\{\Gamma_\alpha, C_\alpha\}$, and $\{\Gamma^o, C^o\} \in \mathcal{FC}_{ad} \times k(X) \subset \mathcal{FC}_{ad} \times k(H)$ such that

$$\Gamma_\alpha \overset{\tau_{wq}}{\longrightarrow} \Gamma^o, \quad \& \quad C_\alpha \overset{\rho_H}{\longrightarrow} C^o.$$

Let $\{\mu^\alpha\} \in \mathcal{R}$ denote the solution of (7.55) corresponding to Γ_α. By Theorem 7.8.4, $\mathcal{R}|_{t,\omega}$ is uniformly tight for a.a $(t,\omega) \in I \times \Omega$ and therefore there exists a $\mu^o \in \mathcal{R}$ such that, along a subnet if necessary,

$$\mu_{t,\omega}^\alpha \overset{w}{\longrightarrow} \mu_{t,\omega}^o \qquad dt \times dP \ a.a \ (t,\omega) \in I \times \Omega.$$

Since $\{C_\alpha, C^o\}$ are compact sets and $\{\mu_{t,\omega}^\alpha\}, \alpha \in \Lambda$, is uniformly tight for almost all $(t,\omega) \in I \times \Omega$, it follows from similar arguments as in the proof of Corollary 7.8.7 that we have

$$\overline{\lim} \, \mu_{t,\omega}^\alpha(C_\alpha) \leq \mu_{t,\omega}^o(C^o) \ dt \times dP \ a.a \ (t,\omega) \in I \times \Omega.$$

From this it is easy to verify that

$$\frac{1}{T} \int_0^T \int_\Omega \mu_{t,\omega}^o(H \setminus C^o) dt dP \leq \underline{\lim} \, \frac{1}{T} \int_0^T \int_\Omega \mu_{t,\omega}^\alpha(H \setminus C_\alpha) dt dP.$$

By lower semicontinuity of ν on $\mathcal{K}(H)$, we have $\nu(C^o) \leq \underline{\lim}_\alpha \nu(C_\alpha)$. Thus it follows from these results that $\Gamma \longrightarrow J(\Gamma)$ is lower semicontinuous in the weak operator topology on $B_1(\mathcal{L}(Y,U)) \equiv \mathcal{FC}_{ad}$. Since $B_1(\mathcal{L}(Y,U))$ is compact in the weak operator topology J attains its infimum on $\mathcal{FC}_{ad}$. This proves the existence of an optimal control law in $\mathcal{FC}_{ad}$. •

For computational purposes, it may be more convenient to replace $\mathcal{K}(H)$ by the class $fcl(H)$ of all finite dimensional closed linear subspaces of H. For more on hyperspace topologies see [Papageorgiou (1997)].

Remark 7.8.9 In order to prove the existence of an optimal control law for the cost functional (7.78) with the Hausdorff dimension function $d_H(\cdot)$ included, it is necessary to develop a technique that does not require the lower semicontinuity condition. We leave it as an open problem.

Next we consider the question of control of explosion or blow up time.

(Control of Blowup Time) Let $B_r(V)$ denote the ball of radius $r > 0$ in V, and V_r its closure in H. Suppose the initial measure ν_0 is supported on a closed set $D_r \subset V \cap V_r$, that is $\nu_0(D_r) = 1$. Define the random time

$$\tau_r(\Gamma) \equiv \inf\{t \geq 0 : \int_0^t \mu_s^\Gamma(H \setminus V_r) \, ds > 0\}.$$

We use the convention $\inf \emptyset = \infty$. Since $\mathcal{F}_t^w$ is right continuous, it is easy to see that the set $\{\tau_r(\Gamma) < t\}$ is $\mathcal{F}_t^w$ measurable for every $t \geq 0$. Thus τ_r is a stopping time and so integrable possibly taking values from $(0, \infty) \cup \{\infty\}$. The problem is to find a feedback control $\Gamma \in \mathcal{FC}_{ad}$ that maximizes

$$J_r(\Gamma) \equiv \mathbb{E}\tau_r(\Gamma).$$

This is solved in the following theorem.

Theorem 7.8.10 *Consider the system (7.54/7.55) with the admissible feedback operators $\mathcal{FC}_{ad} = B_1(\mathcal{L}(Y, U))$ and suppose the assumptions of Theorem 7.8.1 hold and the reachable set $\mathcal{R}$ satisfy the properties (P1) and (P2) of Theorem 7.8.4. Then, there exists a feedback control law $\Gamma^o \in \mathcal{FC}_{ad}$ that maximizes the blowup time in the sense that $J_r(\Gamma^o) \geq J_r(\Gamma) \ \forall \ \Gamma \in \mathcal{FC}_{ad}$.*

Proof Let Λ be any directed set and $\{\Gamma_\alpha\}_{\alpha \in \Lambda} \subset \mathcal{FC}_{ad}$ be a maximizing net. Since $\mathcal{FC}_{ad}$ is compact in the weak operator topology, there is a subnet that converges τ_{wo} to a unique limit (because Hausdorff) $\Gamma^o \in \mathcal{FC}_{ad}$. By virtue of weak convergence (Theorem 7.8.4) of

$$\mu_{t,\omega}^\alpha \equiv \mu_{t,\omega}^{\Gamma^\alpha} \xrightarrow{w} \mu_{t,\omega}^{\Gamma^o} \equiv \mu_{t,\omega}^o \quad dt \times dP \quad a.e.$$

and the fact that $H \setminus V_r$ is an open set we have

$$\varliminf \mu_{s,\omega}^\alpha(H \setminus V_r) \geq \mu_{s,\omega}^o(H \setminus V_r) \ dt \times dP \quad a.e.$$

Hence

$$\varliminf \int_0^t \mu_{s,\omega}^\alpha(H \setminus V_r) \ ds \geq \int_0^t \varliminf \mu_{s,\omega}^\alpha(H \setminus V_r) \ ds$$

$$\geq \int_0^t \mu_{s,\omega}^o(H \setminus V_r) \ ds \quad P - a.s, \tag{7.80}$$

and therefore

$$\inf\{t \geq 0 : \varliminf \int_0^t \mu_{s,\omega}^\alpha(H \setminus V_r) \ ds > 0\}$$

$$\leq \inf\{t \geq 0 : \int_0^t \mu_{s,\omega}^o(H \setminus V_r) \ ds > 0\}. \tag{7.81}$$

It is clear that for every $\varepsilon > 0$, there exists $\alpha_\varepsilon \in \Lambda$ such that

$$\varliminf_{\beta \in \Lambda} \int_0^t \mu_{s,\omega}^\beta(H \setminus V_r) \ ds \leq \varepsilon + \int_0^t \mu_{s,\omega}^\alpha(H \setminus V_r) \ ds$$

for all $\alpha > \alpha_\varepsilon$ uniformly in t on any finite interval $I \subset [0, \infty)$. Then a little reflection reveals the following inequality,

$$\inf\{t \geq 0 : \varliminf_{\beta} \int_0^t \mu_{s,\omega}^\beta (H \setminus V_r) \, ds > 0\}$$

$$\geq \inf\{t \geq 0 : \int_0^t \mu_{s,\omega}^\alpha (H \setminus V_r) \, ds + \varepsilon > 0\}$$

for every $\alpha > \alpha_\varepsilon$. Thus it follows from (7.81) and the above inequality that for every $\varepsilon > 0$,

$$\varlimsup_{\alpha} \left\{ \inf\{t \geq 0 : \int_0^t \mu_{s,\omega}^\alpha (H \setminus V_r) \, ds + \varepsilon > 0\} \right\}$$

$$\leq \inf\{t \geq 0 : \int_0^t \mu_{s,\omega}^o (H \setminus V_r) \, ds > 0\}.$$

Since $\varepsilon > 0$ is arbitrary, this implies that

$$\varlimsup_{\alpha} \tau_r(\Gamma_\alpha) \leq \tau_r(\Gamma^o) \quad P - a.s.$$

Then

$$\varlimsup J_r(\Gamma_\alpha) \equiv \varlimsup \mathbb{E}\{\tau_r(\Gamma_\alpha)\}$$

$$\leq \mathbb{E}\{\varlimsup \tau_r(\Gamma_\alpha)\} \leq \mathbb{E}\{\tau_r(\Gamma^o)\} \equiv J_r(\Gamma^o).$$

Thus we have proved that $\Gamma \longrightarrow J(\Gamma)$ is upper semicontinuous on $B_1(\mathcal{L}(Y, U))$ with respect to the weak operator topology. Since $\mathcal{FC}_{ad}$ is τ_{wo} compact and J is τ_{wo} upper semicontinuous, J attains its supremum on $\mathcal{FC}_{ad}$. •

7.9 Some Problems for Exercise

Here we present some interesting problems which have substantial practical significance. This is evident from their descriptions given below.

P1: Give a detailed proof of Theorem 7.5.4 following the discussion preceding the statement.

P2: Justify the definition of the duality pairing for the pair $\{Z_b, Z_b^*\}$ given by the expression (7.37) and verify that

$$|<z^*, z>| \leq \|z^*\|_{Z_b^*} \|z\|_{Z_b}.$$

Hint: Follow the steps in the construction of these spaces.

P3: Consider the problem stated in Remark 7.7.3. For cumulative filtering error over the time interval I, the proper objective functional is

$$\bar{\varrho}(\Gamma) = \int_\Omega \int_I J(\mu_t^\Gamma)(v)\lambda(dt)W(dv)$$

where λ is a bounded positive measure. Verify that this functional admits an optimal observer (observation operator Γ). Hint: Follow the steps in the proof of Theorem 7.7.2.

P4: Consider the example of the abstract Gelfand triple $V_n \hookrightarrow G_n \hookrightarrow V_n^*$ given after Theorem 7.5.2. Here $V_n = W_0^{m,p}(D^n, M_n), n \geq 1$, with $m \geq (nd/p) + 1$. Justify the necessity of choosing m as stated. Hint: Recall Sobolev embedding theorems.

P5: In certain problems of hydrodynamics it may be desirable to use feedback controls to smooth the flow. According to Theorem 7.8.1, $\mu \in L_1^{wa}(I, L_1(\Omega, M_2(V))) \cap L_\infty^{wa}(I, L_1(\Omega, M_2(H)))$. Thus another natural control problem is to find a $\Gamma \in \mathcal{F}\mathcal{C}_{ad}$ that minimizes the functional

$$J(\Gamma) = \mathbb{E}\left\{ \int_I \| \mu_t^\Gamma \|_{M_2(V)} \, dt \right\} + \mathbb{E}\left\{ ess - sup\{|\mu_t^\Gamma|_{M_2(H)}\} \right\}.$$

Find suitable conditions under which this problem has a solution. Hint: The identity (7.57) and the expression for the feedback control operator given by (7.65) are useful.

P6: Another interesting stopping time problem is given as follows. Let $r > 0$ be a finite number, and define

$$\tau_2(\Gamma) \equiv \inf\{t \geq 0 : \int_0^t \| \mu_s^\Gamma \|_{M_2(V)} \, ds > r\} \in [0, \infty],$$

and $J(\Gamma) \equiv \mathbb{E}\{\tau_2(\Gamma)\}$. Prove the existence of a feedback control law $\Gamma^o \in \mathcal{F}\mathcal{C}_{ad}$ that maximizes $J(\Gamma)$. Hint: Give sufficient conditions so that the map $\Gamma \to J(\Gamma) \equiv \mathbb{E}\tau_2(\Gamma)$ is upper semi continuous on $\mathcal{F}\mathcal{C}_{ad}$ which is assumed to be compact in the weak operator topology.

P7: (Evasion of danger zone) Consider the stochastic NSE (7.54) and suppose the assumptions of Theorem 7.8.4 hold with the admissible feedback controls given by $\mathcal{F}\mathcal{C}_{ad} \equiv B_1(\mathcal{L}(Y, U))$. Let D be an open subset of H and ν

a nonnegative countably additive measure on I having bounded variation. The problem is to find a feedback control law $\Gamma \in \mathcal{FC}_{ad}$ that minimizes the cost functional

$$J(\Gamma) \equiv \int_I \mathbb{E}\{\mu_t^\Gamma(D)\}\nu(dt).$$

Prove the existence of an optimal feedback control law. Hint: $\mathcal{FC}_{ad}$ is compact in the weak operator topology. It suffices to verify that J is weakly lower semicontinuous.

Chapter 8

Some Elements of Malliavin Calculus

8.1 Introduction

Malliavin Calculus is a stochastic calculus of random variables defined on
the Wiener measure space. It was originally introduced by Paul Malliavin
in the mid seventies and since then it has been extensively studied and extended by many authors such as Malliavin himself [Malliavin (1995)] [Malli-
avin (1976)] [Malliavin (1997)], Bell [Bell (1987)], Stroock [Stroock (1981)]
[Stroock (1981)], Watanabe [Watanabe (1984)], Nualart [Nualart (1989)]
[Nualart (2009)] [Nualart (1997)], Shigekawa [Shigekawa (1998/2004)], Da
Prato [Da Prato (2007)], Øksendal [Oksendal (1997)], Friz [Friz (2002)] and
many others.

In this chapter we present some elements of this calculus and its application to stochastic differential equations and central limit theorem. In
particular, we are interested in the Malliavin derivative and its adjoint. For
simplicity, we concentrate on functionals of R^d valued Brownian motion as
introduced in Chapter 2. In fact, these results also hold for infinite dimensional Hilbert space valued Brownian motions with nuclear covariance. We
consider the Ornstein-Uhlenbeck operator given by the negative of the composition of the Malliavin derivative (operator) and its adjoint. Further, we
study the associated semigroup $T_t, t \geq 0$, known as the Ornstein-Uhlenbeck
semigroup. Using the Ornstein-Uhlenbeck operator, we construct Sobolev
spaces on the Wiener measure space. We study the questions of existence
and smoothness of densities of measures induced by the solutions of finite dimensional stochastic differential equations using Hörmander-Malliavin hy-

poellipticity conditions. In the final section, we present some results on the central limit theorem in relation to Wiener chaos taking values from the real line. This result is due to Nualart and his school [Nualart (2009)] (see also the references therein). We extend this result to Wiener chaos taking values from infinite dimensional Hilbert space and conclude the chapter with some comments.

8.2   Abstract Wiener Space

Let $(\Omega, \mathcal{B}, P)$ be a complete probability space and $I \equiv [0, T]$ a closed bounded interval and $\{w(t), t \in I\}$, a d-dimensional standard Brownian motion on $(\Omega, \mathcal{B}, P)$ starting from the origin with probability one. We set $\Omega \equiv C_0(I, R^d)$, the canonical sample space, where $C_0(I, R^d) \equiv \{x \in C(I, R^d) : x(0) = 0\}$, and let $\mathcal{B}$ denote the Borel σ-algebra generated by open or closed subsets of the set Ω and let μ^w denote the canonical Wiener measure. This is the classical Wiener measure space $(\Omega, \mathcal{B}, \mu^w)$ as introduced in Chapter 1. Furnished with the sup-norm topology, $E \equiv C_0(I, R^d)$ is a Banach space. Consider now the subspace $H_1 \equiv \{x \in E : \dot{x} \in L_2(I, R^d) \equiv H\}$ of E. We furnish this with the following scalar product

$$(x, y)_1 \equiv \int_I (\dot{x}(t), \dot{y}(t))_{R^d} dt. \tag{8.1}$$

Completed with respect to this scalar product, H_1 is a Hilbert space with the associated norm given by $\| x \| \equiv \sqrt{(x, x)_1} = \| x \|_1$. Note that for $x \in H_1$, if $\| x \|_1 = 0$, then $x(t) = 0$ for all $t \in I$. From Sobolev embedding theorem, it follows that the inclusion $H_1 \hookrightarrow E$ is continuous and dense. In the context of stochastic process, this space is known as Cameroon-Martin space. The triple (μ^w, H_1, E) is known as the abstract Wiener space. This space was originally introduced by Gross [Gross (1965/66)]. Note also that the Wiener measure of H_1 is zero. This follows from the fact that Brownian motion is nowhere differentiable. And also recall that w is only Hölder continuous exponent α for $0 \le \alpha < (1/2)$.

Some elementary functionals of the Wiener process are as described below. Let $\mathcal{S}_1$ denote the class of functionals of the form

$$\mathcal{S}_1 \equiv \{F : F(w) \equiv f((x_1, w(t_1)), (x_2, w(t_2)), \cdots, (x_n, w(t_n)))\}$$

where w is the d dimensional Wiener process defined on the probability space $(\Omega, \mathcal{B}, P)$, $x_i \in R^d$ for all $1 \leq i \leq n$ and $f : R^n \longrightarrow R$ is a continuous map having at most polynomial growth. Let $\mathcal{S}_2$ denote the class of functionals of the form

$$\mathcal{S}_2 \equiv \{F : F(w) = f(h_1(w), h_2(w), \cdots, h_n(w))\} \qquad (8.2)$$

with f having no more than polynomial growth, where, by $h_i(w)$, we mean a linear functional of w given by the Wiener integral,

$$h_i(w) \equiv \int_I < h_i(t), dw(t) >_{R^d} \equiv \sum_{j=1}^{d} \int_I h_i^j(t) dw_j(t), h_i \in L_2(I, R^d) \equiv H,$$

with $h_i \equiv \{h_i^1, h_i^2, \cdots, h_i^d\}'$ for $1 \leq i \leq n$. The elements of $\mathcal{S}_1$ and $\mathcal{S}_2$ are known as cylindrical functions defined on the Wiener measure space with $\mathcal{S}_1 \subset \mathcal{S}_2$ and that they form an algebra under addition and scalar multiplication.

8.3 Malliavin Derivative and Integration by Parts

Malliavin derivative is a special case of Gâteaux derivative which is a directional derivative. Let $F : E \longrightarrow R$; then its Gâteax derivative at $x \in E$, in the direction $h \in E$, is defined if there exists an $\eta \in E^*$ such that

$$\lim_{\varepsilon \to 0} \left\{ \frac{F(x + \varepsilon h) - F(x)}{\varepsilon} \right\} = (\eta, h)_{E^*, E}$$

giving $DF(x) = \eta$. Note that the dual E^* is the space of vector valued Borel measures $M(I, R^d)$. Often, the derivative may exist only in the direction of a proper subspace of the space E. For functions on the abstract Wiener space, the correct direction is provided by the Cameroon-Martin space H_1 and in this case $\eta = DF(x) \in H_1^*$. Thus the proper duality pairing is $H_1^* - H_1$ product (pairing).

Let us now consider the Brownian functionals from $\mathcal{S}_2$. Since $h_i(w)$ is a Gaussian random variable with mean zero and variance $\| h_i \|_H^2$ and f is a polynomial, every $F \in S_2$ is in $L_2(\Omega)$, that is, $S_2 \subset L_2(\Omega)$. In fact, it is easy to verify that $S_2 \subset L_p(\Omega)$ for all $p \geq 1$ as long as f has polynomial growth. Even exponential growth is admissible. Without loss of generality, assume that $\{h_i\} \in H \equiv L_2(I, R^d)$ are orthonormal. In that case, it is easy to compute and verify that

$$\mathbb{E}|F|^p \equiv \int_{R^n} |f(x)|^p \mu^n(dx) < \infty \qquad (8.3)$$

for every finite $p \geq 1$, where μ^n denotes the standard Gaussian measure on R^n given by

$$\mu^n(dx) = (1/2\pi)^{n/2} \exp\{-(1/2)|x|^2_{R^n}\} \, dx \qquad (8.4)$$

with dx denoting the Lebesgue measure. Let us compute the Malliavin derivative of the functional F given by

$$F(w) \equiv f(h_1(w), \cdots, h_n(w))$$

in the Cameroon-Martin direction $h \in H_1$. This is given by

$$< DF(w), h >= \sum \partial_i f(h_1(w), \cdots, h_n(w)) \, (h_i, h) \qquad (8.5)$$

where $\partial_i f$ denotes the derivative of f with respect to its i-th argument. Hence,

$$DF(w) = \sum \partial_i f(h_1(w), \cdots, h_n(w)) \, h_i. \qquad (8.6)$$

Again, since $\partial_i f$ is a polynomial, $DF \in L_2(\Omega, H)$, in fact, it is in $L_p(\Omega, H)$ for every $p \geq 1$.

For illustration, why the Malliavin derivative has to be computed in the direction of Cameroon-Martin subspace, consider the elementary function

$$F(w) = g(w) = \int_I < g(t), dw(t) >, g \in H.$$

Let $\xi \in H_1$, then

$$\lim_{\varepsilon \to 0} \left\{ \frac{F(w + \varepsilon\xi) - F(w)}{\varepsilon} \right\} = \int_I < g(t), d\xi(t) > \ = \int_I < g(t), \dot{\xi}(t) > dt.$$

Since $\xi \in H_1$, $\dot{\xi} \in H \equiv L_2(I, R^d)$ and so the scalar product and hence the derivative is well defined. Trying with $\xi \in E$, the reader can easily see that, in general, $\dot{\xi} \notin H$ and hence the above scalar product is not defined. The same statement applies for any ξ outside of H_1.

To proceed further we need some elementary identities involving Malliavin derivative. The following two lemmas provide integration by parts formulae.

Lemma 8.3.1 *For every $F \in S_2$ and $h \in H$,*

$$\mathbb{E}(< DF, h >_H) = \mathbb{E}(Fh(w)). \tag{8.7}$$

Proof Let F be given by $F(w) = f(h_1(w), \cdots, h_n(w))$ for any given $n \in N$ where, without loss of generality, we may assume that $\{h_i\}$ is any orthonormal basis of H. Then, clearly

$$DF(w) = \sum (\partial_i f) h_i \in L_2(\Omega, H),$$

and for any $h \in H$,

$$\mathbb{E}\{< DF, h >_H\} = \mathbb{E}\{\sum (\partial_i f)(h_i, h)_H\}.$$

Since $\{x_i \equiv h_i(w)\}_{i=1}^n$ are independent random variables with zero mean and unit variance, we have

$$\mathbb{E}\{< DF, h >_H\} = \int_{R^n} \sum (\partial_i f)(x)\ (h_i, h)_H\ \mu^n(dx)$$

$$= \int_{R^n} f(x)\{\sum (h_i, h)\ x_i\}\ \mu^n(dx)$$

$$= \mathbb{E}\{Fh(w)\}. \tag{8.8}$$

This proves the identity for any finite $n \in N$. Since S_2 is dense in $L_2(\Omega)$ the identity holds for all $F \in L_2(\Omega)$ for which $DF \in L_2(\Omega, H)$. •

An alternative proof is given by the well known Girsanov formula according to which the shifted measure $\mu_h(\cdot) \equiv \mu(\cdot + h)$ is absolutely continuous with respect to the original measure μ without a shift. In this case, the Radon-Nikodym derivative of μ_h with respect μ is given by

$$\rho(h) = \exp\{\int_I < h, dw > -(1/2) \int_I |h|^2 dt\}$$

and $d\mu_h = \rho d\mu$. Recall that unlike Lebesgue measure, Gaussian measure is not translation invariant. Clearly,

$$\int_\Omega F(w) d\mu(w) = \int_\Omega F(w + h) d\mu_{-h}(w),$$

where the positive translation of the argument of the function is compensated by the negative shift of the measure. Now replacing h by εh, and taking the derivative of the resulting expression with respect to ε and letting $\varepsilon \to 0$, while noting that the left hand member of the above identity is independent of ε, we obtain

$$0 = \int_{\Omega} \{< DF(w), h >_H - F(w)h(w)\}d\mu.$$

This leads to the same result. We need another elementary identity.

Lemma 8.3.2 *For every $F, G \in \mathcal{S}_2 \subset L_2(\Omega)$ and $h \in H$, the following identity holds*

$$\mathbb{E}\{F < DG, h >_H\} = \mathbb{E}\{GFh(w)\} - \mathbb{E}\{G < DF, h >_H\}. \quad (8.9)$$

Proof Clearly

$$D(GF) = GDF + FDG.$$

Thus

$$\mathbb{E}\{< D(GF), h >\} = \mathbb{E}\{G < DF, h >_H\} + \mathbb{E}\{F < DG, h >\}.$$

Hence, it follows from Lemma 8.3.1 that

$$\mathbb{E}\{GFh(w)\} = \mathbb{E}\{G < DF, h >_H\} + \mathbb{E}\{F < DG, h >\}$$

and we have

$$\mathbb{E}\{F < DG, h >\} = \mathbb{E}\{GFh(w)\} - \mathbb{E}\{G < DF, h >_H\}.$$

This completes the proof. ●

For symmetric appearance, it is preferable to write the above expression as

$$\mathbb{E}\{GFh(w)\} = \mathbb{E}\{F < DG, h >_H\} + \mathbb{E}\{G < DF, h >_H\}.$$

Remark 8.3.3 It is well known that Lebesgue measure is translation invariant and as a consequence the integration by parts formula is simple and it is given by

$$\int_{R^n} (Df, y)g dx = - \int_{R^n} (Dg, y)f dx$$

for any $y \in R^n$ and for all $f, g \in C_0^1$ (C^1 functions vanishing at infinity). In contrast, if R^n is given the standard Gaussian measure μ^n as introduced earlier, the formula is very different and it is given by

$$\int_{R^n} (Df, y)g \, \mu^n(dx) = \int_{R^n} (y, x)fg \, \mu^n(dx) - \int_{R^n} (Dg, y)f \, \mu^n(dx).$$

Compared with the Lebesgue measure, the Gaussian measure produces an extra term (first term on the right) and this is due to lack of translation invariance of the Gaussian measure.

Also note the similarity of this expression and (8.9) which is proved for Wiener measure space (Ω, μ^w). Evidently, the integration by parts formula given by the expression (8.9) is the infinite dimensional version of the finite dimensional case given above.

For convenience of the reader, we recall some basic definitions. Let X, Y be a pair of Banach spaces and $A : \mathcal{D}(A) \subset X \longrightarrow Y$ a linear operator with domain $\mathcal{D}(A) \subset X$ and range $\mathcal{R}(A) \subset Y$. The set

$$\Gamma(A) \equiv \{(x, y) \in X \times Y : x \in \mathcal{D}(A), y = Ax\}$$

is the graph of the operator A.

Definition 8.3.4 *The operator A is said to be closed if its graph is closed, that is, if $(x_n, y_n) \in \Gamma(A)$ and $x_n \to x$ in X and $y_n \to y$ in Y imply that $(x, y) \in \Gamma(A)$. The operator A is closeable if $(0, y) \in \Gamma(A)$ implies that $y = 0$.*

As seen above, the Malliavin derivative D is well defined on $\mathcal{S}_2$ in the direction H_1. For construction of Sobolev spaces on the Wiener measure space Ω, we need to verify that D has a closed extension with dense domain. In the case of finite dimensional spaces $(\Omega = R^n)$, the differential operator D is a closed densely defined linear operator on the Banach space $L_p(R^n)$ and hence the Sobolev spaces $\{W^{m,p}(R^n), 1 \le p < \infty\}$ are well defined Banach spaces having continuous duals $W^{-m,q}$ $(1/p + 1/q = 1)$. These spaces are absolutely essential and are widely used in the study of partial differential equations. The following result plays a crucial role in this direction.

Lemma 8.3.5 *The operator D defined on $\mathcal{S}_2 \subset L_2(\Omega)$ is closeable from $L_2(\Omega)$ to $L_2(\Omega, H)$.*

Proof Let $\{G_n\} \in \mathcal{D}(D)$ and $G_n \xrightarrow{s} 0$ in $L_2(\Omega)$ and $DG_n \xrightarrow{s} \eta$ in $L_2(\Omega, H)$. We show that $\eta = 0$. Let $F \in \mathcal{S}_2 \subset L_2(\Omega)$ and $h \in H$. Then it follows from Lemma 8.3.2 that

$$\mathbb{E}\{F < DG_n, h >_H\} = \mathbb{E}\{G_n F h(w)\} - \mathbb{E}\{G_n < DF, h >_H\}. \quad (8.10)$$

Letting $n \to \infty$, we arrive at the identity

$$\mathbb{E}\{F < \eta, h >_H\} = 0.$$

This is valid for every $F \in \mathcal{S}_2$ and every $h \in H$. Since $\mathcal{S}_2$ is dense in $L_2(\Omega)$ it is clear that $< \eta, h >= 0$, μ^w almost everywhere. Again, this is true for every $h \in H$ and hence $\eta = 0$, μ^w almost surely. Consequently, the Malliavin differential operator D is closeable. This completes the proof. •

In view of the above Lemma, the operator D, representing the Malliavin differential operator (in the Cameroon Martin direction), admits a unique closed extension with domain which is dense in $L_2(\Omega)$ and it is precisely the Sobolev space $W^{1,2}(\Omega)$ with the norm topology

$$\| F \|_{W^{1,2}} \equiv (\| F \|_{L_2(\Omega)}^2 + \| DF \|_{L_2(\Omega, H)}^2)^{1/2}. \quad (8.11)$$

This is in fact the extension of the weighted Sobolev space for functionals belonging to $\mathcal{S}_2$. Without loss of generality, we may assume that $\{h_i\}$ are orthonormal in H. Then using the standard Gaussian measure μ^n on R^n as mentioned above (see (8.3)) one can construct the family of weighted Sobolev spaces $\{W^{1,2}(R^n, \mu^n)\}_n$ with the norm topology

$$\| f \|_{W^{1,2}(R^n, \mu^n)} = \left(\int_{R^n} |f(x)|^2 \mu^n(dx) + \int_{R^n} |\nabla f|_{R^n}^2 \mu^n(dx) \right)^{1/2}. (8.12)$$

Note that for every $F \in \mathcal{S}_2$ with the representing function f, the two norms coincide giving

$$\| F \|_{W^{1,2}} = \| f \|_{W^{1,2}(R^n, \mu^n)}$$

for every finite integer n. Since the Malliavin differential operator admits extension to a closed operator with dense domain in $L_p(\Omega)$, for all $p > 1$, one can define the entire family of Sobolev spaces on $\Omega \equiv (\Omega, \mu^w)$ as one does on R^n. Thus $W^{1,p}(\Omega)$ is well defined along with its dual $W^{-1,q}(\Omega)$. We discuss this later in the sequel.

8.4 Operator δ the Adjoint of the Operator D

The adjoint of the Malliavin differential operator D, denoted by δ, is very useful for analysis of Wiener functionals. In this section we present some basic properties of this operator. First note that $D : \mathcal{D}(D) \subset L_2(\Omega) \longrightarrow L_2(\Omega, H)$ and hence formally $\delta : \mathcal{D}(\delta) \subset L_2(\Omega, H) \longrightarrow L_2(\Omega)$ and so by definition, for $F \in \mathcal{D}(D)$ and $u \in \mathcal{D}(\delta)$,

$$\mathbb{E}\{(DF, u)_H\} = \mathbb{E}\{F\delta(u)\}. \tag{8.13}$$

We present some basic identities involving the adjoint operator δ.

Lemma 8.4.1 (Identity A) *For $F_j \in \mathcal{S}_2$ and $h_j \in H$, define $u = \sum_{j=1}^n F_j h_j$. Then*

$$\delta(u) = \sum F_j h_j(w) - \sum < DF_j, h_j >_H . \tag{8.14}$$

Proof Starting with the following expression and using Lemma 8.3.2 we obtain

$$\sum < D(F_j G), h_j >_H = \sum G < DF_j, h_j >_H + \sum F_j < DG, h_j >_H$$
$$= \sum G < DF_j, h_j > + < DG, \sum F_j h_j >$$
$$= \sum G < DF_j, h_j > + < DG, u > .$$

By Lemma 8.3.1,

$$\mathbb{E}\{< D(F_j G), h_j >\} = \mathbb{E}\{F_j G h_j(w)\}.$$

Using this on the left hand side of the preceding identity we arrive at the following expression

$$\mathbb{E}\{\sum F_j G h_j(w)\} = \mathbb{E}\{\sum G < DF_j, h_j >\} + \mathbb{E}\{< DG, u >\}$$
$$= \mathbb{E}\{\sum G < DF_j, h_j >\} + \mathbb{E}\{G\delta(u)\}.$$

Hence we have

$$\mathbb{E}\{G \sum F_j h_j(w)\} = \mathbb{E}\{G \sum < DF_j, h_j >\} + \mathbb{E}\{G\delta(u)\}.$$

Since this holds for every $G \in \mathcal{S}_2$ it follows from this that

$$\delta(u) = \sum F_j h_j(w) - \sum < DF_j, h_j > .$$

This completes the proof. •

Remark 8.4.2 (Wiener Integral) *It follows from the above result that if* $u = h \in H = L_2(I, R^d)$, $\delta(u) = h(w) = \int_I < h(t), dw(t) >_{R^d}$. *This is just the Wiener integral.*

Let $D_h(F) \equiv < DF, h >_H$ denote the directional derivative of F in the direction $h \in H_1 \subset H$.

Lemma 8.4.3 (Identity B) *For* $u \equiv \sum F_j h_j$,

$$D_h(\delta(u)) = \delta(D_h u) + < h, u >_H . \tag{8.15}$$

Proof Using the expression (8.14) and basic chain rule, we obtain

$$D_h(\delta(u)) = D_h(\sum F_j h_j(w)) - D_h(\sum < DF_j, h_j >_H)$$
$$= \sum \{< DF_j, h > h_j(w) + F_j < D(h_j(w)), h >\} - \sum D_h < DF_j, h_j >$$
$$= \sum D_h(F_j) h_j(w) + \sum F_j < h_j, h > - \sum < D(D_h F_j), h_j >$$
$$= \sum D_h(F_j) h_j(w) + < u, h >_H - \sum < D(D_h F_j), h_j >$$
$$= \sum D_h(F_j) h_j(w) - \sum < D(D_h F_j), h_j > + < u, h > .$$

Define $v \equiv \sum (D_h F_j) h_j = D_h u$ and note that the first two terms give $\delta(v) = \delta(D_h u)$ and hence follows (8.15). •

Lemma 8.4.4 (Identity C) *For* $F \in \mathcal{D}(D)$ *and* $u \in \mathcal{D}(\delta)$

$$\delta(Fu) = F\delta(u) - < DF, u >_H . \tag{8.16}$$

Proof Taking any $G \in \mathcal{S}_2 \cap L_2$ and using the duality and Lemma 8.3.2, we compute

$$\mathbb{E}\{G\delta(Fu)\} = \mathbb{E}\{F < DG, u >\}$$
$$= \mathbb{E}\{< D(FG), u > -G < DF, u >\}$$
$$= \mathbb{E}\{FG\delta(u) - G < DF, u >\}$$
$$= \mathbb{E}\{G[F\delta(u) - < DF, u >_H]\}.$$

Since this holds for all G in a dense subset of $L_2(\Omega)$, we have

$$\delta(Fu) = F\delta(u) - < DF, u >_H .$$

This completes the proof. •

It is clear from the above result that if the assumptions of the identity C hold and

$$[F\delta(u)- <DF, u>_H] \in L_2(\Omega)$$

then $Fu \in \mathcal{D}(\delta)$.

Lemma 8.4.5 (Identity D) *Let $\{e_i\}$ be a complete orthonormal system for H and suppose $u, v \in \mathcal{S}_H$, the space of H-valued cylindrical functions. Then the following identity holds*

$$\mathbb{E}\{\delta(u)\delta(v)\} = \mathbb{E}\left\{(u,v)_H + \sum_{i,j}^{\infty}(D_{e_j}(u), e_i)_H(D_{e_j}(v), e_i)_H\right\}$$

$$= \mathbb{E}\left\{(u,v)_H + \sum_{j}^{\infty}(D_{e_j}(u), D_{e_j}(v))_H\right\}. \quad (8.17)$$

Proof Using completeness of the family $\{e_i\}$ in H, and the fact that δ is the dual of the operator D, we obtain

$$\mathbb{E}\{\delta(u)\delta(v)\} = \mathbb{E}\{<u, D(\delta(v))>_H\} = \mathbb{E}\left\{\sum(u, e_i)_H(D(\delta(v)), e_i)_H\right\}$$

$$= \mathbb{E}\left\{\sum(u, e_i)D_{e_i}(\delta(v))\right\}. (8.18)$$

Then using the **Identity B** given by (8.15), we obtain

$$\mathbb{E}\{\delta(u)\delta(v)\} = \mathbb{E}\left\{\sum(u, e_i)\{\delta(D_{e_i}(v)) + (v, e_i)\}\right\}$$

$$= \mathbb{E}\left\{(u, v) + \sum(u, e_i)\delta(D_{e_i}(v))\right\}$$

$$= \mathbb{E}\left\{(u, v)_H + \sum <D(u, e_i), D_{e_i}(v)>_H\right\}$$

$$= \mathbb{E}\left\{(u, v)_H + \sum <D_{e_i}(u), D_{e_i}(v)>_H\right\}. \quad (8.19)$$

This is identity D as given by (8.17). •

Identity **(D)** has important implication. This is given in the following corollary. Let $\mathcal{L}_2(H)$ denote the Hilbert space of Hilbert-Schmidt operators on H furnished with the scalar product $<T, S> = Tr(S^*T)$ and norm $\| S \|_{\mathcal{L}_2(H)} = \sqrt{Tr(S^*S)}$.

Corollary 8.4.6 *If $u \in L_2(\Omega, H)$ with $Du \in L_2(\Omega, \mathcal{L}_2(H))$ then $u \in \mathcal{D}(\delta)$ and $\delta(u) \in L_2(\Omega)$.*

Proof This follows from the identity **(D)** given by (8.17). Indeed, for $u = v$ it follows from this that

$$
\begin{aligned}
\mathbb{E}\{(\delta(u))^2\} &= \mathbb{E}\{|u|_H^2 + \sum |D_{e_i} u|_H^2\} \\
&= \mathbb{E}\left\{|u|_H^2 + \sum (Du(e_i), Du(e_i))_H\right\} \\
&= \mathbb{E}\left\{|u|_H^2 + \| Du \|_{\mathcal{L}_2(H)}^2\right\}.
\end{aligned}
\tag{8.20}
$$

This proves the assertion of the corollary. •

In view of the above corollary, we may introduce the Sobolev space $W^{1,2}(\Omega, H)$ as the space of H valued Wiener measurable functions $\{u\}$ on the Wiener measure space Ω such that $u \in L_2(\Omega, H)$ and $Du \in L_2(\Omega, \mathcal{L}_2(H))$. This is furnished with the norm topology given by

$$
\| u \|_{1,2} \equiv \left(\mathbb{E}\left\{|u|_H^2 + \| Du \|_{\mathcal{L}_2(H)}^2\right\} \right)^{1/2}.
$$

Clearly, with respect to this norm topology, $W^{1,2}(\Omega, H)$ is a Banach space. And the adjoint operator δ is a continuous linear operator from $W^{1,2}(\Omega, H)$ to $L_2(\Omega)$ and we may write

$$
\delta \in \mathcal{L}(W^{1,2}(\Omega, H), L_2(\Omega)).
$$

It is also an isometry.

If F and u, appearing in the following identity

$$
\delta(Fu) = F\delta(u) - < DF, u >_H,
$$

are independent the last term vanishes. This is proved in the following Corollary.

Corollary 8.4.7 *If the random elements $F \in \mathcal{D}(D) \subset L_2(\Omega)$ and $u \in L_2(\Omega, H)$ are independent, the last term vanishes leaving $\delta(Fu) = F\delta(u)$.*

Proof Let J be a nonempty subset of $I \equiv [0, T]$ and $J' \equiv I \setminus J$. Let $\mathcal{F}_{J'}$ denote the sigma algebra generated by the increments of Brownian motion

w on the interval J' and suppose F is $\mathcal{F}_{J'}$ measurable and $u = \chi_J(t)h(t)$. Then

$$\delta(Fu) = \int_J < Fh(t), dw(t) >_{R^d} - \int_J < (DF)(t), h(t) >_{R^d} dt. \quad (8.21)$$

Since F is $\mathcal{F}_{J'}$ measurable, $(DF)(t)$ restricted to J is zero. Hence the last term vanishes leaving $\delta(Fu) = \int_J < Fh(t), dw(t) >$. Thus $\delta(Fu) = F\delta(u)$ whenever F is independent of u. •

This is also the fundamental basis of Itô integral as stated in the following result. Let $L_2^a(\Omega, H)$ denote the space of $\mathcal{F}_t$-adapted random processes with values in $H \equiv L_2(I, R^d)$. This is a closed subspace of the Hilbert space $L_2(\Omega, H)$ and so a Hilbert space with the standard norm topology.

Corollary 8.4.8 (Itô Integral) *Let* $u \in L_2^a(\Omega, H)$ *so that* $u \in \mathcal{D}(\delta)$. *Then*

$$\delta(u) = Ito - \int_I < u(t), dw(t) >.$$

Proof Since u is $\mathcal{F}_t$ adapted, there exists a sequence of simple functions u_n such that $u_n \xrightarrow{\ s\ } u$ in $L_2^a(\Omega, H)$ and u_n is given by

$$u_n(t) \equiv \sum_{k=0}^n F_k C_{J_k^n}(t)$$

where $J_k^n \equiv (t_k^n, t_{k+1}^n]$ for $0 \le k \le n$ with $t_0^n = 0$ and $t_n^n = T$; C_σ is the characteristic function of the measurable set σ and F_k is $\mathcal{F}_{t_k^n}$ measurable and belongs to $L_2(\Omega, R^d)$. Then letting $\{v_i\}$ denote the unit vector basis of R^d, it follows from the **Identity C** given by (8.16) that

$$\delta(u_n) = \sum \delta(F_k C_{J_k^n}) = \sum \delta(\sum (F_k, v_i) v_i C_{J_k^n})$$
$$= \sum < F_k, (w(t_{k+1}^n) - w(t_k^n)) >_{R^d} - \sum \int_{J_k^n} \sum_i < D(F_k, v_i), v_i >_{R^d} dt.$$

Since F_k is $\mathcal{F}_{t_k^n}$-measurable, by virtue of Corollary 8.4.7 the last term vanishes and we are left with

$$\delta(u_n) = \sum < F_k, (w(t_{k+1}^n) - w(t_k^n)) >.$$

Letting $n \to \infty$ we obtain

$$\delta(u) = Ito - \int_I < u(t), dw(t) >.$$

This completes the proof. •

In view of the Remark 8.4.2 and Corollary 8.4.8, we have the following important conclusions:

Important Conclusions: *(i): The adjoint operator (divergence operator) δ gives the Wiener integral if u is deterministic, (ii): It gives the Itô integral if $u \in L_2(\Omega, H)$ and adapted, and (iii): It represents a generalization of the above integrals if u is merely an element of $L_2(\Omega, H)$ and this is known as Skorohod integral. This includes anticipating functions.*

From classical analysis we know that the indefinite integral of a function in $L_1(I)$ is absolutely continuous and every absolutely continuous function is an indefinite integral of some function from $L_1(I)$. The following result is the stochastic analog of this classical result. See the concluding remark of this section.

Clark-Ocone Formula *Any $F \in L_2(\Omega)$ having the Malliavin derivative $DF \in L_2(\Omega, H)$, has the representation*

$$F = I\!E(F) + \int_I < \varphi(t), dw(t) > \tag{8.22}$$

where $\varphi \in L_2^a(\Omega, H)$ is uniquely determined by F and it is given by the conditional expectation

$$\varphi(t) = I\!E\{(DF)(t)|\mathcal{F}_t\}.$$

Proof For any $u \in L_2^a(\Omega, H)$ it follows from the duality relation (8.13) that

$$I\!E\{\delta(u)F\} = I\!E\{< u, DF >_H\} = I\!E \int_I < u(t), (DF)(t) >_{R^d} dt$$

$$= I\!E \int_I I\!E\{< u(t), (DF)(t) > |_{\mathcal{F}_t}\} dt$$

$$= I\!E \int_I < u(t), I\!E\{DF(t)|_{\mathcal{F}_t}\} > dt. \tag{8.23}$$

The last line follows from the fact that $u(t)$ is $\mathcal{F}_t$ measurable. Alternatively, since u is adapted, $\delta(u)$ is the Itô integral given by

$$\delta(u) = \int_I < u(t), dw(t) > .$$

Multiplying (8.22) by $\delta(u)$ and taking the expectation of the product we have

$$\mathbb{E}\{\delta(u)F\} = \mathbb{E}\int_I < u(t), \varphi(t) > dt. \qquad (8.24)$$

Evidently, it follows from (8.23) and (8.24) that

$$0 = \mathbb{E}\int_I < u(t), \varphi(t) - \mathbb{E}\{(DF)(t)|_{\mathcal{F}_t}\} > dt.$$

Since this equality holds for all $u \in \mathcal{D}(\delta) \cap L_2^a(\Omega, H)$ and δ is densely defined, we conclude that $\varphi(t) = \mathbb{E}\{(DF)(t)|_{\mathcal{F}_t}\}$ for almost all $t \in I$. In other words, φ is the optional projection of the Malliavin derivative of F. This completes the proof.●

Remark *It is interesting to note that the representation (8.22) is the stochastic analog of the classical representation of absolutely continuous functions by the Lebesgue integral of its derivative*

$$z(t) = z(0) + \int_0^t \dot{z}(s)ds.$$

8.5   Ornstein-Uhlenbeck Operator L

The Ornstein-Uhlenbeck operator, denoted by L, is defined as the composition of the Malliavin differential operator D and the negative of its adjoint δ giving

$$L = -\delta D.$$

We study some basic properties of this operator. Recall that $\delta : \mathcal{D}(\delta) \subset L_2(\Omega, H) \longrightarrow L_2(\Omega)$ and $D : \mathcal{D}(D) \subset L_2(\Omega) \longrightarrow L_2(\Omega, H)$.

Lemma 8.5.1 *For $G, F \in \mathcal{D}(L)$ the following relation holds:*

$$L(GF) = GL(F) + FL(G) + 2 < DG, DF >_H . \qquad (8.25)$$

Proof It follows from Lemma 8.4.4, in particular the identity (8.16), that

$$\delta(G(DF)) = G\delta(DF) - < DG, DF >_H .$$

Thus

$$L(GF) = -\delta(D(GF)) = -\delta\Big(F(DG) + G(DF) \Big)$$
$$= -\delta(FDG) - \delta(GDF)$$
$$= \Big\{ -F\delta(DG) + <DF, DG> \Big\} + \Big\{ -G\delta(DF) + <DG, DF>_H \Big\}$$
$$= FL(G) + GL(F) + 2 <DF, DG>_H .$$

This proves the identity (8.25). •

Let us apply this relation to cylindrical functions $\mathcal{S}_2$ of the form

$$F = f(h_1(w), h_2(w), \cdots, h_n(w))$$

where for each i, $h_i \in H$ and $f \in C^1(R^n)$. We know that the Malliavin derivative of this function is given by

$$DF = \sum (\partial_i f) h_i.$$

Therefore,

$$L(F) = -\delta(DF) = -\delta\Big(\sum (\partial_i f) h_i \Big)$$
$$= -\sum (\partial_i f) h_i(w) + \sum <D(\partial_i f), h_i>_H$$
$$= -\sum (\partial_i f) h_i(w) + \sum <(\partial_j \partial_i f) <h_j, h_i>_H .$$

If the family $\{h_i\}$ is orthonormal in H, this reduces to the form

$$L(F) = \sum \partial_i^2 f - \sum x_i \partial_i f$$
$$= \triangle_n f - x \cdot \bigtriangledown f,$$

where $\triangle_n$ is the n-dimensional Laplacian and ∇ is the gradient operator in R^n. Thus the Ornstein-Uhlenbeck operator on n-dimensional cylindrical subspaces of the infinite dimensional space $L_2(\Omega, \mu^w)$ is given by

$$L_n \equiv \triangle_n - x \cdot \bigtriangledown. \tag{8.26}$$

In the n-dimensional space, L_n is the infinitesimal generator of a Markov process known as the Ornstein-Uhlenbeck process $X_t, t \geq 0$. This process is governed by the SDE

$$dX_t = -X_t dt + \sqrt{2}\ dw(t), X_0 = x, t \geq 0, \tag{8.27}$$

where w is the standard Brownian motion in R^n. Clearly its solution is given by

$$X_t = e^{-t}x + \sqrt{2} \int_0^t e^{-(t-s)}dw(s), t \geq 0.$$

The corresponding semigroup, known as the Ornstein-Uhlenbeck semigroup, is given by the conditional expectation

$$(T(t)\varphi)(x) \equiv E\{\varphi(X_t)|X_0 = x\}.$$

By using Itô differential, the reader can easily verify that

$$d/dt(T(t)\varphi) = L_n T(t)\varphi = T(t)L_n\varphi, t \geq 0,$$

for every $\varphi \in D(L_n)$.

These results also extend to infinite dimensional Ornstein-Uhlenbeck process. Here, we follow Shigekawa [Shigekawa (1998/2004)]. Consider the abstract Wiener measure space (Ω, H_1, μ) where μ is the Wiener measure. For $x \in \Omega$ and $A \in \mathcal{B}_\Omega$, C_A the characteristic function of the set A, the Ornstein-Uhlenbeck transition probability is given by

$$P_t(x, A) \equiv \int_\Omega C_A(e^{-t}x + \sqrt{1 - e^{-2t}}y)\mu(dy).$$

The associated semigroup and the corresponding adjoint semigroup are given by

$$(T_t\varphi)(x) \equiv \int_\Omega P_t(x, dy)\varphi(y) \tag{8.28}$$

and

$$(U_t\mu)(A) \equiv \int_\Omega P_t(x, A)\mu(dx) \tag{8.29}$$

respectively. It is easy to verify that μ is an invariant measure, that is, $U_t\mu = \mu$ for all $t \geq 0$. Indeed, let μ be a measure with mean zero and covariance K and let us denote by μ_a the same measure with mean zero and covariance aK for any $a > 0$. Using this notation, it follows from straightforward computation that

$$U_t\mu(A) = \int_\Omega \int_\Omega C_A(e^{-t}x + \sqrt{1 - e^{-2t}}y)\mu(dy)\mu(dx)$$

$$= \int_\Omega \int_\Omega C_A(e^{-t}x + z)\mu_{1-e^{2t}}(dz)\mu(dx)$$

$$= \int_\Omega \int_\Omega C_A(\xi + z)\mu_{1-e^{2t}}(dz)\mu_{e^{-2t}}(d\xi)$$

$$= \int_\Omega C_A(\eta)\mu_{1-e^{2t}} * \mu_{e^{-2t}}(d\eta) = \mu(A) \tag{8.30}$$

where $\mu * \nu$ denotes the convolution of the measure μ and ν. Thus, the original Wiener measure μ on the Wiener measure space is the invariant measure of the Ornstein-Uhlenbeck process. This property is very useful in the study of the Ornstein-Uhlenbeck semigroup on the space $L_p(\Omega)$, for all $p \geq 1$, as seen below.

Theorem 8.5.2 *The Ornstein-Uhlenbeck semigroup $T_t, t \geq 0$, is a strongly continuous contraction semigroup on $L_p(\Omega) = L_p(\Omega, \mu)$ for all $p \geq 1$, and its infinitesimal generator is $L = -(\delta D)$.*

Proof First we verify that it is a contraction. Clearly, for any $\varphi \in L_1(\Omega)$,

$$|(T_t\varphi)(x)| \leq \int_\Omega |\varphi(y)| P_t(x, dy).$$

Hence it follows from the invariance property of the measure μ that

$$\int_\Omega |(T_t\varphi)(x)| \mu(dx) \leq \int_\Omega \int_\Omega |\varphi(y)| P_t(x, dy) \mu(dx)$$

$$\leq \int_\Omega |\varphi(y)| \mu(dy) = \| \varphi \|_{L_1(\Omega)}.$$

This proves contraction on $L_1(\Omega)$. Now let $p > 1$ and $\varphi \in L_p(\Omega)$. Then, we have

$$|(T_t\varphi)(x)|^p \leq \left(\int_\Omega |\varphi(y)| P_t(x, dy) \right)^p.$$

Now using Hölder inequality corresponding to the conjugate pair $\{p, q\}, (1/p) + (1/q) = 1$, we have

$$\int_\Omega |\varphi(y)| P_t(x, dy) \leq \left(\int_\Omega |\varphi(y)|^p \ P_t(x, dy) \right)^{1/p}$$

μ a.s. Hence,

$$|(T_t\varphi)(x)|^p \leq \int_\Omega |\varphi(y)|^p P_t(x, dy)$$

μ - a.s, and it follows from this that

$$\int_\Omega |(T_t\varphi)(x)|^p \mu(dx) \leq \int_\Omega \int_\Omega |\varphi(y)|^p P_t(x, dy) \mu(dx).$$

Again, by invariance of the measure μ, it follows from this that

$$\int_\Omega |(T_t\varphi)(x)|^p \mu(dx) \leq \int_\Omega |\varphi(y)|^p \mu(dy).$$

Hence we have

$$\| T_t \varphi \|_{L_p(\Omega)} \leq \| \varphi \|_{L_p(\Omega)}$$

proving that $T_t, t \geq 0$, is a contraction semigroup on the Banach space $L_p(\Omega) = L_p(\Omega, \mu)$ for all $p \in [1, \infty)$. Now we verify strong continuity. Since the space of cylindrical functions $\mathcal{S}_2$ is dense in $L_p(\Omega)$, for every $\varphi \in L_p(\Omega)$, there exists a sequence $\varphi_n \in \mathcal{S}_2$ such that $\varphi_n \xrightarrow{s} \varphi$ in $L_p(\Omega)$. Hence, it follows from the following inequality

$$\| T_t \varphi - \varphi \|_{L_p(\Omega)} \leq \| T_t \varphi - T_t \varphi_n \|_{L_p(\Omega)} + \| T_t \varphi_n - \varphi_n \|_{L_p(\Omega)} + \| \varphi_n - \varphi \|_{L_p(\Omega)}$$

and the contraction property that

$$\| T_t \varphi - \varphi \|_{L_p(\Omega)} \leq 2 \| \varphi - \varphi_n \|_{L_p(\Omega)} + \| T_t \varphi_n - \varphi_n \|_{L_p(\Omega)} .$$

Since $\varphi_n \in \mathcal{S}_2$, it is clear that for any given $n \in N$,

$$\| T_t \varphi_n - \varphi_n \|_{L_p(\Omega)} \to 0$$

as $t \downarrow 0$. Then by standard $\{\varepsilon, \delta\}$ argument, we have

$$\lim_{t \downarrow 0} \| T_t \varphi - \varphi \|_{L_p(\Omega)} = 0$$

proving strong continuity. That the operator L is the corresponding infinitesimal generator follows from the discussion following (8.26). This completes the proof. •

Corollary 8.5.3 *The infinitesimal generator* L *of the O.U semigroup* $T_t, t \geq 0$, *is a densely defined closed linear operator (domain and range) in* $L_p(\Omega)$ *with the resolvent* $R(\lambda, L)$ *satisfying*

$$\| R(\lambda, L) \| \equiv \| (\lambda - L)^{-1} \|_{\mathcal{L}(L_p(\Omega))} \leq (1/\lambda) \tag{8.31}$$

for $\lambda \in \rho(L)$ *where the resolvent set* $\rho(L) \supset (0, \infty)$. *Further it is also dissipative.*

Proof The first statement follows from Hille-Yosida theorem [[Ahmed (1991)], Theorem 2.2.8, p27]. The dissipative property follows from Lumar-Phillips theorem [[Ahmed (1991)], Theorem 2.2.14, p33].

8.6   Sobolev Spaces on Wiener Measure Space $\Omega \equiv (\Omega, \mu)$

It is well known that by use of the Laplacian in R^n, one can easily construct the classical Sobolev spaces based on $L_2(R^n)$ as well as $L_p(R^n)$. Let $f \in L_2(R^n)$ with $\partial_i f \in L_2(R^n), 1 \le i \le n$. Then by integration by parts we have

$$< (1 - \triangle)f, f >= |f|^2_{L_2(R^n)} + \| \triangledown f \|^2_{L_2(R^n, R^n)} \equiv \| f \|^2_{W^{1,2}(R^n)},$$

where

$$W^{1,2} \equiv \{ f \in L_2(R^n) : \triangledown f \in L_2(R^n, R^n) \}$$

is the classical Sobolev space which is furnished with the norm topology

$$\| f \|_{W^{1,2}} \equiv \left(|f|^2_{L_2(R^n)} + \| \triangledown f \|^2_{L_2(R^n, R^n)} \right)^{1/2}.$$

Note that the operator $(1 - \triangle)$ is a positive selfadjoint unbounded operator in $L_2(R^n)$ having bounded inverse. Therefore, by use of spectral theory one can define the square root $(1 - \triangle)^{1/2}$ and introduce the scalar product in $W^{1,2}$ through the expression

$$< f, g >_{W^{1,2}} \equiv < (1 - \triangle)^{1/2}f, (1 - \triangle)^{1/2}g >_{L_2(R^n)}$$

for $f, g \in W^{1,2}$ and this also provides the norm as obtained by integration by parts. The topological dual of $W^{1,2}$ is $W^{-1,2}$ which is characterized as follows:

$$W^{-1,2} \equiv \left\{ g : R^n \longrightarrow R : \sup\{| < g, f > | : \| f \|_{W^{1,2}} \le 1\} < \infty \right\}. \quad (8.32)$$

In other words, it is the class of generalized functions $\{g\}$ from R^n to R such that $(1 - \triangle)^{-1/2}g \in L_2$ and

$$| < g, f >_{W^{-1,2}, W^{1,2}} | = | < (1 - \triangle)^{-1/2}g, (1 - \triangle)^{1/2}f >_{L_2(R^n)} | < \infty$$
$$(8.33)$$

for every $f \in W^{1,2}$. This relation also verifies that for $g \in W^{-1,2}$, we have $(1 - \triangle)^{-1/2}g \in L_2(R^n)$.

Similarly, one can construct Sobolev spaces based on $L_p(R^n)$ spaces. It is well known that for any integer $m \in N$ one can construct the spaces

$$W^{m,p} \equiv \{ f \in L_p : D^\alpha f \in L_p, |\alpha| \le m \} \quad (8.34)$$

where α is a multi index given by $\alpha = (\alpha_1, \alpha_2, \cdots, \alpha_n), \alpha_i \in N$ & $|\alpha| = \sum \alpha_i$. With respect to the norm topology,

$$\| f \|_{W^{m,p}} \equiv \left(\sum_{|\alpha| \le m} (\| D^\alpha f \|_{L_p})^p \right)^{1/p},$$

these are Banach spaces and their duals are given by $W^{-m,q}$ where $\{p, q\}$ are the conjugate pairs, $(1/p + 1/q = 1)$ and $1 \le p, q < \infty$.

Similarly, in perfect analogy with the finite dimensional case as discussed above, using the Ornstein-Uhlenbeck operator L one can construct Sobolev spaces on the Wiener measure space (Ω, μ). According to Corollary 8.5.3, we have seen that L is dissipative and so $-L$ is accretive with $1 \in \rho(L)$ and therefore $(1 - L)$ is a positive unbounded operator on $L_2(\Omega)$ having bounded inverse. Indeed, it follows from the inequality (8.31) that

$$\| (1 - L)^{-1} \|_{\mathcal{L}(L_p)} \le 1. \tag{8.35}$$

One can construct Sobolev spaces using either the Malliavin differential operator D or the Ornstein-Uhlenbeck operator L. The later is more convenient since it's domain and range are in the same space $L_p(\Omega)$ while the domain of D is in $L_p(\Omega)$ and range in $L_p(\Omega, H)$. In any case they are equivalent. Once we have the operator L, the construction is similar to that of classical Sobolev spaces on R^n. For $m \in N$, and $\{p, q \ge 1\}$ conjugate pair $(1/p + 1/q = 1)$, we define $W^{m,p}(\Omega)$ as

$$W^{m,p}(\Omega) \equiv \left\{ \varphi \in L_p(\Omega) : (1 - L)^{m/2} \varphi \in L_p(\Omega) \right\}.$$

This is furnished with the norm topology

$$\| \varphi \|_{W^{m,p}} \equiv \| (1 - L)^{m/2} \varphi \|_{L_p(\Omega)} .$$

Completion of this space with respect to this norm topology turns $W^{m,p}(\Omega)$ into a Banach space. To avoid introducing new notations, we continue to use the same notation for the completion. Just like in the classical case, its dual is given by $W^{-m,q}(\Omega)$. Indeed, for any $\psi \in W^{-m,q}(\Omega)$ and $\varphi \in W^{m,p}(\Omega)$ one can introduce the pairing

$$< \psi, \varphi >_{W^{-m,q}(\Omega), W^{m,p}(\Omega)} = < \psi, (1 - L)^{-m/2}(1 - L)^{m/2} \varphi >$$
$$= < (1 - L)^{-m/2} \psi, (1 - L)^{m/2} \varphi >_{L_q, L_p} . \tag{8.36}$$

Using this duality pairing, one can define a norm on $W^{-m,q}(\Omega)$ as follows

$$\| \psi \|_{W^{-m,q}} = \sup\{< \psi, \varphi >_{W^{-m,q}(\Omega), W^{m,p}(\Omega)} : \| \varphi \|_{W^{m,p}} \leq 1\}.$$

Again, with respect to this norm topology, it is a Banach space of (a class of) generalized functionals of Brownian motion. These spaces are defined not only for integers, but also for real numbers giving the family of Banach spaces $\{W^{s,p}, s \in R\}$ with duals given by $\{W^{-s,q}, s \in R\}$.

The reader can easily check the action of the Malliavin derivative and the divergence operators on these Sobolev spaces. For example, for all $p \in (1, \infty)$, we have

$$D : W^{s,p}(\Omega) \longrightarrow W^{s-1,p}(\Omega, H) \quad \text{and} \tag{8.37}$$

$$\delta : W^{s,p}(\Omega, H) \longrightarrow W^{s-1,p}(\Omega) \quad \text{and} \tag{8.38}$$

$$L : W^{s,p}(\Omega) \longrightarrow W^{s-2,p}(\Omega). \tag{8.39}$$

Further, $W^{s,p}(\Omega)$ is invariant under the Ornstein-Uhlenbeck semigroup, $T_t : W^{s,p}(\Omega) \longrightarrow W^{s,p}(\Omega)$. Indeed, using the contraction property, it is easy to see that for every $\varphi \in W^{s,p}(\Omega)$

$$\| T_t\varphi \|_{W^{s,p}(\Omega)} = \| (1-L)^{s/2}T_t\varphi \|_{L_p(\Omega)} = \| T_t(1-L)^{s/2}\varphi \|_{L_p(\Omega)}$$

$$\leq \| (1-L)^{s/2}\varphi \|_{L_p(\Omega)} = \| \varphi \|_{W^{s,p}(\Omega)} .$$

Thus $W^{s,p}(\Omega)$ is invariant under the Ornstein-Uhlenbeck semigroup. Let us verify (8.39); the others are similar. For any $\varphi \in W^{s,p}(\Omega)$ we have

$$\| L\varphi \|_{W^{s-2,p}} = \| L(1-L)^{-1}(1-L)^{s/2}\varphi \|_{L_p}$$

$$= \| -(1-(1-L)^{-1})(1-L)^{s/2}\varphi \|_{L_p} .$$

Clearly, it follows from (8.35) and the above identity that

$$\| L\varphi \|_{W^{s-2,p}} \leq 2 \| (1-L)^{s/2}\varphi \|_{L_p} = 2 \| \varphi \|_{W^{s,p}} . \tag{8.40}$$

Thus we have (8.39).

For more details on Sobolev spaces see the excellent exposition given by Shigekawa [Shigekawa (1998/2004)]. In the following Chapter, we will have more discussions on the Malliavin operators $\{D, \delta, L\}$ as we deal with evolution equations on Fock spaces.

8.7 Smoothness of Probability Measures

In this section we discuss briefly the question of existence and smoothness of Radon-Nikodym derivatives of measures induced by functionals of Brownian motions. One of the major applications of Malliavin Calculus is in the study of existence and smoothness of densities of probability measures induced by the solutions of stochastic (Itô) differential equations in R^n. This is where the Malliavin derivative D plays the crucial role.

The general problem can be stated as follows. Let E be a Banach space and $\mathcal{B}_E$ the sigma algebra of Borel subsets of the set E and μ a Borel measure giving the measure space $(E, \mathcal{B}_E, \mu)$. Similarly let $(\hat{E}, \mathcal{B}_{\hat{E}}, \lambda)$ be another measure space and $T : E \longrightarrow \hat{E}$ a measurable map. Then define the measure

$$\nu(\cdot) = T(\mu)(\cdot) \equiv \mu T^{-1}(\cdot).$$

This is another measure on $\hat{E}$ (strictly on $\mathcal{B}_{\hat{E}}$) induced by the map T. The basic questions are: Is ν absolutely continuous with respect to the measure λ ? Equivalently, does ν have Radon-Nikodym derivative with respect to the existing measure λ on $\hat{E}$ giving $d\nu = gd\lambda$ where $g \in L_1(\hat{E}, \lambda)$. If so, g is called the Radon-Nikodym derivative (RND) of the measure ν with respect to the measure λ. The second question is, if such a density exists, how smooth is it ?

Here, in the study of functionals of Brownian motion, our measure space $(E, \mathcal{B}_E, \mu)$ is the Wiener measure space $(\Omega, \mathcal{B}_\Omega, \mu^w)$ where $\Omega \equiv C_0(I, R)$ or $C_0(I, R^d)$ is the path space of Brownian motions introduced in Chapter 1 and $\mathcal{B}_\Omega$ is the Borel algebra generated by the standard metric topology on Ω and μ^w is the Wiener measure. The target space $\hat{E}$ is the Euclidean space R^n and λ is the Lebesgue measure on $\mathcal{B}_{R^n}$. The map

$$T = F \equiv (F^1, F^2, \cdots, F^n)$$

is the R^n valued functionals of Brownian motion. We wish to study the question of existence and smoothness of g (Radon-Nikodym density) relating

$$d\nu = gd\lambda \qquad (8.41)$$

where $\nu(B) \equiv (\mu^w F^{-1})(B), B \in \mathcal{B}_{R^n}$. As stated earlier, existence of such densities depend largely on the existence of Malliavin derivatives of $\{F^i\}$ and in particular, the Malliavin covariance matrix given by

$$Q^F \equiv \{< DF^i, DF^j >_H, 1 \le i, j \le n\} \qquad (8.42)$$

where $H = L_2(I, R)$ or $L_2(I, R^d)$.

The following theorem due to Bouleau and Hirsch [Bouleau (1991)] provides sufficient conditions under which ν has Radon-Nikodym derivative with respect to the Lebesgue measure λ implying the existence of a density. We present this result without proof. For proof see [Bouleau (1991)].

Theorem 8.7.1 *Suppose the following conditions hold:*

(B1): The random variable $F^i \in W^{1,2}(\Omega)$ for all $1 \le i \le n$.

(B2): The Malliavin covariance matrix Q^F is almost surely invertible.

Then, the measure ν induced by the map F is absolutely continuous with respect to the Lebesgue measure λ.

This result proves the existence of a density g satisfying relation (8.41). But it does not say how smooth the function $g : R^n \longrightarrow R$ is. For smoothness, it is intuitively clear that the Brownian functionals $\{F^i, 1 \le i \le n\}$ must be smooth also in the sense that they have Malliavin derivatives of higher orders. This is transparent in the following definition.

Definition 8.7.2 *The vector F is said to be non degenerate if it satisfies the following two conditions:*

(ND1): Each component $F^i \in W^\infty \equiv \cap_{m \ge 1, p \ge 1} W^{m,p}(\Omega, \mu^w)$

(ND2): The random variable $det(Q^F)$ has the property: $(1/det(Q^F)) \in L_p(\Omega, \mu^w)$ for all $p \ge 1$.

Before we proceed further let us state the following basic result which is well known in the literature. The basic idea of proof is contained in the following lemma.

Lemma 8.7.3 *Let ν be a (probability) measure on R^n and suppose there* *exists a positive number c such that for every* $f \in C_b^1(R^n)$

$$| \int_{R^n} Df(x)\nu(dx)|_{R^n} \le c \parallel f \parallel_\infty \equiv c \, \sup\{|f(x)|, x \in R^n\}.$$

Then, ν *is absolutely continuous with respect to Lebesgue measure* λ *and has a unique density* $\rho \in L_1(R^n, \lambda)$.

Proof For proof see Malliavin [[Malliavin (1976)], Lemma 3.2.10, p250]. Friz [Friz (2002)] gives a simpler proof based on the theory of Fourier transform.

Based on Malliavin calculus, one can prove the following smoothness result.

Theorem 8.7.4 *Suppose F is nondegenerate in the sense of Definition 8.7.2. Then the density of the measure* ν *induced by F, with respect to Lebesgue measure* λ, *is smooth in the sense that* $(d\nu/d\lambda) = g \in C^\infty(R^n)$.

Proof For detailed proof see Nualart [[Nualart (2009)], Theorem 5.7,p35]. We present only a brief outline following Nualart. Let $\eta \in C_0^\infty(R^n)$ arbitrary. Then consider the random variable $\eta(F) \equiv \eta \circ F$. Clearly, its Malliavin derivative is given by

$$D\eta(F) = \sum_{i=1}^n (\partial_i \eta)(F) DF^i.$$

Then take the scalar product of this with the M-derivative DF^j giving

$$< D\eta(F), DF^j >_H = \sum_{i=1}^n \partial_i \eta(F) < DF^i, DF^j >_H \equiv \sum_i Q_{j,i}^F \, \partial_i \eta(F)$$

where $H = L_2(I, R^d)$ and Q^F is the Malliavin covariance matrix of F. Since, by assumption, Q^F is almost surely invertible, it follows from this that

$$\partial_i \eta(F) = \sum_j^n ((Q^F)^{-1})_{j,i} < D\eta(F), DF^j >$$

$$= < D\eta(F), \sum_{j=1}^n ((Q^F)^{-1})_{j,i} DF^j >_H, 1 \le i \le n.$$

Then, taking any $G \in W^\infty$ and computing the scalar product with $\partial_i \eta(F)$ on the Wiener space, we have

$$\mathbb{E}\{\partial_i \eta(F) G\} = \mathbb{E}\{< D\eta(F), G((Q^F)^{-1} DF)_i >\}.$$

Now using the duality relation (8.13) and the divergence operator δ for D^*, we obtain

$$\mathbb{E}\{\partial_i \eta(F) G\} = \mathbb{E}\{\eta(F) \delta(G((Q^F)^{-1} DF)_i)\}. \qquad (8.43)$$

Note that this is an integration by parts formula in the Malliavin calculus. Then, under the assumptions (ND1) and (ND2), $(G(Q^F)^{-1} DF)_i$ is in the domain $\mathcal{D}(\delta)$ for each i and hence the expression on the right hand side is well defined. Define $H_i(F, G) \equiv \delta(G((Q^F)^{-1} DF)_i)$ and note that under the given assumptions, $H_i \in W^\infty$. Hence, there exists a nonnegative constant $c_i \geq 0$, such that $\mathbb{E}|H_i| \leq c_i < \infty$. By choosing $G = 1$ in the above expression we obtain

$$|\mathbb{E}\partial_i \eta(F)| \leq c_i \parallel \eta \parallel_\infty, 1 \leq i \leq n. \qquad (8.44)$$

This means that for the first order differentials, the above inequality holds and hence one can conclude that the probability measure induced by F on R^n has a density p. That is, $d\nu = pd\lambda$. For further smoothness, this estimate must be carried out for derivatives of arbitrary order. For multi index $\alpha = (\alpha_1, \alpha_2, \cdots, \alpha_n)$ with $\alpha_i \in N$, the expression (8.43) is more complex (see [Nualart (2009)], Theorem 5.9, p36; [Bell (1987)], p46) and it is given by

$$\mathbb{E}\{\partial_\alpha \eta(F) G\} = \mathbb{E}\{\eta(F) H_\alpha(F, G)\}. \qquad (8.45)$$

On the basis of this estimate it follows from Lemma 8.7.3 that ν has a density $p \in C^\infty$ giving $d\nu = pd\lambda$. This completes our brief outline of the proof. •

8.7.1   *Smoothness under Ellipticity Condition*

Using Theorem 8.7.1, we prove the existence of a density (RND with respect to Lebesgue measure) of the measure ν induced by the autonomous stochastic differential equation in R^n given by

$$dX_t = b(X_t)dt + \sigma(X_t)dW(t), X_0 = x, t \in I \qquad (8.46)$$

where W is a d dimensional standard Brownian motion.

Theorem 8.7.5 *Suppose the drift and the diffusion parameters b and σ respectively are uniformly Lipschitz with Lipschitz constant $K > 0$,*

$$|b(\xi) - b(\zeta)|^2_{R^n} + \parallel \sigma(\xi) - \sigma(\zeta) \parallel^2_{H.S} \leq K^2 |\xi - \zeta|^2_{R^n}, \forall \, \xi, \zeta \in R^n.$$

Let $X_t \equiv X_t(x)$ denote the solution of equation (8.46) and let $Q^X_t \equiv (DX_t)(DX_t)'$ denote the corresponding Malliavin covariance matrix. Let τ be the stopping time

$$\tau \equiv \inf\{t \in (0, T] : det(Q^X_t) > 0\}.$$

If the set determining the stopping time τ is nonempty, the measure ν_t induced by X_t for $t > \tau$ has a density ρ_t giving $d\nu_t = \rho_t d\lambda$.

Proof Under the Lipschitz assumption, it follows from Theorem 1.3.2 that equation (8.46) has a unique continuous solution. In view of theorem 8.7.1, it suffices to verify the conditions (B1) and (B2) for $t > \tau$. Let us compute the Malliavin covariance matrix. Write equation (8.46) as an integral equation

$$X_t = x + \int_0^t b(X_s)ds + \int_0^t \sigma(X_s)dW(s), t \in I. \tag{8.47}$$

Now, applying the Malliavin differential operator D starting at time $r \in [0, t]$, and recalling that D is a closed operator, we have

$$D_r X_t = \int_r^t D_r b(X_s)ds + \sigma(X_r) + \int_r^t D_r \sigma(X_s)dW(s), t \in I. \tag{8.48}$$

The first term on the right is obvious, the second and the third follow from the following argument. Let $\mathcal{F}_t, t \geq 0$, denote the sigma algebra induced by the Brownian motion W and suppose $F(t)$ is an $\mathcal{F}_t$ adapted (matrix valued) process. Then for any $r \in [0, t]$, the Malliavin derivative of the random element $Z_t \equiv \int_0^t F(s)dW(s)$ evaluated at time $r \in [0, t]$ is given by

$$D_r \left(\int_0^t F(s)dW(s) \right) = F(r) + \int_r^t D_r F(s)dW(s). \tag{8.49}$$

The reader can easily justify this by taking $\varepsilon > 0$ sufficiently small and writing

$$
D_r\left(\int_0^t F(s)dW(s)\right) = D_r\left(\int_0^{r-\varepsilon} F(s)dW(s)\right) + D_r\left(\int_{r-\varepsilon}^{r+\varepsilon} F(s)dW(s)\right)
$$
$$
+ D_r\left(\int_{r+\varepsilon}^t F(s)dW(s)\right)
$$
$$
= 0 + D_r\left(\int_{r-\varepsilon}^{r+\varepsilon} F(s)dW(s)\right) + D_r\left(\int_{r+\varepsilon}^t F(s)dW(s)\right)
$$

and using similar arguments as in Lemma 8.4.7 and letting $\varepsilon \to 0$. Under the differentiability assumption on $\{b, \sigma\}$, equation (8.48) can be written as

$$
D_r X_t = \sigma(X_r) + \int_r^t \partial b(X_s) D_r X_s ds
$$
$$
+ \sum_{i=1}^d \int_r^t \partial \sigma^i(X_s) D_r X_s dW_i(s), \qquad (8.50)
$$

$t \in (r, T]$, where ∂b and $\partial \sigma^i$ are the $n \times n$ Jacobian matrices corresponding to the n-vectors b and σ^i where σ^i denotes the i-th column of σ. Thus the Malliavin derivative of the process X satisfies the above integral equation. Similarly, we can obtain the derivative of the function $x \longrightarrow X_t(x)$ which we may denote by $Y(t) \equiv \partial X_t$. Clearly it follows from equation (8.47) that it satisfies the following linear integral equation in Y

$$
Y(t) = I + \int_0^t \partial b(X_s) Y(s) ds + \sum_{i=1}^d \int_0^t \partial \sigma^i(X_s) Y(s) dW_i(s), \quad (8.51)
$$

$t \in I$. For convenience of notation, we write $\partial b(X_s) = B(s)$ and $\partial \sigma^i(X_s) \equiv \Sigma^i(s)$ and $D_r X_t \equiv V_r(t)$. Using these notations we can rewrite equations (8.50) and (8.51) as linear integral equations given by

$$
V_r(t) = \sigma(X_r) + \int_r^t B(s) V_r(s) ds
$$
$$
+ \sum_{i=1}^d \int_r^t \Sigma^i(s) V_r(s) dW_i(s), t \in (r, T] \quad (8.52)
$$

and

$$
Y(t) = I + \int_0^t B(s) Y(s) ds + \sum_{i=1}^d \int_0^t \Sigma^i(s) Y(s) dW_i(s), t \in I. \quad (8.53)
$$

Since the matrices B and Σ^i are $\mathcal{F}_t$ adapted and uniformly bounded, these equations have unique adapted solutions which are denoted by $\{V_r, Y\}$ respectively. Note that the equations (8.53) and (8.52) are the integral versions of the following SDE's

$$dY(t) = B(t)Y(t)dt + \sum_{i=1}^{d} \Sigma^i(t)Y(t)dW_i(t), Y(0) = I, t \in [0, T] \quad (8.54)$$

$$dV_r(t) = B(t)V_r(t)dt + \sum_{i=1}^{d} \Sigma^i(t)V_r(t)dW_i(t), V_r(r) = \sigma(X_r), \quad (8.55)$$

$t \in (r, T]$ respectively. The adjoint equation corresponding to equation (8.54) is given by

$$dZ(t) = -Z(t)B(t)dt - \sum_{i=1}^{d} Z(t)\Sigma^i(t)dW_i(t) + Z(t)\sum_{i=1}^{d} \Sigma^i(t)(\Sigma^i(t))' dt,$$
$$Z(0) = I, t \in [0, T]. \quad (8.56)$$

Again, under the same assumptions on $\{b, \sigma\}$, this equation has an $\mathcal{F}_t$-adapted solution. We show that $Z(t)$ is the inverse of the matrix $Y(t)$. Since $Z(0) = Y(0) = I$ (identity), it suffices to show that the Itô differential of the product $Z(t)Y(t)$ is zero. The Itô differential is given by

$$d(Z(t)Y(t)) = dZ(t)Y(t) + Z(t)dY(t) + < dZ(t), dY(t) >, t \in I \quad (8.57)$$

where the last term denotes the quadratic variation process. Using the equations (8.54) and (8.56) into the above identity, we obtain $d(Z(t)Y(t)) = 0$. Hence, $Y(t)^{-1}$ exists and it is given by $Z(t)$ for all $t \in I$. Thus the initial value problem (8.55), for the Malliavin derivative $V_r(t), t \in (r, T]$, has a unique solution determined by the transition matrix $\Psi(t, r) \equiv Y(t)Y^{-1}(r)$ and this is given by

$$V_r(t) = \Psi(t, r)\sigma(X_r), t \geq r. \quad (8.58)$$

From this we conclude that the solution process $\{X_t\}$ has Malliavin derivative verifying condition (B1) of Theorem 8.7.1. Now consider the matrix

$$Q_t^X \equiv \int_0^t [V_r(t)V_r(t)'] dr = Y(t)\Gamma(t)Y'(t) \quad (8.59)$$

where

$$\Gamma(t) \equiv \int_0^t Y^{-1}(r)\sigma(X_r)\sigma(X_r)^{'}(Y^{-1}(r))^{'} dr$$

$$\equiv \int_0^t Z(r)\sigma(X_r)\sigma(X_r)^{'} Z(r)^{'} dr$$

and note that it is a symmetric and nonnegative random $n \times n$ matrix. Since Y is nonsingular for all $t \in I$, Q_t^X is almost surely positive definite if and only if $\Gamma(t)$ is positive definite. Thus if the set

$$\{t \in (0,T] : det(Q_t^X) > 0\}$$

is nonempty, and $t > \tau$, the stopping time, the matrix Q_t^X is invertible for all $t > \tau$ proving condition (B2). Hence it follows from Theorem 8.7.1 that ν_t is absolutely continuous with respect to the Lebesgue measure for $t > \tau$ and hence it has a density ρ_t, for all $t > \tau$. This completes the proof. •

Since $\Gamma(t)$ is a nonnegative symmetric matrix, it is clear that if $\Gamma(t)$ is nonsingular for some $t_0 \in [0, \infty)$, then it is nonsingular for all $t \geq t_0$.

Remark 8.7.6 *From the expression (8.59) it is clear that $\Gamma(t)$ is nonsingular whenever the classical uniform ellipticity condition holds. That is, there exists a constant $c > 0$ such that*

$$\sum_{i,j} a_{i,j}(x)\xi_i\xi_j \geq c|\xi|_{R^n}^2 \ \text{for all } x \in R^n$$

where $a = (\sigma\sigma^{'})$. However, as seen in the above theorem, this is not a necessary condition.

The ellipticity question arises in the study of general parabolic partial differential equations, diffusion processes and in particular Markov processes. Diffusion processes are related to the backward and forward Kolmogorov equations, where the backward equation is given by

$$\partial\varphi/\partial t + A\varphi = 0, A\varphi \equiv \sum_{i,j} a_{i,j}(\partial^2\varphi/\partial x_i \partial x_j) + \sum b_i \partial\varphi/\partial x_i,$$

and the forward equation is its adjoint. For any smooth ψ_0, the solution of the forward equation

$$\partial\psi/\partial t = A^*\psi, \psi(0,x) \equiv \psi_0(x), t \geq 0, x \in R^n$$

is given by $\psi(t,x) = E\{\psi_0(X_t(x))\} \equiv (T(t)\psi_0)(x)$ where $X_t(x)$ is the solution of equation (8.47). Here $T(t), t \geq 0$, is the Markov semigroup which is strongly continuous on the Banach space of bounded uniformly continuous functions on R^n denoted by $BUC(R^n)$. Under the ellipticity condition these equations have unique smooth solutions, given that the coefficients $\{a, b\}$ are sufficiently smooth. In the following subsection, we present relaxed (weaker) assumptions due to Hörmander under which the smoothness holds without the uniform ellipticity condition.

8.7.2 Smoothness under Hörmander's Conditions

The pioneering work of Hörmander [Hormander (1967)] demonstrated that the strong ellipticity condition is not necessary. Under much weaker assumptions, smoothness hold. The technique used to prove this is based on the Lie algebra. Let us consider the SDE in R^n written in terms of coordinates

$$dx^i = b^i(x)dt + \sum_\alpha^d \sigma_\alpha^i(x)dw^\alpha, 1 \leq i \leq n, t \geq 0, x^i(0) = x_i \qquad (8.60)$$

where w is a standard d-dimensional Brownian motion. Let φ be any C_b^2 function on R^n. Then, by use of Itô's formula, it is easy to verify that the process $\varphi(x(t))$ satisfies the SDE

$$d\varphi = \sum_\alpha \sigma_\alpha^i \partial_i \varphi dw^\alpha + b^i \partial_i \varphi dt + (1/2)\Big(< \partial_{i,j}\varphi \, dx^i \, dx^j > \Big) \qquad (8.61)$$

where repeated indices are used to mean summation. The last term, which is denoted by $<,>$, stands for the quadratic variation process. Evaluating this we obtain

$$(1/2)\Big(< \partial_{i,j}\varphi \, dx^i \, dx^j > \Big) = (1/2)\Big(< D^2\varphi dx, dx > \Big)$$

$$= (1/2)Tr(D^2\varphi\sigma\sigma^*)dt = (1/2)\Big(\sum_\alpha^d \sum_{i,j=1}^n (\sigma_\alpha^i \sigma_\alpha^j \partial_{i,j}\varphi) \Big)dt$$

$$\equiv (1/2)\Big(\sum_\alpha^d \sigma_\alpha^i \sigma_\alpha^j \partial_{i,j}\varphi \Big)dt.$$

Thus equation (8.61) reduces to

$$d\varphi = \sum_\alpha \sigma_\alpha^i \partial_i \varphi \, dw^\alpha + b^i \partial_i \varphi \, dt + (1/2)\left(\sum_\alpha^d \sigma_\alpha^i \sigma_\alpha^j \partial_{i,j} \varphi\right) dt. \quad (8.62)$$

Using the following notations

$$A_\alpha = \sum_{i=1}^n \sigma_\alpha^i \partial_i \quad A_0 \equiv \sum_{i=1}^n b^i \partial_i, 1 \le \alpha \le d, \quad (8.63)$$

we can rewrite equation (8.62) in the standard form used in differential geometry giving

$$d\varphi = \left(A_0 + (1/2)\sum_{\alpha=1}^d A_\alpha^2\right)\varphi \, dt + \sum_{\alpha=1}^d A_\alpha \varphi \, dw^\alpha. \quad (8.64)$$

It is clear from the expression (8.64) that the differential generator of the associated Markov process is given by

$$\mathcal{A} = \left(A_0 + (1/2)\sum_{\alpha=1}^d A_\alpha^2\right).$$

Stratonovich Model: To convert (8.60) to Stratonovich model, we must adjust the drift parameter from b^i to $\tilde{b}^i$ given by

$$\tilde{b}^i = b^i - c^i, c^i \equiv (1/2)\sum_{\alpha=1}^d \sum_{\ell=1}^n \sigma_\alpha^\ell \partial_\ell(\sigma_\alpha^i).$$

For detailed derivation of this transformation see [Ahmed (1998)]. Thus the Stratonovich model, equivalent to equation (8.60), is given by

$$dx^i = \tilde{b}^i(x)dt + \sum_\alpha \sigma_\alpha^i(x) \circ dw^\alpha, t \ge 0, x^i(0) = x_i; 1 \le i \le n.$$

Using these facts, the reader can easily verify that equation (8.64) reduces to

$$d\varphi = A_0\varphi \, dt + \sum_{\alpha=1}^d A_\alpha \varphi \circ dw^\alpha.$$

This shows that the Stratonovich calculus works like the classical Newtonian calculus unlike Itô calculus.

It is the Lie-algebraic properties of this family of vector fields $\{A_0, A_1, \cdots, A_d\}$ that determine the existence and smoothness of the density of the measure valued function induced by the solution of the SDE (8.60). We state the celebrated Hörmander's condition.

Hörmander's Condition (HO) The vectors generated by the Lie brackets

$$\left\{ A_i, [A_i, A_j], [[A_j, A_k], A_\ell], \cdots, 1 \le i \le d \text{ and } 0 \le j, k, \ell, \cdots \le d \right\}_x$$

evaluated at x span R^n.

Theorem 8.7.7 *Consider the system (8.60) and let ν_t denote the measure induced by the state vector $x(t)$ and suppose $\{A_i, 0 \le i \le d\}$ is a family of C^∞ vector fields corresponding to the drift and diffusion parameters $\{b, \sigma\}$ and that they satisfy the Hörmander's condition (HO). Then, for each $t > 0$, the measure ν_t has a density $\rho_t \in C^\infty(R^n)$ with respect to the Lebesgue measure.*

Proof The proof requires several estimates. We do not present it here. Interested reader is referred to the detailed proof given in Nualart [[Nualart (2009)], Theorem 7.4, p50]. See also Friz [Friz (2002)] where he gives a simpler proof using Stratonovich model. •

This result is the probabilistic counterpart of Hörmander's hypoelliptic result discovered by Malliavin [Malliavin (1976)], [Malliavin (1995)] and is a much weaker assumption than strict ellipticity condition. This is because there are examples of very simple SDE's where σ is degenerate and yet, under Hörmander's conditions, the induced measure has density.

8.7.3 *Some Illustrative Examples*

For illustration we present some simple examples. Some of these examples are taken from Friz [Friz (2002)].

Example 1 Consider the 2-dimensional system

$$dx_1 = x_1 dt + x_1 dw_1$$
$$dx_2 = 0 dt + dw_1 + dw_2. \tag{8.65}$$

The vector fields of the differential generator are given by

$$A_1 = x_1 \partial_1 + \partial_2, \quad A_2 = \partial_2, \quad A_0 = x_1 \partial_1.$$

To determine the Hörmander's conditions, we evaluate the Lie bracket

$$[A_1, A_2] = A_1 A_2 - A_2 A_1 = x_1 \partial_{12} + \partial_{22} - x_1 \partial_{21} - \partial_{22} = 0.$$

Similarly we have $[A_1, A_0] = 0$, $[A_2, A_0] = 0$. Now look at the span

$$Span\{A_1, A_2, [A_1, A_2], [A_1, A_0], [A_2, A_0]\}_{x_1=0}$$
$$= Span\{x_1 \partial_1 + \partial_2, \partial_2, 0\}_{x_1=0} \neq R^2.$$

So the measure induced by the system may not have density. In fact it does not have. Note that the vector field evaluated at $x_1 = 0$ does not span R^2. The implication is: if the process starts on the x_2-axis ($x_1 = 0$), it remains locked there traveling back and forth on the x_2 axis with no possibility of escape. Thus the associated measure cannot be absolutely continuous with respect to the Lebesgue measure.

Example 2 Consider the system

$$dx_1 = dt + x_2 dw_1 \tag{8.66}$$
$$dx_2 = 0dt + dw_1 + dw_2.$$

By inspection, one can identify the vector fields as

$$A_0 = \partial_1, A_1 = x_2 \partial_1 + \partial_2, \quad A_2 = \partial_2.$$

Computing the Lie bracket $[A_1, A_2]$ we have $[A_1, A_2] = -\partial_1$ and $[A_1, A_0] = [A_2, A_0] = 0$ and hence the span

$$Span\{A_1, A_2, [A_1, A_2]\} = Span\{x_2 \partial_1 + \partial_2, \partial_2, -\partial_1\} = R^2.$$

Thus the probability measure induced by the system (8.66) has smooth density.

Example 3 (degenerate case) Consider the system

$$dx_1 = 0dt + dw_1 \tag{8.67}$$
$$dx_2 = x_1 dt.$$

Here the drift is $b = (0, x_1)'$ and the dispersion σ equals

$$\sigma = \begin{pmatrix} 1 & 0 \\ 0 & 0 \end{pmatrix}.$$

Clearly, $\sigma\sigma' = \sigma$ is degenerate. On the other hand $A_0 = x_1\partial_2$, $A_1 = \partial_1$ and $A_2 = 0$. Thus the

$$Span\{A_1, A_2, [A_1, A_2], [A_1, A_0], [A_2, A_0]\} = Span\{\partial_1, \partial_2\} = R^2.$$

So Hörmander's condition holds and the corresponding measure has density. It is also easy to see that, for initial condition $x_1(0) = x_1$, $x_2(0) = x_2$,

$$x_1(t) = x_1 + w_1(t), x_2(t) = x_2 + x_1 t + \int_0^t w_1(s)ds.$$

The process is gaussian with mean $\bar{x}_1(t) = x_1$, $\bar{x}_2(t) = x_2 + x_1 t$ and covariance

$$Q(t) = \begin{pmatrix} t & t^2/2 \\ t^2/2 & t^3/3 \end{pmatrix}$$

which is nonsingular for $t > 0$. The measure induced by this process is Gaussian and the corresponding density is given by

$$\rho_t(x) \equiv (\sqrt{3}/\pi t^2)\exp\{-(1/2)(Q^{-1}(t)(x - \bar{x}(t)), (x - \bar{x}(t)))\}.$$

This shows the power of Hörmander's condition, it covers degenerate diffusion predicting existence of smooth density.

Example 4 Consider the system

$$dx_1 = x_2 dt \tag{8.68}$$

$$dx_2 = x_1 dt + x_2 dw_2.$$

Clearly this system is degenerate. The vector fields are

$$A_0 = x_2\partial_1 + x_1\partial_2, \quad A_1 = 0 \quad A_2 = x_2\partial_2.$$

The Lie bracket $[A_0, A_2] = x_1\partial_2 - x_2\partial_2$. So the

$$Span\{A_2, [A_0, A_2]\}_{x_2=0} = Span\{x_1\partial_2\} \neq R^2.$$

However, we observe that the process cannot remain locked in the degenerate region $\{x \in R^2 : x_2 = 0\}$ because of the presence of the term $x_1\partial_2$. It leaves the set immediately. Similarly

$$Span\{A_2, [A_0, A_2]\}_{x_1=0} = Span\{x_2\partial_2\} \neq R^2.$$

In this case the process can remain locked in the set $\{x_1 = 0\}$ with no possibility of escape and hence the associated measure does not have a density.

Example 5 This is a slightly modified example of Friz. Consider the system

$$dx_1 = x_2 dt + dw_1 \qquad (8.69)$$

$$dx_2 = f(x_1)dw_2$$

where f and all its derivatives vanish at $x_1 = 0$. The vector fields are

$$\{A_0 = x_2\partial_1, A_1 = \partial_1, A_2 = f(x_1)\partial_2\}.$$

The corresponding Lie brackets are $[A_0, A_1] = 0$ and $[A_0, A_2] = x_2 f'(x_1)\partial_2 - f(x_1)\partial_1$ and $[A_1, A_2] = f'(x_1)\partial_2$. Then the span of these vector fields is

$$Span\{A_1, A_2, [A_0, A_1], [A_0, A_2], [A_1, A_2]\}_{x_1=0}$$

$$= Span\{A_1\} = Span\{\partial_1\} \neq R^2.$$

Due to the presence of A_1 the process cannot stay on the set $\{x_1 = 0\}$; it will shoot out of x_2-axis instantly. Hence the density exists even though Hörmander's condition does not hold. Thus, Hörmander's condition is also a sufficient condition, better than strict ellipticity condition.

8.7.4   *Some Comments on Degeneracy Condition*

As seen in the Definition 8.7.2, for C^∞ smoothness of the density corresponding to the measure induced by a Wiener functional F, one needs very strong conditions. One of them is that $F \in L_p(\Omega) \equiv L_p(\Omega, \mu^w)$ for all $p \geq 1$. We present here two examples to indicate the severity of restriction imposed by this demand for smoothness. Consider the n-th degree homogeneous chaos

$$F(w) \equiv \tilde{F}(w, K) \equiv \int_{I^n} K(\tau_1, \tau_2, \cdots, \tau_n) dw(\tau_1) dw(\tau_2) \cdots dw(\tau_n)$$

where $K \in \hat{L}_2(I^n)$. If K merely belongs to $\hat{L}_2$, we have $F \in L_2(\Omega)$. In order for F to be in $L_p(\Omega)$ for every $p \geq 1$, we need a very strong condition. Let $\{\varphi_i\}$ be a complete orthonormal basis of $L_2(I)$ and define the set

$$\mathcal{K}_n \equiv \Big\{ K \in \hat{L}_2(I^n) : K(t_1, t_2, \cdots, t_n)$$

$$= \sum_{i_1, i_2, \cdots, i_n} a_{i_1, i_2, \cdots, i_n}\, \varphi_{i_1}(t_1) \cdots \varphi_{i_n}(t_n) \text{ and } \sum_{i_1, \cdots, i_n} |a_{i_1, i_2, \cdots, i_n}| < \infty \Big\},$$

where the coefficients $\{a_{i_1,i_2,\cdots,i_n}, i_k \geq 1, k = 1, 2, \cdots n\}$ are symmetric in their arguments. Define the set

$$M_n \equiv \{F : F(w) = \tilde{F}(w, K), K \in \mathcal{K}_n\}.$$

We show that $M_n \subset L_p(\Omega)$ for all $p \geq 1$. Let $K \in \mathcal{K}_n$ and define the random variable

$$Z(w) \equiv \int_{I^n} K(t_1, t_2, \cdots, t_n) dw(t_1) \cdots dw(t_n)$$

$$= \sum_{i_1,i_2,\cdots,i_n \geq 1} a_{i_1,i_2,\cdots,i_n} \varphi_{i_1}(w) \cdots \varphi_{i_n}(w)$$

where $\varphi_{i_k}(w) \equiv \int_I \varphi_{i_k}(t) \, dw(t)$. Let $Y \in L_q(\Omega)$ for q so that $((1/p)+(1/q) = 1)$. Then

$$\mathbb{E}\{ZY\} = \sum_{i_1,i_2,\cdots,i_n \geq 1} a_{i_1,i_2,\cdots,i_n} \mathbb{E}\{\varphi_{i_1}(w) \cdots \varphi_{i_n}(w) Y(w)\}.$$

By Hölder inequality, we obtain

$$|\mathbb{E}\{ZY\}|$$

$$\leq \sum_{i_1,i_2,\cdots,i_n \geq 1} |a_{i_1,i_2,\cdots,i_n}| \left(\mathbb{E}|\varphi_{i_1}(w) \cdots \varphi_{i_n}(w)|^p \right)^{1/p} \left(\mathbb{E}|Y(w)|^q \right)^{1/q}.$$

Since $\{\varphi_{i_k}(w), k = 1, 2, \cdots, n\}$ are independent Gaussian random variables with zero mean and unit variance, the reader can easily compute and find that

$$\mathbb{E}|\varphi_{i_1}(w) \cdots \varphi_{i_n}(w)|^p = C_p^n \equiv \left((1/\sqrt{\pi}) 2^{p/2} \Gamma((p+1)/2) \right)^n.$$

Hence

$$|\mathbb{E}(ZY)| \leq \left(\sum_{i_1,i_2,\cdots,i_n \geq 1} |a_{i_1,i_2,\cdots,i_n}| \right) \left(C_p^n \right)^{1/p} \| Y \|_{L_q(\Omega)}$$

for every $Y \in L_q(\Omega)$. Since, by definition of the set $\mathcal{K}_n$, the infinite series is bounded, it is evident that $Z \in L_p(\Omega)$ for every $p \geq 1$. Thus

$$M_n \subset \bigcap_{p \geq 1} L_p(\Omega).$$

Another class of functionals that satisfy similar properties is the class of cylindrical functionals of Brownian motion given by continuous functions having at most polynomial growth. We denote this class by $\mathcal{P}_n$ as defined below

$$\mathcal{P}_n \equiv \Big\{ Z : Z(w) \equiv f(\varphi_1(w), \varphi_2(w), \cdots, \varphi_n(w)) \quad \text{for some } f \text{ satisfying ;}$$

$$f \in C(R^n), \text{ and } |f(x)| \leq C\{1 + |x|^r\}, C \geq 0, r \in [0, \infty) \Big\}.$$

Without loss of generality, we may assume that the set $\{\varphi_i\}$ is orthonormal. By direct computation, the reader can readily verify that every $Z \in \mathcal{P}_n$ is in $L_p(\Omega)$ for all $p \in (0, \infty)$. Indeed, we have the following estimate

$$\mathbb{E}|Z|^p \equiv \int_\Omega |Z|^p d\mu^w \leq 2^p C^p \Big\{ 1 + (2n)^{rp/2}(1/\sqrt{\pi})\Gamma((1 + rp)/2) \Big\},$$

where $\Gamma(\cdot)$ is the standard gamma function.

Remark 8.7.8 It is important to recall that for the existence of smooth densities, the vector fields must be sufficiently smooth. In many practical applications, such smoothness conditions are rarely satisfied.

8.8   Central Limit Theorem for Wiener-Itô Functionals

In his monograph [Nualart (2009)], Nualart has presented interesting applications of Malliavin calculus to central limit theorem and finance. It is shown in [Nualart (2009)] that under certain assumptions, Wiener's homogeneous chaos obeys central limit theorem. This is then extended to functionals admitting Wiener-Itô expansion. Interested reader is referred to Nualart for many details. Here we present a simple but interesting result originally given in [Nualart (2009)] and also some extensions thereof. Also we prove a central limit theorem for Wiener chaos taking values in an infinite dimensional Hilbert space generalizing Nualart's result and conclude the section with some brief comments.

Denote the Hilbert space $L_2(I)$ by $\hat{H}$ and the inner product by $<, >_{\hat{H}}$. Let $\mathcal{H}_n$ denote the space of Wiener's homogeneous chaos of degree n. This is a closed linear subspace of the space $L_2(\Omega)$. We present the following central limit theorem in $\mathcal{H}_n$. The proof is due to Nualart [Nualart (2009)] and we present it in details in order to be able to see the distinction in the proof for the infinite dimensional case to follow in Theorem 8.8.4.

Theorem 8.8.1 *Let* $\{Z_m\} \subset \mathcal{H}_n$ *satisfy the following properties:*

(P1): $\mathbb{E}|Z_m|^2 \longrightarrow \sigma^2$ *as* $m \to \infty$

(P2): $|DZ_m|^2_{\hat{H}} \xrightarrow{s} n\sigma^2$ *in* $L_1(\Omega)$.

Then, there exists a subsequence of $\{Z_m\}$, *relabeled as* $\{Z_m\}$, *and an element* $Z_0 \in \mathcal{H}_n$ *such that* $Z_m \xrightarrow{\mathcal{L}} Z_0$ *(in law) and* $\mathcal{L}(Z_0) = N(0, \sigma^2)$, *Gaussian with zero mean and variance* σ^2.

Proof Let $K_m \in \hat{L}_2(I^n)$ correspond to $Z_m = g_n(K_m)$. Clearly it follows from (P1) that the sequence $\{Z_m\}$ is contained in a bounded subset of $\mathcal{H}_n \subset L_2(\Omega)$ and therefore it is tight and by Prohorov's theorem it contains a subsequence along which it converges in law to a random variable $Z_0 \in \mathcal{H}_n$. In other words, $\mathcal{L}(Z_m) \equiv \mu^m \xrightarrow{w} \mu^0$. We show that $\mu^0 \equiv \mathcal{L}(Z_0) = N(0, \sigma^2)$. Consider the characteristic functional $\psi_m(s) \equiv \mathbb{E}\{e^{isZ_m}\} = \int e^{isx} \mu^m(dx)$. Clearly, it follows from weak convergence of the sequence $\{\mu^m\}$ that

$$\psi_m(s) \longrightarrow \psi_0(s) \equiv \mathbb{E}\{e^{isZ_0}\} = \int e^{isx} \mu^0(dx)$$

for every $s \in R$. Since the sequence $\{Z_m\}$ is bounded in $L_2(\Omega)$, one can easily verify that for every $s \in R$,

$$\dot{\psi}_m(s) \equiv \mathbb{E}\{iZ_m e^{isZ_m}\} = i \int x e^{isx} \mu^m(dx) \longrightarrow$$

$$i \int x e^{isx} \mu^0(dx) = \mathbb{E}\{iZ_0 e^{isZ_0}\} = \dot{\psi}_0(s)$$

as $m \to \infty$. On the other hand, using the Ornstein-Uhlenbeck operator $L = -\delta D$ we have $LZ_m = -nZ_m$. Hence we have

$$\dot{\psi}_m(s) = -(i/n)\mathbb{E}\{LZ_m e^{isZ_m}\} = -(i/n)\mathbb{E}\{-(\delta DZ_m)e^{isZ_m}\}$$

$$= (i/n)\mathbb{E}\{< DZ_m, D(e^{isZ_m}) >_{\hat{H}}\} = -(s/n)\mathbb{E}\{|DZ_m|^2_{\hat{H}} e^{isZ_m}\}.$$

Note that the last integral is a duality pairing between $L_1(\Omega)$ and $L_\infty(\Omega)$ with the first member of the integrand converging strongly by assumption (P2) and the second member converging in the weak star sense as seen above. Thus using the assumption (P2) and the fact that $Z_m \xrightarrow{\mathcal{L}} Z_0$ and letting $m \to \infty$ we obtain

$$\lim_{m \to \infty} \dot{\psi}_m(s) = -(s/n)(n\sigma^2)\psi_0(s).$$

It follows from the above two limits that

$$\dot{\psi}_0(s) = -s\sigma^2 \psi_0(s), \psi_0(0) = 1.$$

Solving this equation we have $\psi_0(s) = e^{-(s^2/2)\sigma^2}$. This is the characteristic functional of a normal distribution with mean zero and variance σ^2. Thus $\mathcal{L}(Z_0) = N(0, \sigma^2)$. This completes the proof. •

In view of Theorem 8.8.1, it is evident that, under the given assumptions, each subspace $\mathcal{H}_n \subset L_2(\Omega) = \mathcal{G}$ satisfies the central limit theorem. It follows from section 2.2 that the product space $\prod_{n \geq 0} \mathcal{H}_n$ coincides with the Fréchet space $\Psi_\alpha(\mathcal{K}_\alpha)$ which is the isomorphic image of the Fréchet space $\mathcal{K}_\alpha$. This is an algebraic isomorphism (not topological). Thus, in view of the above theorem we may conclude that this space also satisfies the central limit theorem. Let

$$\pi_n : \prod_{k \geq 0} \mathcal{H}_k \longrightarrow \mathcal{H}_n$$

denote the projection of the product space to its n-th coordinate space. Let $\{F_k\} \in \prod_{n \geq 0} \mathcal{H}_n$ and suppose that it converges in the product topology to F_0 and that for each $n \in N$, $\pi_n(F_k) \xrightarrow{\mathcal{L}} \pi_n(F_0)$ which has the normal distribution $N(0, \sigma_n^2)$. Let $R^\infty \equiv \prod^\infty R$ denote the Cartesian product of infinite copies of the real line R. This is furnished with the Tychonoff product topology turning it into a topological space. Suppose that it is also endowed with the corresponding topological Borel field $\mathcal{B}_\infty$ generated by open sets so that $(R^\infty, \mathcal{B}_\infty)$ is a measurable space. We may then consider the sequence of random variables $\{F_k\}$ as well as F_0 as measurable functions from Ω to $(R^\infty, \mathcal{B}_\infty)$. Thus F_0 is an infinite dimensional random variable which induces a Gaussian measure $\nu^0 \equiv \mu^w F_0^{-1}$ on $(R^\infty, \mathcal{B}_\infty)$ and it is given by the product measure $\prod N(0, \sigma_n^2)$ with covariance operator $Q_0 \equiv diag\{\sigma_n^2\}$. This is generally a finitely additive cylinder set measure. If $\sigma_n^2 = 1$ for all $n \in N$, F_0 is a cylindrical Gaussian random variable with values in R^∞. Nualart [Nualart (2009)] calls it chaotic central limit theorem. Thus in general, the random variable $\tilde{F}_0$ given by the infinite series $\tilde{F}_0 \equiv \sum_{n \geq 0} \pi_n(F_0)$ may not belong to any of the Hilbert spaces $L_2(\Omega) \equiv \mathcal{G}_\alpha \subset \mathcal{H}_\alpha \subset \mathcal{G}_\alpha^*$.

For the scalar valued random variable $\tilde{F}_0$ to be an element of $L_2(\Omega) = \mathcal{G}_\alpha$, it is necessary that the covariance operator $Q_0 \equiv Cov(F_0)$ is nuclear. That is,

$$Tr_{\mathcal{G}_\alpha} Q_0 = \sum_{n \geq 0} \mathbb{E}(\pi_n(F_0))^2 = \sum_{n \geq 0} \sigma_n^2 < \infty.$$

In this case the law of $\tilde{F}_0$ is the Gaussian measure $N(0, Tr_{\mathcal{G}_\alpha}(Q_0))$ with mean zero and variance $Tr_{\mathcal{G}_\alpha}(Q_0)$. Thus, if Q_0 is nuclear, it follows from the well known Minlos-Sazanov theorem that the measure ν^0 is countably additive and its support is the Hilbert space $X_1 \equiv \ell_2 \subset R^\infty$.

It is possible, as seen in the example with $\sigma_n^2 = 1$ for $n \geq 1$, that the preceding infinite series diverges to infinity. In other words, $\tilde{F}_0 \notin \mathcal{G}_\alpha = L_2(\Omega)$. In that case we may use one of the larger spaces such as $\mathcal{H}_\alpha$ or $\mathcal{G}_\alpha^*$. We may caution the reader not to confuse $\mathcal{H}_\alpha$ with the Wiener homogeneous chaos which is denoted by $\mathcal{H}_n$ for any integer $n \in N$. If the trace of Q_0 is measured with reference to either of the two larger spaces, we have

$$Tr_{\mathcal{H}_\alpha} Q_0 = \sum_{n \geq 0} (1/n!) \mathbb{E}(\pi_n(F_0))^2 = \sum_{n \geq 0} \sigma_n^2 / n!, \quad \text{for } \mathcal{H}_\alpha$$

and

$$Tr_{\mathcal{G}_\alpha^*} Q_0 = \sum_{n \geq 0} (1/(n!)^2) \mathbb{E}(\pi_n(F_0))^2 = \sum_{n \geq 0} \sigma_n^2 / (n!)^2, \quad \text{for } \mathcal{G}_\alpha^*.$$

If $Tr_{\mathcal{H}_\alpha} Q_0 < \infty$, $\tilde{F}_0 \in \mathcal{H}_\alpha$ and similarly, if $Tr_{\mathcal{G}_\alpha^*} Q_0 < \infty$, $\tilde{F}_0 \in \mathcal{G}_\alpha^*$. In the first case, the measure ν_0 corresponding to F_0 is supported on the Hilbert space X_2 given by

$$X_2 \equiv \{x \in R^\infty : \| x \|_{X_2}^2 \equiv \sum_{n=1}^\infty (1/n!) |x_n|^2 < \infty\}$$

and in the second case it is supported on the Hilbert space X_3 given by

$$X_3 \equiv \{x \in R^\infty : \| x \|_{X_3}^2 \equiv \sum_{n=1}^\infty (1/n!)^2 |x_n|^2 < \infty\}.$$

Clearly these are separable Hilbert spaces embedded in the Tychonoff space R^∞. The reader can easily check that the embeddings $X_1 \hookrightarrow X_2 \hookrightarrow X_3$ are continuous and dense. Clearly, this expands the scope of application of central limit theorem to a larger class of Wiener-Itô functionals.

In view of the above analysis we have the following result.

Theorem 8.8.2 *Consider a sequence of random variables* $\{Z_m, m \in N\}$, *where each member is given by* $Z_m \equiv \sum_{n \geq 0} g_n(K_{n,m})$, $K_{n,m} \in \hat{L}_2(I^n)$ *and suppose that it's projection* $\pi_n(Z_m) \equiv g_n(K_{n,m})$ *satisfies the following properties:*

(Q1): For each $n \in N$, $\lim_{m \to \infty} \mathbb{E}(\pi_n(Z_m))^2 = \sigma_n^2$

(Q2): For each $n \in N$, $\lim_{m \to \infty} |D\pi_n(Z_m)|_{\hat{H}}^2 = n\sigma_n^2$ in the $L_1(\Omega)$ sense.

Then, (A): If $Tr_{\mathcal{G}_\alpha} Q_0 = \sum_{n \geq 0} \sigma_n^2 \equiv \sigma^2 < \infty$, the sequence of random variables $\{Z_m\} \in \mathcal{G}_\alpha$ and $Z_m \xrightarrow{\mathcal{L}} Z_0 \in \mathcal{G}_\alpha$ and it has the normal distribution $N(0, \sigma^2)$.

(B): If $Tr_{\mathcal{H}_\alpha} Q_0 = \sum_{n \geq 0} \sigma_n^2/n! \equiv \gamma^2 < \infty$, the sequence of random variables $\{Z_m\} \in \mathcal{H}_\alpha$ and $Z_m \xrightarrow{\mathcal{L}} Z_0 \in \mathcal{H}_\alpha$ and it has the normal distribution $N(0, \gamma^2)$.

(C): If $Tr_{\mathcal{G}_\alpha^*} Q_0 = \sum_{n \geq 0} \sigma_n^2/(n!)^2 \equiv \varrho^2 < \infty$, the sequence of random variables $\{Z_m\} \in \mathcal{G}_\alpha^*$ and $Z_m \xrightarrow{\mathcal{L}} Z_0 \in \mathcal{G}_\alpha^*$ and it has the normal distribution $N(0, \varrho^2)$. •

Remark 8.8.3 Since the Hilbert spaces $\{\mathcal{G}_\beta, \mathcal{G}_\gamma, \mathcal{G}_\delta\}$ are functionals of Brownian motions or fields and have finite dimensional range, the results of Theorem 8.8.1 and Theorem 8.8.2 also hold for the family of spaces $\mathcal{G}_\beta \hookrightarrow \mathcal{H}_\beta \hookrightarrow \mathcal{G}_\beta^*$, $\mathcal{G}_\gamma \hookrightarrow \mathcal{H}_\gamma \hookrightarrow \mathcal{G}_\gamma^*$, and $\mathcal{G}_\delta \hookrightarrow \mathcal{H}_\delta \hookrightarrow \mathcal{G}_\delta^*$ (see Chapter 3). The reader may find it interesting to verify if this is true for the spaces $\{\mathcal{G}_\varphi, \mathcal{H}_\varphi, \mathcal{G}_\varphi^*\}$ and $\{\mathcal{G}_\Pi, \mathcal{H}_\Pi, \mathcal{G}_\Pi^8\}$ related to functionals of fractional Brownian motion and the Lêvy process respectively.

Consider now the spaces $\{\mathcal{G}_\varpi, \mathcal{H}_\varpi, \mathcal{G}_\varpi^*\}$ as introduced in sections 2.6 and 3.6. These are functionals of infinite dimensional Brownian motion and they take values in an abstract Hilbert space H. The results presented above do not hold for these spaces since the elements of these spaces have range in infinite dimensional Hilbert space H. However, with an additional assumption we can extend Nualart's central limit theorem also to these spaces.

Let H be a separable Hilbert space with a complete orthonormal basis $\{h_i\}$. Recall the Wiener space $L_2(\Omega, \mu; H)$ as seen in Corollary 2.6.2 with the sum decomposition

$$L_2(\Omega, \mu; H) \equiv \mathcal{G}_\varpi = \sum \oplus \mathcal{G}_{\varpi,n}.$$

The following result generalizes the central limit Theorem 8.8.1.

Theorem 8.8.4 *For a fixed $n \in N$, consider the sequence of random variables $\{Z_m\} \subset \mathcal{G}_{\varpi,n} \subset L_2(\Omega, \mu; H)$ with covariance $P_m \in \mathcal{L}_1^+(H)$ given* by $\mathbb{E}(Z_m, h)^2_H \equiv (P_m h, h)$, $h \in H$, and suppose it satisfies the following properties:

(P1): There exists a $P \in \mathcal{L}^+_1(H)$ such that as $m \to \infty$,

$$\mathbb{E}|Z_m|^2_H = \sum_{i \geq 1}(P_m h_i, h_i) \longrightarrow \sum_{i \geq 1}(P h_i, h_i) \equiv Tr(P) \equiv \sigma^2(< \infty)$$

(P2): $\lim_{r \to \infty} \sum_{i \geq r} \mathbb{E}(Z_m, h_i)^2_H \equiv \lim_{r \to \infty} \sum_{i \geq r}(P_m h_i, h_i)_H = 0$ uniformly with respect to $m \in N$.

(P3): For every $\eta, \zeta \in H$, $\{< (DZ_m, \eta)_H, (DZ_m, \zeta)_H >_{\hat{H}}\} \xrightarrow{s} n(P\eta, \zeta)$ in $L_1(\Omega)$.

Then, there exists a subsequence of $\{Z_m\}$, relabeled as $\{Z_m\}$, and an element $Z_0 \in \mathcal{G}_{\varpi,n}$ such that $Z_m \xrightarrow{\mathcal{L}} Z_0$ (in law) and $\mathcal{L}(Z_0) = N(0, P)$.

Proof Essentially the proof follows from arguments similar to those of Theorem 8.8.1 with some exceptions. We present an outline. In the case of random variables with range in finite dimensional spaces, the assumption (P1) is sufficient to guarantee tightness and hence the weak convergence of the sequence of measures $\{\nu_m\} \equiv \{\mathcal{L}(Z_m)\}$ by Prohorov's theorem. For the infinite dimensional state space, this is no longer true. Here we need both (P1) and (P2) to guarantee the tightness. This follows from Theorem 2 given in [Gihman and Skorohod, [Gihman (1971)], Chapter 6, Theorem 2, p377]. Thus, under the assumptions (P1) and (P2) there exists a subsequence of the sequence $\{Z_m\}$, relabeled as the original sequence, and an element $Z_0 \in \mathcal{G}_{\varpi,n}$ such that $Z_m \xrightarrow{\mathcal{L}} Z_0$. Equivalently, $\mathcal{L}(Z_m) = \nu_m \xrightarrow{w} \nu_0 = \mathcal{L}(Z_0)$. Following similar arguments as in Theorem 8.8.1, we prove that ν_0 is a Gaussian measure on H with mean zero and covariance P. We use characteristic functionals and the Ornstein-Uhlenbeck operator $L = -\delta D$. For convenience of notation, let $D_F \varphi$ denote the Fréchet derivative of any $\varphi \in C(H)$ whenever it exists. By definition the characteristic functionals of the random variables $\{Z_m, Z_0\}$ (or equivalently the measures $\{\nu_m, \nu_0\}$) are given by

$$\Psi_m(\xi) \equiv \mathbb{E}\{e^{i(Z_m,\xi)}\} \equiv \int_H e^{i(x,\xi)}\nu_m(dx), \xi \in H$$

$$\Psi_0(\xi) \equiv \mathbb{E}\{e^{i(Z_0,\xi)}\} \equiv \int_H e^{i(x,\xi)}\nu_0(dx), \xi \in H.$$

Since $Z_m \in L_2(\Omega, \mu; H)$ and $Z_m \xrightarrow{\mathcal{L}} Z_0$, it is easy to verify that, for every $\eta \in H$,

$$(D_F \Psi_m(\xi), \eta)_H = i \int_h (x, \eta) e^{i(x, \xi)} \nu_m(dx)$$

$$\longrightarrow i \int_H (x, \eta) e^{i(x, \xi)} \nu_0(dx) = (D_F \Psi_0(\xi), \eta)_H \quad (8.70)$$

as $m \to \infty$. Alternatively, using Ornstein-Uhlenbeck operator, we have

$$(D_F \Psi_m(\xi), \eta) = i\mathbb{E}\{(Z_m, \eta) e^{i(\xi, Z_m)}\}$$

$$= (-i/n)\mathbb{E}\{(LZ_m, \eta) e^{i(\xi, Z_m)}\}$$

$$= (-1/n)\mathbb{E}\{< (DZ_m, \eta)_H, (DZ_m, \xi)_H >_{\hat{H}} e^{i(\xi, Z_m)}\}. \quad (8.71)$$

Note that $< (DZ_m, \eta)_H, (DZ_m, \xi)_H >_{\hat{H}} \in L_1(\Omega, \mu)$ and $e^{i(\xi, Z_m)} \in L_\infty(\Omega, \mu)$. Thus, using the assumption (P3) and letting $m \to \infty$ in the expression (8.71), we obtain

$$\lim_{m \to \infty} (D_F \Psi_m(\xi), \eta) \longrightarrow -(P\eta, \xi)\Psi_0(\xi), \xi \in H. \quad (8.72)$$

Equating (8.70) with (8.72) we arrive at the following identity

$$(D_F \Psi_0(\xi), \eta)_H = -(P\eta, \xi)\Psi_0(\xi), \ \forall \ \eta \in H. \quad (8.73)$$

Since $\eta \in H$ is arbitrary, and P is symmetric, it follows from this that

$$D_F \Psi_0(\xi) = -\Psi_0(\xi) \, P\xi, \quad \text{for} \ \xi \in H.$$

Combining this with the fact that $\Psi(0) = 1$, we have

$$\Psi_0(\xi) = e^{-(1/2)(P\xi, \xi)_H}.$$

This is the characteristic functional of a Gaussian measure on H with mean zero and covariance operator P and so $\mathcal{L}(Z_0) = N(0, P)$. This completes the outline of our proof. •

Using the above result the reader can easily prove the following corollary.

Corollary 8.8.5 Let $\{Z^m\} \in \mathcal{G}_\varpi$ and suppose $Z^m \xrightarrow{s} Z^o$ in $\mathcal{G}_\varpi$ with $P^o \in \mathcal{L}_1^+(H)$ being the covariance operator corresponding to Z^o. Further suppose that for each $n \in N$, the sequence $\{\pi_n(Z^m)\}_{m \geq 1}$ satisfies the assumptions of the central limit theorem given by Theorem 8.8.4. Then

$\mathcal{L}(Z^o) = N(0, P^o)$, a Gaussian measure with mean zero and covariance $P^o \in \mathcal{L}_1^+(H)$.

Proof Hint: $\{\pi_n(Z^m)\}_{n \geq 1}$ are independent H valued square integrable random variables with covariance $\{P_n^m\} \in \mathcal{L}_1^+(H)$. For each $n \in N$, $P_n^m \xrightarrow{s} P_n^o$ in the Banach space $\mathcal{L}_1(H)$. By virtue of independence of the random variables $\{\pi_n(Z^o)\}, n \in N$, $P^o = \sum P_n^o$.

Remark 8.8.6 If the sum $\sum_{n=1}^{\infty} P_n^o$ fails to converge in $\mathcal{L}_1(H)$ to an element $P^o \in \mathcal{L}_1^+(H)$ we have to consider larger spaces. Here also one may consider Nualart's chaotic central limit theorem. For the infinite dimensional case, we need the product space $\mathcal{T} \equiv \prod^{\infty} H$ (product of infinite copies of the Hilbert space H) endowed with the Tychnoff product topology. For any finite integer n, define the index set

$$\Lambda_n \equiv \{(\alpha_1, \alpha_2, \cdots, \alpha_n) : \alpha_i \in N, \alpha_i \neq \alpha_j \text{ for } i \neq j\}$$

and let $\mathcal{T}_n \equiv \mathcal{P}_{\Lambda_n}(\mathcal{T})$ denote the projection of $\mathcal{T}$ to the product of any n copies of H determined by the index set $\Lambda_n \subset N^n$. Following similar arguments as in Theorem 8.8.2, we obtain a Gaussian measure on the Tychonoff space $\mathcal{T}$ given by $N(0, Q)$ where $Q \equiv diag P_n^o$. In general, this is a finitely additive cylindrical Gaussian measure well defined on the algebra of cylinder sets in $\mathcal{T}$. Since, for each $n \in N$, $P_n^o \in \mathcal{L}_1^+(H)$, the measure $N(0, Q)$ is countably additive on the sigma algebra generated by cylinder sets with base determined by $\mathcal{T}_n$ only for finite n. In other words, $N(0, Q)$ is only a finitely additive Gaussian cylinder set measure on $\mathcal{T}$. For a cylinder set Γ based on $\mathcal{T}_n$, the family of measures $\{\mu_{\Lambda_n}, n \in N\}$ given by

$$\mu_{\Lambda_n}(\Gamma) \equiv \prod_{i=1}^{n} N(0, P_{\alpha_i}^o)(\mathcal{P}_{\alpha_i}^{-1}(\Gamma))$$

is well defined. Clearly, this family does not have a countably additive extension to the sigma algebra generated by the cylinder sets.

Remark 8.8.7 Parallel to the discussion preceded by Theorem 8.8.2, the sum $\sum_{n=1}^{\infty} P_n^o$ may fail to converge in $\mathcal{L}_1(H)$ to an element $P^o \in \mathcal{L}_1^+(H)$. We leave it to the interested reader to verify that under the assumptions of Theorem 8.8.4, and some assumptions similar to those of Theorem 8.8.2, one can readily extend the results of Theorem 8.8.2 also to the case of

Gelfand triple $\{\mathcal{G}_{\varpi}, \mathcal{H}_{\varpi}, \mathcal{G}_{\varpi}^*\}$. In this case the corresponding Hilbert spaces are given by similar expressions with R replaced by the Hilbert space H. We use π_n to denote the projection map taking $\mathcal{T}$ to its n-th component space. Consider the vector space given by $X_1 \equiv \ell_2(\prod^{\infty} H) \equiv \ell_2(H)$ and let it be endowed with the norm topology $\| x \|_{X_1} = \left(\sum_{n \geq 1} |\pi_n(x)|_H^2 \right)^{1/2}$. Clearly, this is a separable Hilbert space and it is a rather narrow space and the measure $N(0, P^o)$ may not be supported on this. We consider larger spaces X_2 and X_3 which are given by

$$X_2 \equiv \{x \in \mathcal{T} : \quad \| x \|_{X_2} \equiv \left(\sum_{n=1}^{\infty} (1/n!) |\pi_n(x)|_H^2 \right)^{1/2} < \infty\}$$

and

$$X_3 \equiv \{x \in \mathcal{T} : \quad \| x \|_{X_3}^2 \equiv \left(\sum_{n=1}^{\infty} (1/n!)^2 |\pi_n(x)|_H^2 \right)^{1/2} < \infty\}$$

respectively. Again the embeddings $X_1 \hookrightarrow X_2 \hookrightarrow X_3$ are continuous and dense. The spaces X_2 and X_3 are much larger, and therefore the measure $N(0, P^o)$ is more likely to be supported on one of these spaces rather than X_1.

8.9   Malliavin Calculus for Fr-Brownian Motion

Before we close this Chapter, we mention some recent development towards Malliavin calculus. In a recent paper [Duncan (2000)], Duncan and his colleagues have developed a calculus similar to that of Malliavin on the space of functionals of fractional Brownian motion.

It is well known that the class of exponential functionals of Brownian motion is dense in $L_2(\Omega)$. A similar result was proved by Duncan and his school [Duncan (2000)] for fractional Brownian motion. We present it here briefly. Let $\mathcal{E}$ denote the class of functionals given by

$$\mathcal{E} \equiv \left\{ e(\psi) : e(\psi) \equiv \exp\left\{ \int_I \psi(t) dB_H(t) - (1/2) \| \psi \|_{L_2^\varphi(I)}^2 \right\}, \psi \in L_2^\varphi(I) \right\},$$
(8.74)

where, as seen in Chapter 2,

$$\| \psi \|_{L_2^\varphi(I)}^2 \equiv \int_{I^2} \varphi(t - s) \psi(t) \psi(s) dt ds.$$

It was shown in [Duncan (2000)] that for any linearly independent set $\{\psi_i \in L_2^{\varphi}(I)\}$, the functionals $\{e(\psi_i)\}$ are linearly independent and that this set is dense in $L_2^{\varphi}(\Omega)$. Therefore one can use them as the basis (functions) for the Hilbert space $L_2^{\varphi}(\Omega)$. These functionals are also used to define what is known as wick products $e(\psi) \diamond e(\phi) \equiv e(\psi + \phi)$. Using this notion of wick product Duncan et al. constructed a differential calculus very similar to the Malliavin Calculus. The Malliavin derivatives are first defined for functionals belonging to the class $\mathcal{E}$ and then extended to $L_2^{\varphi}(\Omega)$ by continuity and density arguments. For details see Duncan [Duncan (2000)].

8.10   Some Problems for Exercise

P1: Refer to the discussion following Lemma 8.3.1.

(a): Verify that the shifted measure $\mu_h(\cdot) \equiv \mu(h + \cdot)$ on the Wiener measure space $(C_0(I, R^n), \mathcal{B}_0, \mu)$ is absolutely continuous with respect to the original measure μ and that its RND (Radon Nikodym derivative) is given by

$$\rho(h) \equiv \exp\Big\{ \int (h, dW) - (1/2) \int |h|^2 dt \Big\}$$

for $h \in L_2(I, R^n)$.

(b): Repeat the problem for the fractional Brownian motion B_H. The RND of the shifted measure with respect to the measure induced by the fractional Brownian motion on $C_0(I, R^n)$ is given by

$$\rho_H(h) \equiv \exp\Big\{ \int (h, dB_H) - (1/2) \int |h|_{L_2^{\varphi}}^2 dt \Big\}$$

for $h \in L_2^{\varphi}(I, R^n)$. Show that $\lim_{H \downarrow 1/2} \rho_H = \rho$. Hint: Recall section 2.7 and the fact that $\lim_{H \downarrow 1/2} \varphi(t - s) = \delta(t - s)$.

P2: Verify the sequence of identities in the expression (8.8).

P3: Let Ω denote the Wiener measure space. Prove that the Malliavin differential operator D defined on $\mathcal{S}_2 \subset L_p(\Omega), p \geq 1$, admits a closed extension on $L_p(\Omega)$ for every $p > 1$. Hint: Follow the procedure used in the proof of Lemma 8.3.5.

P4: Prove that the operator δ, adjoint of the Malliavin differential operator D, is a continuous linear map from $W^{1,2}(\Omega, H)$ to $L_2(\Omega)$ and that it is an isometric isomorphism. Hint: Follows from definition (section 8.6).

P5: Let Ω denote the Wiener measure space and consider the Sobolev space $W^{1,2}(\Omega, H)$. Compare this with the classical Sobolev spaces $W^{1,2}(R^n, H)$ (similarities and differences) (see section 8.6).

P6: Justify the validity of Clark-Ocone formula (8.22) for the homogeneous chaos

$$F(w) = \int_{I^n} K_n(t_1, t_2, \cdots, t_n) dw(t_1) dw(t_2) \cdots dw(t_n), K_n \in \hat{L}_2(I^n).$$

Identify the process $\varphi \in L_2^a(\Omega, H)$ in terms of F. Hint: Follow the proof leading to the equation (8.22).

P7: Verify that the Ornstein-Uhlenbeck operator $L = -\delta D$ is a dissipative unbounded linear operator on $L_p(\Omega)$ for $p \geq 1$. Hint: Follows from section 8.6.

P8: Give a detailed proof of Lemma 8.7.3; and prove the estimate (8.45). Hint: (see [Nualart (2009)], Theorem 5.9, p36; [Bell (1987)], p46)

P9: Consider the Lotka-Volterra model for prey-predator system:

$$dx_1 = (\alpha_1 x_1 + \beta_1 x_1 x_2) dt + \sigma_1 dw_1$$
$$dx_2 = (-\alpha_2 x_2 + \beta_2 x_1 x_2) dt + \sigma_2 dw_2$$

with all the coefficients nonnegative. Use Hörmander's conditions and determine if the measure on R^2 induced by the process $x(t), t \geq 0$, possesses density for the following cases: (1) All the coefficients are positive; (2) $\sigma_2 = 0$; (3) $\sigma_1 = 0$; (4): The diffusion parameters are functions of state: $\sigma_1 = \sigma_1(x_1), \sigma_2 = \sigma_2(x_2)$, satisfying $\sigma_1(0) = \sigma_2(0) = 0$.

P10: Extend the results of Theorem 8.8.2 to the case of Gelfand triple $\{\mathcal{G}_\varpi, \mathcal{H}_\varpi, \mathcal{G}_\varpi^*\}$.
Hints: Use the assumptions of Theorem 8.8.4, and introduce assumptions very similar to those of Theorem 8.8.2.

Chapter 9

Evolution Equations on Fock Spaces

9.1 Introduction

In this Chapter we study evolution equations on Fock spaces. We prove existence, uniqueness and regularity properties of solutions. In section 9.2, we study the basic properties of the Malliavin operators on the space $\mathcal{G}_\alpha$. We prove that these Malliavin operators, restricted to $\mathcal{G}_\alpha$, are unbounded linear operators while they are bounded operators from $\mathcal{G}_\alpha$ to its dual $\mathcal{G}_\alpha^*$. In fact, this holds for all the Fock spaces $\{\mathcal{G}_\alpha, \mathcal{G}_\beta, \mathcal{G}_\gamma, \mathcal{G}_\delta, \mathcal{G}_\varpi\}$ which we may denote by $\mathcal{G}$ without any subscript. We prove existence and uniqueness of mild solutions for linear and semilinear evolution equation on the Fock spaces mentioned above. In section 9.3, we go beyond the Ornstein-Uhlenbeck operator and the corresponding semigroup. We show that there exist general linear unbounded operators on Fock spaces that generate a broader class of semigroups. In section 9.4, we use the Gelfand triple $\mathcal{G} \hookrightarrow \mathcal{H} \hookrightarrow \mathcal{G}^*$ and consider linear operators mapping $\mathcal{G}$ to $\mathcal{G}^*$ and evolution equations based on such operators. We prove existence of weak solutions both for linear and semilinear problems. In section 9.5 an example involving Gaussian random field is presented. This involves perturbation in the Fréchet space $\mathcal{K}$ of the kernels determining the unperturbed elements of the space $L_2(\mathcal{M}, \mu^w)$. In section 9.6, we construct nonlinear, monotone and coercive operators on Wiener-Sobolev spaces using Ornstein-Uhlenbeck operator as the basis and a duality map. And then we consider evolution equations based on such operators. We consider linear, semilinear and nonlinear evolution equations on these Fock spaces and present several results on existence, uniqueness and regularity properties of their solutions.

9.2   Malliavin Operators on Fock Spaces

In this section we want to study Malliavin operators on Fock spaces introduced in Chapter 2. For simplicity of presentation, we consider the Gelfand triple $G_\alpha \hookrightarrow H_\alpha \hookrightarrow G_\alpha^*$ and their isomorphic images $\mathcal{G}_\alpha \hookrightarrow \mathcal{H}_\alpha \hookrightarrow \mathcal{G}_\alpha^*$ (see Chapter 3). We know that

$$\mathcal{G}_\alpha = \sum \oplus \mathcal{G}_{\alpha,n}, \quad \mathcal{H}_\alpha = \sum \oplus \mathcal{H}_{\alpha,n}.$$

First, let us consider the Malliavin derivative of an element $f_n \in \mathcal{G}_{\alpha,n}$, the space of n-th order homogeneous chaos. Clearly, f_n has the representation

$$f_n(w) = g_n(K_n, w), K_n \in \hat{L}_2(I^n),$$

where $\{g_n\}$ are the orthogonal functionals of Brownian motion as introduced in Chapter 2. We recall that the Malliavin derivative makes sense only in the direction of the Cameroon-Martin subspace $H_1 \subset \Omega$. Thus, for any element $h \in H_1 \subset H \equiv L_2(I)$, we have

$$\lim_{\varepsilon \to 0}(1/\varepsilon)\{f_n(w+\varepsilon h) - f_n(w)\} = <Df_n(w), \dot{h}> = n\int_I (g_{n-1}(K_n(t, \cdot)), \dot{h})_H dt.$$

The last identity follows from the symmetry of the kernel K_n. Hence, for almost all $t \in I$,

$$(Df_n)(w)(t) \equiv (Df_n)(t) = ng_{n-1}(K_n(t, \cdot)). \tag{9.1}$$

For convenience of notation, we omit w. Note that this is only a $(n-1)$ fold Wiener integral and it is a stochastic process belonging to $L_2(\Omega, H)$ and

$$\mathbb{E}\int_I |(Df_n)(t)|^2 dt = n^2(n-1)! \parallel K_n \parallel^2_{L_2(I^n)} = n(n! \parallel K_n \parallel^2). \tag{9.2}$$

Lemma 9.2.1   *The Malliavin operator D is an unbounded operator on $L_2(\Omega) = \mathcal{G}_\alpha$, and it maps $\mathcal{D}(D) \subset \mathcal{G}_\alpha$ to $L_2(\Omega, H)$ or equivalently $L_2(I, \mathcal{G}_\alpha)$. The domain is dense in $\mathcal{G}_\alpha$. Further, it is a continuous linear operator from $\mathcal{G}_\alpha$ to the larger spaces $L_2(I, \mathcal{H}_\alpha) \subset L_2(I, \mathcal{G}_\alpha^*)$.*

Proof The domain of the operator is given by

$$\mathcal{D}(D) \equiv \{f \in \mathcal{G}_\alpha : Df \in L_2(I, \mathcal{G}_\alpha)\}.$$

Let $\{K_n\}$ denote the Fourier-Wiener kernels of f. Then it follows from (9.1) that

$$(Df)(t) = \sum_{n \geq 1} n g_{n-1}(K_n(t, \cdot)).$$

Hence,

$$\int_I |(Df)(t)|_{\mathcal{G}_\alpha}^2 \, dt = \sum_{n \geq 1} n(n! \parallel K_n \parallel_{L_2(I^n)}^2) \geq \parallel K \parallel_{G_\alpha}^2 = \parallel f \parallel_{\mathcal{G}_\alpha}^2.$$

For each integer n define $C_n = \frac{n! \parallel K_n \parallel^2}{\sum_{n=1}^\infty n! \parallel K_n \parallel^2}$. Clearly $C_n \geq 0$ and $\sum_{n=1}^\infty C_n = 1$. Since G_α is isometrically isomorphic to $\mathcal{G}_\alpha$, $\parallel f \parallel_{\mathcal{G}_\alpha} = \parallel K \parallel_{G_\alpha}$ and hence it follows from the first identity of the above expression that

$$\left(\frac{\parallel Df \parallel_{L_2(I, \mathcal{G}_\alpha)}}{\parallel f \parallel_{\mathcal{G}_\alpha}} \right)^2 = \sum_{n=1}^\infty n C_n.$$

Examining the series, it is clear that there exists a sequence $\{C_n^o\}$ satisfying the required properties, $C_n^o \geq 0$, $\sum_{n=1}^\infty C_n^o = 1$, such that $\sum_{n=1}^\infty n C_n^o = +\infty$. Hence by the isometric isomorphism, there exists an $f^o \in B_1(\mathcal{G}_\alpha)$, unit ball of the Hilbert space $\mathcal{G}_\alpha$, such that $\parallel Df^o \parallel_{L_2(I, \mathcal{G}_\alpha)} = +\infty$. This shows that the Malliavin operator D is an unbounded operator from $\mathcal{G}_\alpha$ to $L_2(I, \mathcal{G}_\alpha)$. However, if D is restricted to the subspace $M_\alpha \subset \mathcal{G}_\alpha$ consisting of those $f \in \mathcal{G}_\alpha$ for which the Fourier-Wiener Kernels $\{K_n\}$ are such that $R \equiv \{R_n \equiv \sqrt{n} K_n, n \geq 0\} \in G_\alpha$, then

$$\int_I |(Df)(t)|_{\mathcal{G}_\alpha}^2 = \sum_{n \geq 1} n! \parallel \sqrt{n} K_n \parallel_{L_2(I^n)}^2$$

$$= \sum_{n \geq 1} n! \parallel R_n \parallel_{L_2(I^n)}^2 = \parallel R \parallel_{G_\alpha}^2 < \infty.$$

Thus the domain $\mathcal{D}(D) \neq \emptyset$. Now we show that it is dense in $\mathcal{G}_\alpha$. Consider the subspace $G_{\alpha,m} \subset G_\alpha$ where

$$G_{\alpha,m} \equiv \{K \in G_\alpha : K_n = 0 \, \forall \, n \geq m + 1\}.$$

Then, it is clear that the isomorphic image $\Phi_\alpha(G_{\alpha,m})(\subset \mathcal{G}_\alpha)$ is in the domain of the operator D for every finite $m \in N$. From this it is easy to see that the domain $\mathcal{D}(D)$ is dense in $\mathcal{G}_\alpha$. Now considering the larger space, we prove continuity of the map

$$D : \mathcal{G}_\alpha \longrightarrow L_2(I, \mathcal{H}_\alpha).$$

Note that for almost all $t \in I$, $(Df)(t) \in \mathcal{H}_\alpha$. Indeed, using the expression for Df and the duality pairing in $\mathcal{H}_\alpha$, we obtain

$$
\begin{aligned}
|(Df)(t)|^2_{\mathcal{H}_\alpha} &= \sum_{n \geq 1} (1/n!)n^2 < g_{n-1}(K_n(t, \cdot)), g_{n-1}(K_n(t, \cdot)) > \\
&= \sum (n/n!)n! \parallel K_n(t, \cdot) \parallel^2_{L_2(I^{n-1})} \\
&\leq \sum_{n \geq 1} n! \parallel K_n(t, \cdot) \parallel^2_{L_2(I^{n-1})},
\end{aligned}
\tag{9.3}
$$

for almost all $t \in I$. Integrating this over the interval I, it follows from the isometric isomorphism between G_α and $\mathcal{G}_\alpha$ that

$$\int_I |(Df)(t)|^2_{\mathcal{H}_\alpha} dt \leq \sum_{n \geq 1} n! \parallel K_n \parallel^2_{L_2(I^n)} \leq \parallel f \parallel^2_{\mathcal{G}_\alpha}.$$

In other words, $Df \in L_2(I, \mathcal{H}_\alpha)$ for every $f \in \mathcal{G}_\alpha$ and hence $D \in \mathcal{L}(\mathcal{G}_\alpha, L_2(I, \mathcal{H}_\alpha)) \subset \mathcal{L}(\mathcal{G}_\alpha, L_2(I, \mathcal{G}^*_\alpha))$ with bound equal or less than 1, that is, it is a nonexpansive operator. This completes the proof. •

In general we can show that the higher order Malliavin derivatives exist and they are unbounded operators on $\mathcal{G}_\alpha$.

Proposition 9.2.2 *For $f \in \mathcal{G}_\alpha$, having the Wiener-Itô expansion*

$$f = \mathbb{E}(f) + \sum_{n \geq 1} g_n(K_n),$$

the k-th order Malliavin derivative is given by

$$(D^k f)(t_1, t_2, \cdots, t_k) = \sum_{n \geq k} \frac{n!}{(n-k)!} \, g_{n-k}(K_n(t_1, t_2, \cdots, t_k; \cdot)),$$

$\forall \, k \leq n, t_k \in I$. *The operator D^k is an unbounded operator from $\mathcal{G}_\alpha$ to $L_2(I^k, \mathcal{G}_\alpha)$ while it is a bounded linear operator from $\mathcal{G}_\alpha$ to $L_2(I^k, \mathcal{H}_\alpha)$.*

Proof The expression for $D^k f$ follows from direct computation. It is not difficult to show that for $f \in \mathcal{D}(D^k)$,

$$
\begin{aligned}
\int_{I^k} \parallel D^k f \parallel^2_{\mathcal{G}_\alpha} dt_1 \cdots dt_k &= \int_{I^k} \mathbb{E}|(D^k f)(t_1 t_2 \cdots t_k)|^2 dt_1 dt_2 \cdots dt_k \\
&= \sum_{n \geq k} \frac{(n!)^2}{(n-k)!} \parallel K_n \parallel^2_{L_2(I^n)}.
\end{aligned}
$$

Thus the domain of D^k is given by

$$\mathcal{D}(D^k) \equiv \{ f \in \mathcal{G}_\alpha : \| D^k f \|^2_{L_2(I^k, \mathcal{G}_\alpha)} = \sum_{n \geq k} \frac{(n!)^2}{(n-k)!} \| K_n \|^2_{L_2(I^n)} < \infty \}.$$

Noting that $\| f \|^2_{\mathcal{G}_\alpha} = \sum_{n \geq 0} n! \| K_n \|^2_{L_2(I^n)}$, it follows from the above expression that, in general, $D^k f \notin L_2(I^k, \mathcal{G}_\alpha)$. Thus the operator D^k is an unbounded operator from $\mathcal{G}_\alpha$ to $L_2(I^k, \mathcal{G}_\alpha)$. This proves the first statement. For the proof of the second statement, let us consider the larger space $\mathcal{H}_\alpha$ and compute the corresponding norm. Accordingly, we have

$$\int_{I^k} < D^k f, D^k f >_{\mathcal{H}_\alpha} dt_1 \cdots dt_k = \sum_{n \geq k} \left(\frac{n!}{(n-k)!} \right)^2 \| K_n \|^2_{L_2(I^n)}$$

$$= \sum_{n \geq k} \left(\frac{n!}{((n-k)!)^2} \right) n! \| K_n \|^2_{L_2(I^n)} .$$

Define $\xi(n) = n!/((n-k)!)^2$ for $n \geq k$. The reader can easily verify that there exists a finite positive number M_k such that

$$\sup\{\xi(n), n \geq k\} \leq M_k.$$

Hence, it follows from the previous equality that

$$\int_{I^k} < D^k f, D^k f >_{\mathcal{H}_\alpha} dt_1 \cdots dt_k \leq M_k \sum_{n \geq k} n! \| K_n \|^2_{L_2(I^n)} \leq M_k \| f \|^2_{\mathcal{G}_\alpha} .$$

Clearly, this shows that the operator

$$D^k : \mathcal{G}_\alpha \longrightarrow L_2(I^k, \mathcal{H}_\alpha)$$

is a bounded linear operator. This completes the proof. •

Next, we consider the divergence operator δ. In the following lemma we show that it is an unbounded operator from $L_2(I, \mathcal{G}_\alpha)$ to $\mathcal{G}_\alpha$. On the other hand, considering the larger space $\mathcal{H}_\alpha \supset \mathcal{G}_\alpha$, it is a bounded operator from $L_2(I, \mathcal{G}_\alpha)$ to $\mathcal{H}_\alpha$. Thus the divergence operator δ has similar behavior as the Malliavin differential operator D.

Lemma 9.2.3 *The divergence operator δ is an unbounded linear operator from $L_2(I, \mathcal{G}_\alpha)$ to $\mathcal{G}_\alpha$. Its domain is dense and is given by*

$$\mathcal{D}(\delta) \equiv \Big\{ \eta \in L_2(I, \mathcal{G}_\alpha) : \eta(t) = \sum_{n \geq 0} g_n(K_n(t, \cdot)) \ \ with \ \{K_n\} \ satisfying$$

$$\sum_{n \geq 0} (n+1)! \int_I \| K_n(t, \cdot) \|^2_{L_2(I^n)} dt < \infty \Big\}.$$

Further, δ is a bounded linear operator from $L_2(I, \mathcal{G}_\alpha)$ to $\mathcal{H}_\alpha$.

Proof First we consider an elementary functional of the form

$$\eta_n(t) \equiv g_n(K_n(t, \cdot)), K_n \in \hat{L}_2(I^{n+1}), n \in N, \tag{9.4}$$

where $\{g_n\}$ are the orthogonal functionals of Brownian motion as seen in Chapter 2, section 2.2. For convenience of the reader, we repeat

$$\eta_n(t) \equiv g_n(K_n(t, \cdot)) \equiv \int_{I^n} K_n(t; \tau_1, \tau_2, \cdots, \tau_n) dw(\tau_1) \cdots dw(\tau_n), t \in I.$$

Clearly,

$$\| \eta_n \|^2_{L_2(I, \mathcal{G}_\alpha)} = n! \| K_n \|^2_{L_2(I^{n+1})}.$$

Let g be any element of $\mathcal{G}_\alpha$ having the kernel representation $g \equiv \sum_{r=0}^\infty g_r(L_r)$ for some $L \in G_\alpha$. Then by duality, we have

$$< \delta\eta_n, g >_{\mathcal{G}_\alpha} = < \eta_n, Dg > = < \eta_n, \sum_{r \geq 1} Dg_r(L_r) >$$

$$= \int_I < g_n(K_n(t, \cdot)), \sum_{r \geq 1} r g_{r-1}(L_r(t, \cdot)) > dt$$

$$= \int_I < g_n(K_n(t, \cdot)), (n+1) g_n(L_{n+1}(t, \cdot)) > dt$$

$$= \int_I (n+1) n! < K_n(t, \cdot), L_{n+1}(t, \cdot) >_{L_2(I^n)} dt$$

$$= \int_I < K_n(t, \cdot), ((n+1)!) L_{n+1}(t, \cdot) >_{L_2(I^n)} dt. \tag{9.5}$$

The fourth line follows from the fact that the set $\{g_n\}$ is orthogonal. Set $K_n \equiv \tilde{K}_{n+1}$ and note that this is a function of $n+1$ arguments. Using this notation in equation (9.5) and the orthogonality property of $\{g_n\}$, the reader can easily verify that

$$< \delta\eta_n, g >_{\mathcal{G}_\alpha} = (n+1)! \int_I < \tilde{K}_{n+1}(t, \cdot), L_{n+1}(t, \cdot) > dt$$

$$= < g_{n+1}(\tilde{K}_{n+1}), g_{n+1}(L_{n+1}) >$$

$$= < g_{n+1}(\tilde{K}_{n+1}), \sum_{r \geq 0} g_r(L_r) > = < g_{n+1}(\tilde{K}_{n+1}), g > \tag{9.6}$$

This is true for all $g \in \mathcal{G}_\alpha = L_2(\Omega)$ and hence

$$\delta\eta_n = g_{n+1}(\tilde{K}_{n+1}). \tag{9.7}$$

Thus, the divergence operator δ transforms an n-th order chaos into $n+1$-st order chaos. Clearly,

$$\| \delta\eta_n \|_{\mathcal{G}_\alpha}^2 = (n+1)! \, \| \tilde{K}_{n+1} \|_{L_2(I^{n+1})}^2 = (n+1)! \, \| K_n \|_{L_2(I^{n+1})}^2 \, .$$

Now consider the functional

$$\eta(t) = \sum_{n\geq 0} \eta_n(t) \equiv \sum_{n\geq 0} g_n(K_n(t,\cdot)) \tag{9.8}$$

with $K_n(t,\cdot) \in \hat{L}_2(I^n)$ for almost all $t \in I$ and $\int_I \| K_n(t,\cdot) \|_{L_2(I^n)}^2 \, dt < \infty$. In other words, $K_n \in L_2(I^{n+1})$. Clearly,

$$\| \eta(t) \|_{\mathcal{G}_\alpha}^2 = \sum_{n\geq 0} n! \, \| K_n(t,\cdot) \|_{L_2(I^n)}^2$$

and

$$\int_I \| \eta(t) \|_{\mathcal{G}_\alpha}^2 \, dt = \sum_{n\geq 0} n! \int_I \| K_n(t,\cdot) \|_{L_2(I^n)}^2 \, dt$$

$$= \sum_{n\geq 0} n! \, \| \tilde{K}_{n+1} \|_{L_2(I^{n+1})}^2 < \infty. \tag{9.9}$$

In other words, $\eta \in L_2(I,\mathcal{G}_\alpha) = L_2(\Omega, H)$. Now applying the divergence operator δ on the random variable η given by (9.8), it follows from (9.7) that

$$\delta\eta = \sum_{n\geq 0} g_{n+1}(\tilde{K}_{n+1}). \tag{9.10}$$

Then evaluating its $\mathcal{G}_\alpha$ norm we obtain

$$\| \delta\eta \|_{\mathcal{G}_\alpha}^2 = \sum_{n\geq 0} < g_{n+1}(\tilde{K}_{n+1}), g_{n+1}(\tilde{K}_{n+1}) >$$

$$= \sum_{n\geq 0} (n+1)! \, \| \tilde{K}_{n+1} \|_{L_2(I^{n+1})}^2 \, . \tag{9.11}$$

In general the operator δ is unbounded. This follows from the fact that, for any given positive number r, one can construct an $\eta \in L_2(I,\mathcal{G}_\alpha)$ of norm one satisfying $\pi_n(\eta(t)) = 0$, for all $t \in I$, and all $n \leq [r]$, such that

$$\| \delta\eta \|_{\mathcal{G}_\alpha}^2 \geq (1 + [r])$$

where $[r]$ denotes the largest integer not exceeding r. Thus, the operator δ is bounded on $L_2(I,\mathcal{G}_\alpha)$ if and only if the series (9.11) converges. That the domain is dense, follows from the facts that every $\eta \in L_2(I, \mathcal{G}_\alpha)$ has the representation as given in the definition of the domain $\mathcal{D}(\delta)$ and, for every finite integer N, $\eta_N(t) \equiv \sum_{n \geq 0}^{N} g_n(K_n(t, \cdot)) \in D(\delta)$ and that it converges to an element of $L_2(I, \mathcal{G}_\alpha)$. This proves the first assertion. For the second, we compute the $\mathcal{H}_\alpha$ norm of $\delta(\eta)$ giving

$$\| \delta(\eta) \|_{\mathcal{H}_\alpha}^2 = \sum_{n \geq 0} (1/(n+1)!) < g_{n+1}(\tilde{K}_{n+1}), g_{n+1}(\tilde{K}_{n+1}) >$$

$$= \sum_{n \geq 0} (1/n!)n! \int_I \| K_n(t, \cdot) \|_{L_2(I^n)}^2 \, dt$$

$$\leq (1/\Gamma(1)) \| \eta \|_{L_2(I, \mathcal{G}_\alpha)}^2 = \| \eta \|_{L_2(I, \mathcal{G}_\alpha)}^2 < \infty. \tag{9.12}$$

This shows that $\delta \in \mathcal{L}(L_2(I, \mathcal{G}_\alpha), \mathcal{H}_\alpha)$ proving the second assertion. •

It is interesting to observe that δ is in fact a nonexpansive operator from $L_2(I, \mathcal{G}_\alpha)$ to $\mathcal{H}_\alpha$.

Now we consider the Ornstein-Uhlenbeck operator $L = -\delta D$. This is an unbounded linear operator in $\mathcal{G}_\alpha$, but it is a bounded operator from $\mathcal{G}_\alpha$ to the larger space $\mathcal{H}_\alpha$ as shown in the following lemma.

Let π_n denote the projection of $\mathcal{G}_\alpha$ into the space $\mathcal{G}_{\alpha,n}$ of homogeneous chaos of degree n.

Theorem 9.2.4   *The operator L with domain $\mathcal{D}(L) \subset \mathcal{G}_\alpha$ has the simple structure*

$$Lf = \sum_{n \geq 1} -n \, \pi_n(f).$$

It is an unbounded linear operator in $\mathcal{G}_\alpha$ with dense domain. In contrast, it is a bounded linear operator from $\mathcal{G}_\alpha$ to $\mathcal{H}_\alpha$.

Proof Let $f \in \mathcal{G}_\alpha$. Then by virtue of the isometric isomorphism Φ_α, there exists a $K \in G_\alpha$ such that $f = \Phi_\alpha(K)$. For any integer m define

$$h_m \equiv \sum_{n=0}^{m} \pi_n(f) = \sum_{n=0}^{m} g_n(K_n).$$

Clearly,

$$(Dh_m)(t) = \sum_{n=1}^{m} n g_{n-1}(K_n(t, \cdot))$$

and hence it follows from Lemma 9.2.3, in particular the expression (9.7), that

$$L(h_m) = -\delta D(h_m) = \sum_{n=1}^{m}(-n)\delta(g_{n-1}(K_n(t,\cdot)))$$

$$= \sum_{n=1}^{m}(-n)g_n(K_n) = \sum_{n=1}^{m}(-n)\pi_n(f).$$

From this it is clear that the operator L has the structure as stated. It follows from this that the domain of the operator L is given by

$$\mathcal{D}(L) \equiv \left\{ f \in \mathcal{G}_\alpha : \| L(f) \|_{\mathcal{G}_\alpha}^2 \right.$$

$$\left. = \sum_{n\geq 1} n^2 (n!) \| K_n \|_{L_2(I^n)}^2 = \sum_{n\geq 1} n^2 \| \pi_n(f) \|_{\mathcal{G}_{\alpha,n}}^2 < \infty \right\}.$$

$$(9.13)$$

By the isometry Φ_α, $\| f \|_{\mathcal{G}_\alpha}^2 = \| K \|_{G_\alpha}^2 = \sum(n!) \| K_n \|_{L_2(I^n)}^2 < \infty$. Clearly, it follows from this that the infinite series in the above expression will not converge for every $f \in \mathcal{G}_\alpha$. This shows that $\mathcal{D}(L)$ is a proper subspace of $\mathcal{G}_\alpha$ implying that L is an unbounded operator. On the other hand, it is clear that $h_m \in \mathcal{D}(L)$ for every integer m and that $h_m \xrightarrow{s} f$ in $\mathcal{G}_\alpha$. Hence the domain of L is a nonempty dense subset of $\mathcal{G}_\alpha$. This proves the first part. For the second part, consider the expression

$$L(f) = \sum_{n\geq 1}(-n)\pi_n(f) = \sum_{n\geq 1} g_n(-nK_n).$$

By use of the norm topology of $\mathcal{H}_\alpha$ and the isometric isomorphism Φ_α, we obtain

$$\| L(f) \|_{\mathcal{H}_\alpha}^2 = \sum(1/n!) < g_n(-nK_n), g_n(-nK_n) >$$

$$= \sum_{n\geq 1}(n^2/n!)n! \| K_n \|_{L_2(I^n)}^2$$

$$\leq \sup\{(n^2/n!), n \in N\}\left(\sum_{n\geq 1} n! \| K_n \|_{L_2(I^n)}^2\right)$$

$$\leq \sup\{(n^2/n!), n \in N\} \| K \|_{G_\alpha}^2$$

$$= \sup\{(n^2/n!), n \in N\} \| f \|_{\mathcal{G}_\alpha}^2.$$

$$(9.14)$$

Hence,

$$\| L(f) \|_{\mathcal{H}_\alpha} \leq \sqrt{2} \, \| f \|_{\mathcal{G}_\alpha} \tag{9.15}$$

for all $f \in \mathcal{G}_\alpha$ and therefore $L \in \mathcal{L}(\mathcal{G}_\alpha, \mathcal{H}_\alpha)$, the space of bounded linear operators from $\mathcal{G}_\alpha$ to $\mathcal{H}_\alpha$. This completes the proof. •

We have already seen in Chapter 8 that L generates a C_0-semigroup of contractions on $L_p(\Omega)$ spaces for all $p \geq 1$. Here, in the space $\mathcal{G}_\alpha \equiv L_2(\Omega)$ the semi group has a very simple structure. We state this in the following corollary. For convenience and in conformity with standard notations used in general semigroup theory [Ahmed (1991)], we use the notation $T(t)$ instead of T_t.

Corollary 9.2.5 *The Ornstein-Uhlenbeck operator L generates a contraction semigroup in the Hilbert space $\mathcal{G}_\alpha$ and it is exponentially stable.*

Proof Define

$$T(t)f \equiv \sum_{n \geq 1} e^{-nt} \pi_n(f).$$

For every $f \in \mathcal{D}(L)$, it follows from the above expression that

$$\lim_{t \downarrow 0} \frac{T(t)f - f}{t} = Lf$$

in the norm topology of $\mathcal{G}_\alpha$. It is now easy to verify that

$$d/dt(T(t)f) = T(t)(Lf) = L(T(t)f)$$

for all $f \in D(L)$. Further, for every $f \in \mathcal{G}_\alpha$ we have

$$\| T(t)f \|_{\mathcal{G}_\alpha} \leq e^{-t} \| f \|_{\mathcal{G}_\alpha}, t \geq 0.$$

This completes the proof. •

Using the Ornstein-Uhlenbeck operator L we can study very simple evolution equations on $\mathcal{G}_\alpha$.

Lemma 9.2.6 *Consider the following differential equation on the Fock space $\mathcal{G}_\alpha$*

$$(d/dt)\varphi = L\varphi + h(t), \varphi(0) = \varphi_0, t \in I \equiv [0, b], b < \infty, \tag{9.16}$$

where L is the Ornstein-Uhlenbeck operator. Then, for each $\varphi_0 \in \mathcal{G}_\alpha$ and $h \in L_1(I, \mathcal{G}_\alpha)$, the evolution equation has a unique mild solution $\varphi \in C(I, \mathcal{G}_\alpha)$.

Proof Since the Fock space $\mathcal{G}_\alpha$ is a Hilbert space, the proof follows from general semigroup theory [Ahmed (1991)] on Banach spaces. The solution is given by the well known variation of constants formula,

$$\varphi(t) = T(t)\varphi_0 + \int_0^t T(t-s)h(s)ds, t \in I,$$

where $T(t), t \geq 0$, is the semigroup generated by the operator L. The fact that $t \longrightarrow \varphi(t)$ is continuous follows from the Bochner integrability of h and the strong continuity of the semigroup T. This completes the proof. ●

This result can be easily extended to semilinear evolution equations on $\mathcal{G}_\alpha$.

Theorem 9.2.7 *Consider the following semilinear evolution equation on $\mathcal{G}_\alpha$*

$$(d/dt)\varphi = L\varphi + h(t, \varphi), \varphi(0) = \varphi_0, t \in I \equiv [0, b] \tag{9.17}$$

where L is the Ornstein-Uhlenbeck operator and $h : I \times \mathcal{G}_\alpha \longrightarrow \mathcal{G}_\alpha$ is measurable in the first variable and locally Lipschitz in the second argument and there exists $\ell \in L_1^+(I)$ such that

$$|h(t,z)|_{\mathcal{G}_\alpha} \leq \ell(t)\big(1 + |z|_{\mathcal{G}_\alpha}\big).$$

Then, for each $\varphi_0 \in \mathcal{G}_\alpha$, the evolution equation (9.17) has a unique mild solution $\varphi \in C(I, \mathcal{G}_\alpha)$.

Proof The proof follows from general semigroup theory [see Ahmed [Ahmed (1991)], Theorem 5.2.3]. We present only an outline. Let $T(t), t \geq 0$, denote the Ornstein-Uhlenbeck semigroup. Under the given assumption, using the integral equation

$$\varphi(t) = T(t)\varphi_0 + \int_0^t T(t-s)h(s, \varphi(s))ds, t \in I, \tag{9.18}$$

one can easily derive an a-priori estimate and using this estimate one can then construct a closed bounded set $\Gamma \subset C(I, \mathcal{G}_\alpha)$. Then, define the operator Υ by

$$\Upsilon(\varphi)(t) \equiv T(t)\varphi_0 + \int_0^t T(t-s)h(s, \varphi(s))ds, t \in I,$$

and consider the fixed point problem $\varphi = \Upsilon(\varphi)$ on the set Γ. Using this operator one can prove that, for sufficiently large integer n, the n-th iterate of Υ is a contraction and hence by Banach fixed point theorem it has a unique fixed point on Γ. It is then shown that this is also the unique fixed point of the operator Υ itself. This fixed point is also the unique solution of the integral equation (9.18) and hence the unique mild solution of the evolution equation (9.17). This completes our outline. •

9.3   Evolution Equations on Abstract Fock Spaces

In Chapters 2 and 3 we have constructed various Fock spaces, such as $\{\mathcal{G}_\alpha, \mathcal{G}_\beta, \mathcal{G}_\gamma, \mathcal{G}_\delta, \mathcal{G}_\varpi, \mathcal{G}_\varphi, \mathcal{G}_\Pi\}$ which are isometrically isomorphic to the Hilbert spaces $\{G_\alpha, G_\beta, G_\gamma, G_\delta, G_\varpi, G_\varphi, G_\Pi\}$. We have also considered their duals $\{\mathcal{G}_\alpha^*, \mathcal{G}_\beta^*, \mathcal{G}_\gamma^*, \mathcal{G}_\delta^*, \mathcal{G}_\varpi^*, \mathcal{G}_\varphi^*, \mathcal{G}_\Pi^*\}$ which are isomorphic (isometric) images of the duals $\{G_\alpha^*, G_\beta^*, G_\gamma^*, G_\delta^*, G_\varpi^*, G_\varphi, G_\Pi\}$. These spaces are constructed around the pivot spaces $\{H_\alpha, H_\beta, H_\gamma, H_\delta, H_\varpi, H_\varphi, H_\Pi\}$ and their isomorphic images $\{\mathcal{H}_\alpha, \mathcal{H}_\beta, \mathcal{H}_\gamma, \mathcal{H}_\delta, \mathcal{H}_\varpi, \mathcal{H}_\varphi, \mathcal{H}_\Pi\}$.

We wish to consider evolution equations on these spaces and also the spaces introduced in Chapter 7 such as $\{\mathcal{Y}_p, \mathcal{Y}_q\}$ and $\mathcal{Z}_b, \mathcal{Z}_b^*$.

To cover the first set of spaces we remove the indices and denote them by $\{G, H, G^*\}$ and their isomorphic images by $\{\mathcal{G}, \mathcal{H}, \mathcal{G}^*\}$, and recall the following results from Chapter 3. Also, we remove the indices of the isometric maps introduced there by one symbol Φ. We recall that Φ has continuous extension from G to H as well as G^*. Thus, we shall freely use Φ also for the extension without further notice. According to Theorem 3.2.2 (See also Theorems 3.3.2, 3.4.2, 3.5.2), Φ is an isometric isomorphism between G and $\mathcal{G}$.

Lemma 9.3.1 *The triple G, H, G^*, as defined above, are all Hilbert spaces and the embeddings $G \hookrightarrow H \hookrightarrow G^*$ are continuous and dense.*

Writing $\mathcal{H} \equiv \Phi(H)$ and $\mathcal{G}^* \equiv \Phi(G^*)$ we have the following result.

Theorem 9.3.2 *The class $\mathcal{G}$ is the space of regular functionals on Wiener measure space while $\mathcal{H}$ and $\mathcal{G}^*$ are generalized functionals with $\mathcal{G}$ being the* *test functionals and they satisfy the following diagram:*

$$G \hookrightarrow H \hookrightarrow G^*$$

$$\updownarrow \qquad \updownarrow \qquad \updownarrow$$

$$\mathcal{G} \hookrightarrow \mathcal{H} \hookrightarrow \mathcal{G}^*, \tag{9.19}$$

with continuous and dense embeddings where $\hookrightarrow$ stands for inclusion $\updownarrow$ for the isometric isomorphism $\cong$.

Naturally, for $\phi^* \in \mathcal{G}^*$, it's norm is defined by

$$\| \phi^* \|_{\mathcal{G}^*} \equiv \operatorname{Sup}\{< \phi^*, \phi >_{\mathcal{G}^*, \mathcal{G}}, \| \phi \|_{\mathcal{G}} = 1\}.$$

We recall the practical significance of this result as follows. In case a given functional f on the Wiener measure space fails to have finite energy, that is, $\| f \|_{\mathcal{G}} = +\infty$, it means, f does not have standard Wiener-Itô decomposition. It may very well be an element of the dual $\mathcal{G}^*$ having an elevated energy measured in terms of its Hilbertian norm. In the preceding section, we have seen this phenomenon in the case of the Malliavin operators D and δ.

We are interested in the linear evolution equations of the form :

$$(d/dt)\phi = \mathcal{A}\phi + f \quad \text{for} \ \ t \geq 0, \tag{9.20}$$

on the Fock spaces $\mathcal{G}, \mathcal{H}, \mathcal{G}^*$ as discussed above. Let a be a mapping satisfying $a : N \longrightarrow \overline{R}$ and consider the operator $\mathcal{A}$ given by

$$\mathcal{A}\phi \equiv \sum_{n=0}^{n=\infty} a(n) \ \pi_n(\phi). \tag{9.21}$$

The domain of $\mathcal{A}$ is given by

$$\mathcal{D}(\mathcal{A}) \equiv \{\phi \in \mathcal{G} : \sum_{n=0}^{n=\infty} (a(n))^2 \ \| \pi_n\phi \|_{\mathcal{G}}^2 < \infty\}.$$

For $a(n) = -(n)$, this operator coincides with the Ornstein-Uhlenbeck operator L and it also appears in Quantum physics. And there it is known as the number operator [see Bell, [Bell (1987)] ,p21]. In Bell [Bell (1987)]

$a(n) = -(n/2)$ instead of $-n$ because the SDE that he chooses for the Ornstein-Uhlenbeck process is given by

$$dX_t = -(1/2)X_t dt + dw(t).$$

We show that under very general assumptions on $a(n), n \geq 0$, this operator generates a C_0-semigroup in $\mathcal{G}$.

Theorem 9.3.3 *Suppose there exists a finite (possibly nonnegative) real number ω_0 such that the sequence $\{a(n)\}$ satisfies*

$$\sup_{n \geq 0} a(n) \leq \omega_0$$
$$\lim_{n \to \infty} a(n) = -\infty, \qquad (9.22)$$

and that it is finite for each finite n and assumes positive values at most for finitely many n. Then $\mathcal{A}$ generates a C_0-semigroup $\{T(t), t \geq 0\}$ of quasi contractions in $\mathcal{G}$ in the sense that for $\omega > \omega_0$,

$$\| T(t) \|_{\mathcal{L}(\mathcal{G})} \leq e^{\omega t} \quad \text{for all } t \geq 0.$$

Proof For detailed proof see Ahmed [Ahmed (1994)]. We present a brief outline. In the proof we show that (i): $\overline{\mathcal{D}(\mathcal{A})} = \mathcal{G}$, (ii): $\mathcal{A}$ is closed, and (iii): $\| (\lambda I - \mathcal{A})^{-1} \|_{\mathcal{L}(\mathcal{G})} \leq (1/(\lambda - \omega))$ for all $\lambda > \omega$. Hence, the conclusion follows from Hille-Yosida theorem see [Pazy (1983)]], [Ahmed (1991)], Corollary 2.2.11, p30]. •

Remark 9.3.4 For the Quantum mechanical number operator we have $a(n) = -(n)$. Hence, in this case $T(t), t \geq 0$, is a contraction semigroup and it is the Ornstein-Uhlenbeck semigroup we have already seen. In other words, the semigroup generated by $\mathcal{A}$ covers the Ornstein-Uhlenbeck semigroup as a special case.

The following result is an immediate consequence of Theorem 9.3.3 and it generalizes the result of Lemma 9.2.6.

Theorem 9.3.5 *Consider the Cauchy problem on $\mathcal{G}$:*

$$(d/dt)\phi = \mathcal{A}\phi + f \quad \text{for } t \in I \equiv [0, T],$$
$$\phi(0) = \phi_0, \qquad (9.23)$$

where the operator $\mathcal{A}$ satisfies the assumptions of Theorem 9.3.3. Then, for every $\phi_0 \in \mathcal{G}$ and $f \in L_1(I, \mathcal{G})$, the Cauchy problem (9.23) has a unique mild solution $\phi \in C(I, \mathcal{G})$ and it is given by :

$$\phi(t) = T(t)\phi_0 + \int_0^t T(t-s)f(s)ds, t \in I. \tag{9.24}$$

Proof Follows from Theorem 9.3.3 and the variation of constants formula.

9.4 Evolution Equations Determined by Coercive Operators on Fock Spaces

In this section we introduce a general class of evolution equations on the abstract spaces $\{\mathcal{G}, \mathcal{H}, \mathcal{G}^*\}$. Recall that $\mathcal{G}$ is the Hilbert space of regular functionals on the Wiener measure space, while $\mathcal{H}$ and $\mathcal{G}^*$ are the spaces of generalized functionals covering $\mathcal{G}$.

For $g \in \mathcal{G}^*$ and $f \in \mathcal{G}$ we write the duality product as $< g, f >_{\mathcal{G}^*,\mathcal{G}}$. In case $g \in \mathcal{H}$, this duality product reduces to the scalar product in $\mathcal{H}$ and it is denoted by $(g, f)_{\mathcal{H}}$. For convenience of notation, the norm in $\mathcal{H}$ will be denoted by $|\cdot|_{\mathcal{H}}$.

Let $\mathcal{A}$ be a linear operator from $\mathcal{G}$ to $\mathcal{G}^*$ satisfying the following properties:

(P1): there exists a constant $c > 0$ such that

$$|< \mathcal{A}\varphi, \psi >_{\mathcal{G}^*,\mathcal{G}}| \le c \parallel \varphi \parallel_{\mathcal{G}} \parallel \psi \parallel_{\mathcal{G}} \quad \forall \ \varphi, \psi \in \mathcal{G}.$$

(P2): there exist $\lambda \ge 0$ and $\nu > 0$ such that

$$< \mathcal{A}\varphi, \varphi >_{\mathcal{G}^*,\mathcal{G}} + \lambda |\varphi|_{\mathcal{H}}^2 \ge \nu \parallel \varphi \parallel_{\mathcal{G}}^2, \quad \forall \ \varphi \in \mathcal{G}.$$

In general, operators satisfying the property **(P2)** are said to be coercive.

We wish to consider evolution equations described by

$$(d/dt)\phi + \mathcal{A}\phi = f \quad \text{for} \ t \in I \equiv [0, T],$$

$$\phi(0) = \phi_0, \tag{9.25}$$

with initial state $\phi_0 \in \mathcal{H}$ and $f \in L_2(I, \mathcal{G}^*)$. We introduce the following definition.

Definition 9.4.1 *An element $\phi \in L_2(I, \mathcal{G}) \cap C(I, \mathcal{H})$ is said to be a weak solution (or distribution solution) of the Cauchy problem (9.25) if, for each $\zeta \in \mathcal{G}$, the following identity holds*

$$< \dot{\phi}(t) + \mathcal{A}\phi(t) - f(t), \zeta >_{\mathcal{G}^*, \mathcal{G}} = 0$$

for almost all $t \in I$ and $\phi(0) = \phi_0$.

We present the following general result.

Theorem 9.4.2 *Consider the evolution equation given by (9.25) and suppose the operator $\mathcal{A}$ satisfy the properties (P1) and (P2). Then, for every $\phi_0 \in \mathcal{H}$ and $f \in L_2(I, \mathcal{G}^*)$, equation (9.25) has a unique weak solution ϕ satisfying the following properties:*

(1): $\phi \in L_\infty(I, \mathcal{H}) \cap L_2(I, \mathcal{G}) \cap C(I, \mathcal{H})$; (2): $\dot{\phi} \in L_2(I, \mathcal{G}^*)$; and (3): the map $(\phi_0, f) \longrightarrow \phi$ is continuous (even Lipschitz).

Proof The basic technique of proof is based on Galerkin approach [see Ahmed and Teo, [Ahmed (1981)], Theorem 5.1.1, p278]. For convenience of the reader we present it briefly. We use the complete orthogonal system $\{g_n\}$ or any other system that is complete in the class $L_2(\Omega, \mu^w) = \mathcal{G}$ and project the infinite dimensional system to an increasing family of finite dimensional systems of ordinary differential equations determined by the finite dimensional subspaces $\mathcal{G}_n \equiv span\{g_k, 0 \leq k \leq n\}$. This is the well known Galerkin projection method. Existence and regularity properties of solutions for the finite dimensional systems follow from classical results. This sequence of solutions is then shown to converge in various weak topologies to a weak solution of the problem. This requires a-priori bounds. We present these and the necessary steps leading to the proof of existence and regularity properties of the solutions. By scalar multiplying equation (9.25) on both sides by ϕ and considering $\mathcal{G}^* - \mathcal{G}$ duality pairing and integrating by parts we obtain

$$|\phi(t)|_{\mathcal{H}}^2 + 2\nu \int_0^t \| \phi(s) \|_{\mathcal{G}}^2 \, ds \leq |\phi_0|_{\mathcal{H}}^2 + 2\lambda \int_0^t |\phi(s)|_{\mathcal{H}}^2 ds$$

$$+ 2 \int_0^t \| f(s) \|_{\mathcal{G}^*} \| \phi(s) \|_{\mathcal{G}} \, ds. \tag{9.26}$$

Using Cauchy-Schwartz inequality for the last term, it is easy to verify that for each $t \in I$,

$$|\phi(t)|_{\mathcal{H}}^2 + \nu \int_0^t \parallel \phi(s) \parallel_{\mathcal{G}}^2 ds \leq |\phi_0|_{\mathcal{H}}^2 + 2\lambda \int_0^t |\phi(s)|_{\mathcal{H}}^2 ds$$
$$+ (1/\nu) \int_0^t \parallel f(s) \parallel_{\mathcal{G}^*}^2 ds. \qquad (9.27)$$

Then, it follows from Gronwall inequality that for all $t \in I$

$$|\phi(t)|_{\mathcal{H}}^2 \leq \left\{ |\phi_0|_{\mathcal{H}}^2 + (1/\nu) \int_I \parallel f(s) \parallel_{\mathcal{G}^*}^2 ds \right\} \exp\{2\lambda T\} \equiv C_T, \qquad (9.28)$$

with $C_T < \infty$. Using this estimate in (9.27) we obtain

$$\int_0^T \parallel \phi(s) \parallel_{\mathcal{G}}^2 ds \leq (1/\nu) \left(1 + 2\lambda T \exp 2\lambda T \right) \times$$
$$\left\{ |\phi_0|_{\mathcal{H}}^2 + (1/\nu) \int_0^T \parallel f(s) \parallel_{\mathcal{G}^*}^2 ds \right\}. \qquad (9.29)$$

It follows from the property (P1) that the operator $\mathcal{A} \in \mathcal{L}(\mathcal{G}, \mathcal{G}^*)$ with the bound given by the number c. Hence, the distributional derivative of ϕ satisfies the following estimate

$$\parallel \dot{\phi} \parallel_{L_2(I,\mathcal{G}^*)} \leq c \parallel \phi \parallel_{L_2(I,\mathcal{G})} + \parallel f \parallel_{L_2(I,\mathcal{G}^*)}. \qquad (9.30)$$

Thus every ϕ, that satisfies the first identity of equation (9.25) in the distribution sense, must necessarily satisfy the (a-priori) estimates (9.28), (9.29) and (9.30). Let $\mathcal{W}$ denote the vector space described by

$$\mathcal{W} \equiv \{ \varphi : \varphi \in L_2(I,\mathcal{G}) \text{ and } \dot{\varphi} \in L_2(I,\mathcal{G}^*) \}.$$

Endowed with the norm topology,

$$\parallel \varphi \parallel_{\mathcal{W}} \equiv \left(\parallel \varphi \parallel_{L_2(I, \mathcal{G})}^2 + \parallel \dot{\varphi} \parallel_{L_2(I, \mathcal{G}^*)}^2 \right)^{1/2},$$

this is a Hilbert space and it follows from a well known result [[Ahmed (1981)], Theorem 1.2.15, p27] that the embedding $\mathcal{W} \hookrightarrow C(I, \mathcal{H})$ is continuous. Thus, the (weak) solution, if one exists, must satisfy the regularities: $\phi \in L_\infty(I, \mathcal{H}) \cap L_2(I, \mathcal{G}) \cap C(I, \mathcal{H}), \dot{\phi} \in L_2(I, \mathcal{G}^*)$. Now we are prepared to use the Galerkin projection technique. By use of the finite dimensional projections, as mentioned above, one can prove the existence of a sequence $\{\phi_n\}$ solving the finite dimensional (n-dimensional) problems:

$$< \dot{\phi}_n(t), g_r > + < \mathcal{A}\phi_n(t), g_r > = < f(t), g_r >, 1 \leq r \leq n, t \in I,$$

$$\phi_n(0) = \sum_{r=1}^{n} < \phi_0, g_r > g_r.$$

It follows from the a-priori estimates that the sequence $\{\phi_n, \dot{\phi}_n\}$ is contained in a bounded subset of $\mathcal{W}$. Thus there exists a subsequence, relabeled as the original sequence, and an element $\phi^o \in \mathcal{W}$ such that

$$\phi_n \xrightarrow{w*} \phi^o \quad \text{in} \quad L_\infty(I, \mathcal{H}) \tag{9.31}$$

$$\phi_n \xrightarrow{w} \phi^o \quad \text{in} \quad L_2(I, \mathcal{G}) \tag{9.32}$$

$$\dot{\phi}_n \xrightarrow{w} \dot{\phi}^o \quad \text{in} \quad L_2(I, \mathcal{G}^*) \tag{9.33}$$

and $\phi_n(0) \xrightarrow{s} \phi_0$ in $\mathcal{H}$. Then multiplying the finite dimensional system as presented above by any $\eta \in C_0^1((0, T))$, C^1 functions with compact supports in $(0, T)$, and integrating over I we obtain

$$\int_I < \dot{\phi}_n(t), g_r >_{\mathcal{G}^*, \mathcal{G}} \eta(t)dt + \int_I < \mathcal{A}\phi_n(t), g_r >_{\mathcal{G}^*, \mathcal{G}} \eta(t)dt$$

$$= \int_I < f(t), g_r >_{\mathcal{G}^*, \mathcal{G}} \eta(t)dt, 1 \leq r \leq n. \tag{9.34}$$

By virtue of (9.32) and the property (P1), it is clear that $\mathcal{A}\phi_n \xrightarrow{w*} \mathcal{A}\phi^o$ in $L_2(I, \mathcal{G}^*)$. Now letting $n \to \infty$ it follows from this and (9.31)-(9.33) that

$$\int_I < \dot{\phi}^o(t), g_r >_{\mathcal{G}^*, \mathcal{G}} \eta(t)dt + \int_I < \mathcal{A}\phi^o(t), g_r >_{\mathcal{G}^*, \mathcal{G}} \eta(t)dt$$

$$= \int_I < f(t), g_r >_{\mathcal{G}^*, \mathcal{G}} \eta(t)dt, 1 \leq r < \infty. \tag{9.35}$$

Since $\eta \in C_0^1$ is arbitrary and $\{g_r\}$ is complete in $\{\mathcal{G}, \mathcal{H}, \mathcal{G}^*\}$, it follows from the above identity that

$$< \dot{\phi}^o(t) + \mathcal{A}\phi^o(t) - f(t), \zeta >_{\mathcal{G}^*, \mathcal{G}} = 0, a.e \ t \in (0, T)$$

for every $\zeta \in \mathcal{G}$. In other words, the identity

$$\dot{\phi}^o(t) + \mathcal{A}\phi^o(t) = f(t), t \in (0, T)$$

holds in the distribution sense. It remains to verify that $\phi^o(0) = \phi_0$. Choosing any C^1 function η satisfying $\eta(T) = 0$, and subtracting equation (9.34) from equation (9.35) and integrating by parts, we obtain

$$(\phi^o(0) - \phi_n(0), g_r)_{\mathcal{H}} \eta(0) = \int_I < \phi^o - \phi_n, \mathcal{A}^* g_r >_{\mathcal{G}, \mathcal{G}^*} \eta(t) dt$$

$$- \int (\phi^o(t) - \phi_n(t), g_r)_{\mathcal{H}} \dot\eta(t) dt. \quad (9.36)$$

Letting $n \to \infty$, it follows from (9.31)-(9.32) that the expression on the right hand side of (9.36) converges to zero and hence

$$\lim_{n \to \infty} (\phi^o(0) - \phi_n(0), g_r)_{\mathcal{H}} \eta(0) = 0.$$

Thus, it follows from completeness of the set $\{g_r\}$ in $\mathcal{G}$ and the density of the embeddings $\mathcal{G} \hookrightarrow \mathcal{H} \hookrightarrow \mathcal{G}^*$ and arbitrariness of $\eta(0)$, that $\phi_n(0)$ converges weakly to $\phi^o(0)$. But we have seen above that $\phi_n(0) \xrightarrow{s} \phi_0$ in $\mathcal{H}$. Hence we must have $\phi^o(0) = \phi_0$. This proves that ϕ^o is the weak solution of the evolution equation (9.25) in the sense of Definition 9.4.1 satisfying the regularities (1) and (2). The reader can easily verify uniqueness. For the statement (3), note that it follows from the estimates (9.29) and (9.30) that $(\phi_0, f) \longrightarrow \phi^o$ is continuous from $\mathcal{H} \times L_2(I, \mathcal{G}^*) \longrightarrow \mathcal{W}$. This completes our proof. •

We can also prove a semilinear version of Theorem 9.4.2.

Theorem 9.4.3 *Consider the semilinear evolution equation*

$$(d/dt)\phi + \mathcal{A}\phi = \mathcal{B}(\phi) + f$$

$$\phi(0) = \phi_0.$$

Suppose $\mathcal{A}$ satisfies the hypothesis of Theorem 9.4.2, and $\mathcal{B}$ is dissipative and hemicontinuous in $\mathcal{H}$ and satisfies a linear growth condition. Then, for each $\phi_0 \in \mathcal{H}$ and $f \in L_2(I, \mathcal{H})$, the Cauchy problem has a unique weak solution $\phi \in L_\infty(I, \mathcal{H}) \cap C(I, \mathcal{H}_w)$.

Proof For detailed proof see [Ahmed (1994)], [Ahmed (1995)].

9.5 An Example

Consider the Gaussian random fields of Chapters 2 and 3, sections 2.4 and section 3.4. For each integer $n \geq 1$, let $\mathcal{L}_n(\hat{L}_2(D^n))$ denote the space of bounded linear operators in $\hat{L}_2(D^n)$. Consider the infinite product space $\mathcal{L} \equiv \prod_{n \geq 1} \mathcal{L}_n$. We introduce an operator $\mathcal{A}$ as follows. Let $B \in \mathcal{L}$ with $B_n \in \mathcal{L}_n(L_2(D^n))$ such that for each $K \in G$,

$$BK \equiv \{B_n K_n, n \geq 1\} \in \mathcal{K},$$

where $\mathcal{K}$ is the Fréchet space introduced in Chapter 2, section 2.4. We know that by virtue of Theorem 2.4.1, each $\phi \in \mathcal{G}$ has the orthogonal representation $\phi \equiv \sum_{n \geq 1} e_n(K_n) \equiv \Phi(K)$ for some $K \in G$. Define

$$\mathcal{A}\phi \equiv \Phi(BK) \equiv \sum_{n \geq 1} e_n(B_n K_n), \quad \phi = \Phi(K), K \in G. \qquad (9.37)$$

Since

$$\| \mathcal{A}\phi \|_{\mathcal{G}}^2 = \sum_{n \geq 1} n! \, \| B_n K_n \|_{\hat{L}_2(D^n)}^2$$

$$\leq \sum_{n \geq 1} n! \, \| B_n \|_{\mathcal{L}_n(\hat{L}_2(D^n))}^2 \| K_n \|_{\hat{L}_2(D^n)}^2, \qquad (9.38)$$

unless $sup\{\| B_n \|^2, n \geq 1\}$ is finite, the operator $\mathcal{A}$ is generally unbounded in the Hilbert space $\mathcal{G}$. However, under suitable assumptions, it is a bounded linear operator from $\mathcal{G}$ to its dual $\mathcal{G}^*$ which is a larger space. This is given in the following result.

Lemma 9.5.1 *Suppose there exists a non negative constant C such that*

$$sup\{(\| B_n \|_{\mathcal{L}_n} /n!), n \geq 1\} \leq C.$$

Then, $\mathcal{A}$ is a bounded linear operator from $\mathcal{G}$ to its dual $\mathcal{G}^*$.

Proof (outline only) By virtue of our assumption on B, one can verify that

$$| < \mathcal{A}\phi, \psi >_{\mathcal{G}^*, \mathcal{G}} | \leq C \, \| \phi \|_{\mathcal{G}} \| \psi \|_{\mathcal{G}}$$

for all $\phi, \psi \in \mathcal{G}$. Thus we have $\mathcal{A}\phi \in \mathcal{G}^*$; that is, $\mathcal{A} \in \mathcal{L}(\mathcal{G}, \mathcal{G}^*)$. This completes the proof. •

Note that the Quantum mechanical number operator is a very special case of this general operator. For instance, taking $B_n \equiv nI_n$ one has the number operator where $I_n \in \mathcal{L}_n$. Note that even $B_n \equiv e^{\kappa n} I_n$ is admissible for any $\kappa \in R$.

Consider the evolution equation :

$$(d/dt)\phi + \mathcal{A}\phi = f, t \in I \equiv [0, T];$$

$$\phi(0) = \phi_0, \qquad (9.39)$$

in the state space $\mathcal{H}$ where the operator $\mathcal{A}$ is as defined by (9.37).

Definition 9.5.2 The evolution equation (9.39) is said to have a weak solution if there exists a $\phi \in L_\infty(I, \mathcal{H})$ so that for each $\phi^0 \in \mathcal{H}$ for which $\mathcal{A}^*\phi^0 \in \mathcal{H}$, the following identity holds

$$(d/dt) < \phi(t), \phi^0 > + < \phi(t), \mathcal{A}^*\phi^0 > = < f(t), \phi^0 >, \qquad (9.40)$$

for almost all $t \in I$ and $\phi(0) = \phi_0$.

We shall prove the following result. Let $C(I, \mathcal{H}_w)$ denote the space of weakly continuous functions on I with values in $\mathcal{H}$ ($\mathcal{H}$ furnished with the weak topology).

Theorem 9.5.3 *Suppose the assumption of Lemma 9.5.1 hold and there exists a constant $c > 0$ such that*

$$< \mathcal{A}\varphi, \varphi >_{\mathcal{G}^*, \mathcal{G}} \geq -c \parallel \varphi \parallel_\mathcal{H}^2.$$

Then, for each $\phi_0 \in \mathcal{H}$ and $f \in L_2(I, \mathcal{H})$, the Cauchy problem (9.39) has a unique weak solution $\phi \in L_\infty(I, \mathcal{H}) \cap C(I, \mathcal{H}_w)$.

Proof See Ahmed [[Ahmed (1994)], Theorem 4.3], and [[Ahmed (1995)], Theorem 4.6].

Remark 9.5.4 The hypothesis, $< \mathcal{A}\phi, \phi >\geq -c \parallel \phi \parallel_\mathcal{H}^2$ for some $c \geq 0$, is satisfied if there exists a constant $c > 0$ such that $((B_n + cI)K_n, K_n)_{\hat{L}_2(D^n)} \geq 0$ for all $n \in N$. This is not a big restriction; the operator introduced in the previous section trivially satisfies this. Note that under this condition, even if $\phi_0 \in \mathcal{G}$ and $f \in L_1(I, \mathcal{G})$, there may not exist a strong solution.

Under a stronger hypothesis on $\mathcal{A}$, we can prove the existence of strong solutions even with milder assumptions on f and ϕ_0. This is stated below.

Theorem 9.5.5 *Suppose the assumptions of Lemma 9.5.1 hold and there exists a constant $\nu > 0$ such that*

$$Inf_{n \geq 1}\{((B_n e, e)/n!), \parallel e \parallel_{\hat{L}_2(D^n)} = 1\} \geq \nu > 0.$$

Then, for each $\phi_0 \in \mathcal{H}$ and $f \in L_2(I, \mathcal{G}^*)$, the Cauchy problem (9.39) has a unique solution $\phi \in L_\infty(I, \mathcal{H}) \cap L_2(I, \mathcal{G}) \cap C(I, \mathcal{H})$, in the sense that the first equation in (9.39) holds in $\mathcal{G}^*$ for almost all $t \in I$.

Proof Under the given assumption it is easy to verify that

$$< \mathcal{A}\varphi, \varphi >_{\mathcal{G}^*,\mathcal{G}} \geq \nu \parallel \varphi \parallel_{\mathcal{G}}^2$$

for all $\varphi \in \mathcal{G}$. Thus the conclusion follows from Theorem 9.4.2. •

Remark 9.5.6 Coercivity condition is a strong requirement compared to the condition assumed in Theorem 9.5.3. In any case under the assumptions of Theorem 9.5.5, $-\mathcal{A}$ generates a C_0-semigroup $S(t), t \geq 0$, in $\mathcal{H}$. Thus, for each $\phi_0 \in \mathcal{H}$ and $f \in L_1(I, \mathcal{H})$, equation (9.39) has a unique mild solution given by

$$\phi(t) = S(t)\phi_0 + \int_0^t S(t - r)f(r)dr, t \in I.$$

9.6   Evolution Equations on Wiener-Sobolev Spaces

Using the Ornstein-Uhlenbeck operator L we can construct interesting evolution operators. Let $\{p, q\}$ be the conjugate pair with $p \geq 1$ and take any real number $r > 0$ and define the operator C by $C \equiv (1 - L)^{r/2}$. Clearly, C maps $W^{r,p}(\Omega)$ to $L_p(\Omega)$, that is,

$$C : W^{r,p}(\Omega) \equiv W^{r,p}(\Omega, \mu^w) \longrightarrow L_p(\Omega) \equiv L_p(\Omega, \mu^w)$$

and that it is a bounded linear operator in these spaces. Its dual, denoted by C^*, is also a bounded linear operator

$$C^* : L_q(\Omega) \longrightarrow W^{-r,q}(\Omega)$$

where $\{p, q\}$ is the conjugate pair. Let $J : L_p \longrightarrow L_q$ denote the duality map given by

$$J(x) \equiv \{y \in L_q : (y, x)_{L_q, L_p} =\parallel x \parallel_{L_p}^2 =\parallel y \parallel_{L_q}^2\}.$$

If $\{p, q\} \in (1, \infty)$, the spaces $\{L_p, L_q\}$ are strictly convex and the duality map J is single valued. Using the basic operator L leading to the operator

C, we can now construct an operator $\mathcal{A}$ given by $\mathcal{A} \equiv C^* JC$. Generally, this is a nonlinear operator

$$\mathcal{A} : W^{r,p} \longrightarrow W^{-r,q}.$$

If $p = q = 2$, the duality map J is simply the identity map and $\mathcal{A} = C^* C$ is a bounded linear map from $W^{r,2}$ to $W^{-r,2}$. For simplicity of notation, we set $W^{r,p} \equiv X$ and its dual $W^{-r,q} \equiv X^*$.

Lemma 9.6.1 Suppose the conjugate pair $\{p, q\}$ is such that the duality map $J : L_p(\Omega) \longrightarrow L_q(\Omega)$ is single valued. Then the operator $\mathcal{A} \equiv C^* JC$ mapping X to X^* is single valued and it satisfies the following properties:

(P1): $\mathcal{A} : X \longrightarrow X^*$ is nonlinear (for $p \neq q$), bounded and continuous.

(P2): It is coercive in the sense that there exists a number $\beta > 0$ such that

$$< \mathcal{A}\varphi, \varphi >_{X^*, X} \geq \beta \parallel \varphi \parallel_X^2 \ \ \forall \ \varphi \in X.$$

(P3): It is monotone

$$< \mathcal{A}\varphi - \mathcal{A}\psi, \varphi - \psi >_{X^*, X} \geq 0 \ \ \forall \ \varphi, \psi \in X.$$

(P4): For $p = q = 2$, the operator $\mathcal{A} \in \mathcal{L}(X, X^*)$.

Proof For any $\varphi, \psi \in X$ we have

$$| < \mathcal{A}\varphi, \psi >_{X^*, X} | = |(J(C\varphi), C\psi)_{L_q, L_p}|$$
$$\leq \parallel J(C\varphi) \parallel_{L_q} \parallel C\psi \parallel_{L_p} = \parallel C\varphi \parallel_{L_p} \parallel C\psi \parallel_{L_p}.$$

Since C is a bounded linear operator from X to L_p, there exists a finite positive number b such that for all $\psi \in X$, $\parallel C\psi \parallel_{L_p} \leq b \parallel \psi \parallel_X$ and consequently

$$| < \mathcal{A}\varphi, \psi >_{X^*, X} | \leq b^2 \parallel \varphi \parallel_X \parallel \psi \parallel_X.$$

Hence $\mathcal{A}$ is a bounded operator proving (P1). For coercivity, note that the operator C has a bounded inverse [Shigekawa (1998/2004)] which is injective. Thus, there exists a finite positive number $\beta > 0$ such that

$$< \mathcal{A}\varphi, \varphi > \equiv (J(C\varphi), C\varphi)_{L_q, L_p} = \parallel C\varphi \parallel_{L_p}^2 \geq \beta \parallel \varphi \parallel_X^2.$$

This proves (P2). For the property (P3), note that, for any $\varphi, \psi \in X$, we have

$$< \mathcal{A}\varphi - \mathcal{A}\psi, \varphi - \psi >_{X^*, X} = (J(C\varphi) - J(C\psi), C\varphi - C\psi)_{L_q, L_p}. \quad (9.41)$$

One can easily verify that the duality map J is monotone. Indeed, for $x, y \in L_p$,

$$
\begin{aligned}
(J(x) &- J(y), x - y)_{L_q, L_p} \\
&= \| x \|_{L_p}^2 + \| y \|_{L_p}^2 - (J(x), y)_{L_q, L_p} - (J(y), x)_{L_q, L_p} \\
&\geq \left(\| x \|_{L_p} - \| y \|_{L_p} \right)^2 \geq 0.
\end{aligned}
\tag{9.42}
$$

Hence, combining (9.41) and (9.42) we obtain

$$
< \mathcal{A}\varphi - \mathcal{A}\psi, \varphi - \psi >_{X^*, X} \geq 0
\tag{9.43}
$$

proving (P3). In case $p = q = 2$, by the classical Riesz representation theorem, J is the identity map (linear) and consequently $\mathcal{A}$ is a bounded linear operator from X to its dual X^*. This proves (P4). •

Remark 9.6.2 In general, for conjugate pairs $\{p, q\}$ satisfying $1 \leq p, q \leq \infty$, the duality map J may be multi valued and hence the operator $\mathcal{A}$ is multi valued. This is particularly true for $p = 1, q = \infty$.

Now we consider the following evolution equation

$$
\dot{\varphi} + \mathcal{A}\varphi = f, \varphi(0) = \varphi_0, t \in I.
\tag{9.44}
$$

For simplicity, we let $p = q = 2$ implying linearity of the operator $\mathcal{A}$. We return to the nonlinear problem later in the sequel.

Theorem 9.6.3 Consider the evolution equation (9.44) and suppose the operator $\mathcal{A}$ satisfies Lemma 9.6.1 with $p = q = 2$. Then, for every $\varphi_0 \in \mathcal{H} \equiv L_2(\Omega)$ and $f \in L_2(I, X^*)$, the evolution equation has a unique solution $\varphi \in L_\infty(I, \mathcal{H}) \cap L_2(I, X)$ and $\dot{\varphi} \in L_2(I, X^*)$. Further, the solution $\varphi \in C(I, \mathcal{H})$ and it is continuously dependent on the data $\{\varphi_0, f\}$.

Proof The proof is based on Galerkin projection method and a-priori bounds. First note that the embeddings $X \hookrightarrow \mathcal{H} \hookrightarrow X^*$ are continuous and dense. Scalar multiplying the equation (9.44) by φ and using the property (P2) of Lemma 9.6.1 and Schwartz inequality, it is easy to verify that for each $t \in I$,

$$
|\varphi(t)|_{\mathcal{H}}^2 + \beta \int_0^t \| \varphi(s) \|_X^2 \, ds \leq |\varphi_0|_{\mathcal{H}^2} + (1/\beta) \int_0^t \| f(s) \|_{X^*}^2 \, ds.
\tag{9.45}
$$

Further, using Gronwall inequality and the fact that $A \in \mathcal{L}(X, X^*)$ one can deduce that there exists a constant $\gamma > 0$ such that

$$\int_I \| \dot{\varphi}(t) \|_{X^*}^2 \leq \gamma \left\{ |\varphi_0|_{\mathcal{H}}^2 + \int_I \| f(t) \|_{X^*}^2 \, dt \right\}. \tag{9.46}$$

This shows that if φ is any solution of equation (9.44), it must satisfy the a-priori bounds as given above. That is, $\varphi \in L_\infty(I, \mathcal{H}) \cap L_2(I, X)$ and $\dot{\varphi} \in L_2(I, X^*)$. Now let $\{v_i\}$ be any complete orthonormal basis of $\mathcal{H}$ which is orthogonal in X and X^*. Let M_n denote the linear span $Span\{v_i, 1 \leq i \leq n\}$ and P^n the corresponding projection. Projecting the system (9.44) to M_n, we construct the finite dimensional system (system of ordinary differential equations)

$$\dot{X}(t) + AX(t) = F(t), t \in I, X(0) = X_0 \tag{9.47}$$

where $X(t)$ and $X(0)$ are the n-vectors

$$X(t) = \{x_i^n(t), 1 \leq i \leq n\} \quad \text{and} \quad X_0 = \{x_{i,0}^n, 1 \leq i \leq n\}$$

respectively with $\{x_{i,0}^n\}$ being the fourier coefficients in the expansion of the initial state φ_0 given by

$$P^n(\varphi_0) \equiv \varphi_n(0) = \sum_{i=1}^n x_{i,0}^n v_i.$$

Clearly, $\varphi_n(0) \xrightarrow{s} \varphi_0$ in $\mathcal{H}$. The matrix A is given by $\{a_{i,j} \equiv < Av_j, v_i >_{X^*,X}, 1 \leq i, j \leq n\}$ and the vector $F(t) \equiv \{f_i(t), 1 \leq i \leq n\}$ with $f_i(t) = < f(t), v_i >_{X^*,X}$. It follows from the theory of linear ordinary differential equations with constant coefficients that equation (9.47) has a unique continuous solution $X \in C(I, R^n)$. Using this solution, define

$$\varphi_n(t) \equiv \sum_{i=1}^n x_i^n(t) v_i$$

and note that φ_n satisfies the following identity

$$(\dot{\varphi}_n(t), v_i)_{\mathcal{H}} + < A\varphi_n(t), v_i >_{X^*,X} = < f(t), v_i >_{X^*,X}, 1 \leq i \leq n. \tag{9.48}$$

Multiplying this equation by $x_i^n(t)$ and summing over $i \in [1, 2, \cdots, n]$, one can verify that the sequence $\{\varphi_n\}$ and its derivative $\{\dot{\varphi}_n\}$ satisfy the same a-prior bounds as given by (9.45) and (9.46). In other words, $\{\varphi_n\}$ and $\{\dot{\varphi}_n\}$ are contained in a bounded subset of $L_\infty(I, \mathcal{H}) \cap L_2(I, X)$ and $L_2(I, X^*)$

respectively. Note that $L_2(I, X)$ and $L_2(I, X^*)$ are reflexive Banach spaces. Thus, considering the weak star topology on $L_\infty(I, \mathcal{H})$ and weak topologies on $L_2(I, X)$ and $L_2(I, X^*)$, we conclude that there exists a $\varphi \in L_\infty(I, \mathcal{H}) \cap L_2(I, X)$ satisfying $\dot\varphi \in L_2(I, X^*)$ such that along a subsequence, relabeled as the original sequence,

$$
\begin{aligned}
\varphi_n &\xrightarrow{w^*} \varphi &&\text{in}\quad L_\infty(I, \mathcal{H}) \\
\varphi_n &\xrightarrow{w} \varphi &&\text{in}\quad L_2(I, X) \\
\dot\varphi_n &\xrightarrow{w} \dot\varphi &&\text{in}\quad L_2(I, X^*) \\
\mathcal{A}\varphi_n &\xrightarrow{w} \mathcal{A}\varphi &&\text{in}\quad L_2(I, X^*).
\end{aligned}
\tag{9.49}
$$

Hence, it suffices to verify that the limit φ as indicated above is the weak solution of our problem. Let $\eta \in C_0^1(0, T)$ (C^1 functions with compact supports). Multiplying equation (9.48) by η and integrating over the interval I, we have

$$
\int_I < \dot\varphi_n(t), \eta(t)v_i >_{X^*,X} dt + \int_I < \mathcal{A}\varphi_n(t), \eta(t)v_i >_{X^*,X} dt
$$
$$
= \int_I < f(t), \eta(t)v_i >_{X^*,X} dt
\tag{9.50}
$$

for all $1 \le i \le n$. Letting $n \to \infty$, it follows from this and (9.49) that φ satisfies the following identity

$$
\int_I < \dot\varphi(t), \eta(t)v_i >_{X^*,X} dt + \int_I < \mathcal{A}\varphi(t), \eta(t)v_i >_{X^*,X} dt
$$
$$
= \int_I < f(t), \eta(t)v_i >_{X^*,X} dt
\tag{9.51}
$$

for each $i \in N$. Since $\eta \in C_0^1$ is arbitrary and $\{v_i\}$ is a basis, the above identity holds for every $v \in X$ giving

$$
\int_I < \dot\varphi(t), \eta(t)v >_{X^*,X} dt + \int_I < \mathcal{A}\varphi(t), \eta(t)v >_{X^*,X} dt
$$
$$
= \int_I < f(t), \eta(t)v >_{X^*,X} dt.
$$

Hence the first identity in equation (9.44) holds in the distribution (weak) sense. We prove that $\varphi(0) = \varphi_0$. Now choosing $\eta \in C^1$ with $\eta(T) = 0$, and integrating the first term on the left of equation (9.50) and (9.51) and taking the limit we obtain

$$
\lim_{n\to\infty} (\varphi_n(0), \eta(0)v_i)_\mathcal{H} = (\varphi(0), \eta(0)v_i)_\mathcal{H}.
\tag{9.52}
$$

But we know that $\varphi_n(0)$ converges strongly to φ_0. Since $\{v_i\}$ is a basis and $\eta \in C^1$ is otherwise arbitrary we have $\varphi(0) = \varphi_0$ as desired. The continuity of the solution $t \longrightarrow \varphi(t)$ with values in $\mathcal{H}$ follows from [[Ahmed (1981)], Theorem 1.2.15, p27]. Since the system is linear, continuity of solution with respect to the data $\{\varphi_0, f\}$ follows from the estimates (9.45) and (9.46). This completes the proof. ●

We can also consider nonlinear monotone operators and evolution equations based on such operators. We present here a basic frame work for such operators and an important result on the existence and regularity of solutions of abstract evolution equations on Winer-Sobolev (Fock Spaces) spaces.

Let $\{p, q\}$ be the conjugate pair satisfying $1 < q \leq 2 \leq p < \infty$, and let $X \equiv W^{r,p}(\Omega) \equiv W^{r,p}(\Omega, \mu^w)$ with the topological dual given by $X^* \equiv W^{-r,q}(\Omega)$ and $\mathcal{H} \equiv L_2(\Omega)$. Then we have the following inclusions

$$X \hookrightarrow \mathcal{H} \hookrightarrow X^*$$

with continuous and dense embeddings. Let $\mathcal{A}$ denote a nonlinear operator possibly more general than $C^* JC$ and suppose it satisfies the following properties:

(A1): There exist constants $c_1 \geq 0, c_2 > 0$ such that

$$< \mathcal{A}(\varphi), \varphi >_{X^*, X} \geq c_1 + c_2 \parallel \varphi \parallel_X^p \quad \forall \ \varphi \in X.$$

(A2): There exists a constant $c_3 > 0$ such that

$$\parallel \mathcal{A}(\varphi) \parallel_{X^*} \leq c_3(1 + \parallel \varphi \parallel_X^{p-1}) \quad \forall \ \varphi \in X.$$

(A3): $\mathcal{A}$ is monotone and hemicontinuous in the sense that

$$< \mathcal{A}\varphi - \mathcal{A}\psi, \varphi - \psi >_{X^*, X} \geq 0, \quad \forall \ \varphi, \psi \in X,$$

and for every $\varphi, \zeta \in X$, $\mathcal{A}(\varphi + \theta\zeta) \xrightarrow{w} \mathcal{A}(\varphi)$ as $\theta \to 0$.

For simplicity of presentation, here we have assumed the operator $\mathcal{A}$ to be time invariant. Extension to time varying case is fairly straightforward and can be carried out exactly in the same way as in [[Ahmed (1988)], Theorem 4.1, p96]. For study of nonlinear evolution equations involving such operators, we need the following function spaces $L_p(I, X)$, its dual $L_q(I, X^*)$, and $L_\infty(I, \mathcal{H})$. Further, we use the function space

$$\mathcal{W} \equiv \{\varphi : \varphi \in L_p(I, X) \text{ and } \dot\varphi \in L_q(I, X^*)\}$$

where $\dot\varphi$ denotes the distributional derivative of the vector valued function $t \longrightarrow \varphi(t)$. With respect to the norm topology given by

$$\| \varphi \|_\mathcal{W} \equiv \| \varphi \|_{L_p(I,X)} + \| \dot\varphi \|_{L_q(I,X^*)}$$

$\mathcal{W}$ is a Banach space and further, the embedding $\mathcal{W} \hookrightarrow C(I, \mathcal{H})$ is continuous. Strictly speaking, the elements of $\mathcal{W}$ have continuous versions with values in $\mathcal{H}$. The proof of this statement is similar to that of Theorem 1.2.15, p27 [Ahmed (1981)].

Now we can present the following general result.

Theorem 9.6.4 Consider the following nonlinear evolution equation

$$\dot\varphi + \mathcal{A}(\varphi) = f, \varphi(0) = \varphi_0, t \in I, \tag{9.53}$$

with respect to the Gelfand triple $X \hookrightarrow \mathcal{H} \hookrightarrow X^*$ as introduced above. Suppose the operator $\mathcal{A}$ satisfies the assumptions (A1)-(A3), and $f \in L_q(I, X^*)$ and $\varphi_0 \in \mathcal{H}$. Then the system (9.53) has a weak solution $\varphi \in L_\infty(I, \mathcal{H}) \cap L_p(I, X)$ with $\dot\varphi \in L_q(I, X^*)$. The solution is unique if $\mathcal{A}$ is strictly monotone.

Proof The proof is again based on the technique of finite dimensional projection giving a system of nonlinear ordinary differential equations and limiting arguments. Existence and regularity of solutions of finite dimensional nonlinear systems are proved using classical finite dimensional techniques with nonlinear monotone vector fields. Then, one uses the a-priori bounds to prove the boundedness of the approximating solutions $\{\varphi_n\}$. It is shown that this sequence is contained in a bounded subset of the Banach spaces $L_\infty(I, \mathcal{H}) \cap L_p(I, X)$. The distributional derivatives $\{\dot\varphi_n\}$, and $\{\mathcal{A}(\varphi_n)\}$ are proved to be contained in a bounded subset of $L_q(I, X^*)$. Note that, under our assumptions, the spaces $L_p(I, X)$ and its dual $L_q(I, X^*)$ are reflexive. Then, using the weak star topology for $L_\infty(I, \mathcal{H})$ and the weak topology for $L_p(I, X)$ and $L_q(I, X^*)$ one concludes that there exists a $\varphi \in \mathcal{W}$ such that along a subsequence, relabeled as the original sequence,

$$\varphi_n \xrightarrow{w^*} \varphi \quad \text{in} \quad L_\infty(I, \mathcal{H})$$

$$\varphi_n \xrightarrow{w} \varphi \quad \text{in} \quad L_p(I, X) \tag{9.54}$$

$$\dot{\varphi}_n \xrightarrow{w} \dot{\varphi} \quad \text{in} \quad L_q(I, X^*)$$

and that there exists an $\vartheta \in L_q(I, X^*)$ such that

$$\mathcal{A}(\varphi_n) \xrightarrow{w} \vartheta \quad \text{in} \quad L_q(I, X^*). \tag{9.55}$$

Following similar steps as in the linear case (Theorem 9.6.3), we obtain an identity similar to (9.51). This is given by

$$\int_I < \dot{\varphi}(t), \eta(t)v_i >_{X^*,X} dt + \int_I < \vartheta, \eta(t)v_i >_{X^*,X} dt$$

$$= \int_I < f(t), \eta(t)v_i >_{X^*,X} dt. \tag{9.56}$$

Since $\eta \in C_0^1(I)$ is arbitrary and $\{v_i\}$ is a complete basis, this leads to the following identity

$$\dot{\varphi} + \vartheta = f \quad \text{in} \quad L_q(I, X^*) \tag{9.57}$$

which holds in the sense of distribution with $\varphi(0) = \varphi_0$. The problem then reduces to demonstrating that $\vartheta = \mathcal{A}(\varphi)$. This is easily done by use of the assumption (A3) stating that $\mathcal{A}$ is monotone and hemicontinuous. Indeed, due to monotonicity we have

$$\int_I < \mathcal{A}(\psi) - \mathcal{A}(\varphi_n), \psi - \varphi_n >_{X^*,X} dt \geq 0, \tag{9.58}$$

for all $\psi \in L_p(I, X)$. Since $\varphi_n \xrightarrow{w} \varphi$ in $L_p(I, X)$, by use of Mazur's theorem we can construct a sequence using an appropriate convex combination such as $\tilde{\varphi}_n \equiv \sum_{i=1}^n a_i^n \varphi_i$ with $\{a_i^n\} \geq 0$ and $\sum_{i=1}^n a_i^n = 1$ for all $n \in N$, such that

$$\tilde{\varphi}_n \xrightarrow{s} \varphi \quad \text{in} \quad L_p(I, X).$$

Clearly, using this sequence in the expression (9.58) we have

$$\int_I < \mathcal{A}(\psi) - \mathcal{A}(\tilde{\varphi}_n), \psi - \tilde{\varphi}_n >_{X^*,X} dt \geq 0, \tag{9.59}$$

for all $\psi \in L_p(I, X)$. Letting $n \to \infty$ it follows from this that

$$\int_I < \mathcal{A}(\psi) - \vartheta, \psi - \varphi >_{X^*, X} dt \geq 0, \tag{9.60}$$

for all $\psi \in L_p(I, X)$. Take $\psi = \varphi + \tau\rho$ for any $\tau > 0$ and $\rho \in L_p(I, X)$. Substituting this in the preceding expression we obtain

$$\int_I < \mathcal{A}(\varphi + \tau\rho) - \vartheta, \rho >_{X^*, X} dt \geq 0, \tag{9.61}$$

for all $\rho \in L_p(I, X)$. Letting $\tau \to 0$, it follows from the hemicontinuity of $\mathcal{A}$ that

$$\int_I < \mathcal{A}(\varphi) - \vartheta, \rho >_{X^*, X} dt \geq 0, \tag{9.62}$$

for all $\rho \in L_p(I, X)$. This is possible if, and only if, $\vartheta = \mathcal{A}(\varphi)$. Thus, the expression (9.57) is actually

$$\dot\varphi + \mathcal{A}(\varphi) = f \quad \text{in} \quad L_q(I, X^*) \tag{9.63}$$

with $\varphi(0) = \varphi_0$. We prove uniqueness by contradiction. Suppose there are two solutions, φ and ψ corresponding to the same f and initial condition $\varphi(0) = \psi(0) = \varphi_0$. Then they must satisfy

$$\int_I < \dot\varphi - \dot\psi, \varphi - \psi >_{X, X^*} dt + \int_I < \mathcal{A}(\varphi) - \mathcal{A}(\psi), \varphi - \psi >_{X^*, X} dt = 0.$$

By virtue of strict monotonicity of the operator $\mathcal{A}$, it follows from the above identity that

$$\int_I < \dot\varphi - \dot\psi, \varphi - \psi > dt < 0.$$

Since $\varphi, \psi \in \mathcal{W}$ and $\mathcal{W}$ is continuously embedded in $C(I, \mathcal{H})$, these functions are continuous with values in $\mathcal{H}$. Thus the point values are well defined. Then using integration by parts, it follows from the preceding inequality that

$$|\varphi(T) - \psi(T)|^2_{\mathcal{H}} < 0.$$

This is impossible and so the contradiction proves uniqueness. •

9.7 Some Examples for Exercise

P1: Prove that the Malliavin operator δ is a nonexpansive map from $L_2(I, \mathcal{G}_\alpha)$ to $\mathcal{H}_\alpha$.

P2: Justify that the semigroup $T(t), t \geq 0$, as defined in Corollary 9.2.5, is the Ornstein-Uhlenbeck semigroup.

P3: Consider the operator $\mathcal{A}$ given by the expression (9.21). Give a detailed proof of Theorem 9.3.3.

P4: Consider the operator $\mathcal{A}$ as defined by the expression (9.37) and give a detailed proof of Lemma 9.5.1.

P5: Consider the operator $\mathcal{A} \equiv C^* J C$ as defined in Lemma 9.6.1. Verify the estimate (9.46). Hints: Use (9.45) and the evolution equation (9.44) in the sense of distribution.

P6: Consider the operator $\mathcal{A} \equiv C^* J C \in \mathcal{L}(X, X^*)$ as defined in Lemma 9.6.1. For $p = q = 2$, J is the identity map and the operator $\mathcal{A} \in \mathcal{L}(X, X^*)$. Recall the Hilbert space $\mathcal{H} = L_2(\Omega)$. Clearly the operator $\mathcal{A}$ considered as a linear operator on $\mathcal{H}$ is an unbounded operator $\mathcal{A} : D(\mathcal{A}) \subset \mathcal{H} \longrightarrow \mathcal{H}$ with domain dense in $\mathcal{H}$. Prove that $-\mathcal{A}$ generates a C_0-semigroup of contractions (operators) in $\mathcal{H}$. Hints: Using Lemma 9.6.1, verify that

$$\| (\lambda I + \mathcal{A}) \|_{\mathcal{H}} \leq (1/\lambda), \forall \, \lambda > 0,$$

and then use Hille-Yosida theorem.

P7: Prove that the above result also follows directly from Lemma 9.6.3; and then the (mild) solution of equation (9.44) is given by

$$\varphi(t) = S(t)\varphi_0 + \int_0^t S(t - r) f(r) dr, t \geq 0,$$

for every pair $(\varphi_0, f) \in \mathcal{H} \times L_1^{\ell oc}([0, \infty), \mathcal{H})$.

Bibliography

(1). R.A.Adams (1975) *Sobolev Spaces*, Academic Press, New York

(2). N.U.Ahmed, *Fourier Analysis on Wiener Measure Space*, Journal of the Franklin Institute, 286(2),(1968), 143-151.

(3). N.U.Ahmed, *Strong and Weak Synthesis of Nonlinear Systems with Constraints on the System Space*, Information and Control, 23(1),(1973), 71-85.

(4). N.U.Ahmed, Some Novel Properties of Wiener's Canonical Expansion, I.E.E.E. Trans. Systems Sci. Cybernetics, 5, (1969),140-144.

(5). N.U.Ahmed, Closure and Completeness of Wiener's Orthogonal Set G_n in the Class $L_2(\Omega, \mathcal{B}, \mu)$ and it's Applications to Stochastic Hereditary Differential Systems, J. Information and Control, 17,2,(1970),161-174.

(6). N.U.Ahmed, A New Class of Generalized Nonlinear Functionals of White Noise with Applications to Random Integral Equations, Stochastic Analysis and Applications,1(2),(1983),139-158.

(7). T.Hida, Analysis of Brownian Functionals, Carleton University Math. Lect. Notes., 13,(1978).

(8). T.Hida, Generalized Multiple Wiener Integrals, Proc. Japan. Acad., 54, ser.A, (1978), 55-58.

(9). T.Hida, Causal Calculus of Brownian Functionals, and it's Applications. Int. Symp. on Stochastic and Related Topics, Carleton University, Ottawa, (1980).

(10). T.Hida,(1980), Brownian Motion, Springer-Verlag, New York, Heidelberg, Berlin.

(11). T.Hida, and S. Si, (2008), Lectures on White Noise functionals, World Scientific, New Jersey, London, singapore, Beijing, Shanghai,Hong Kong, Taipei, Chennai.

(12). T.Hida, H.H. Kuo, J. Potthoff and L. Streit, (1993), White Noise , Kluwer Academic, Dordrecht, Boston, London.

(13). D.Nualart, M.Zakai, Generalized Brownian Functionals and the Solution to a Stochastic Partial Differential Equation, Journal of Functional Analysis, 84,(1989), 1-18.

(14). D.R.Bell, (1987), The Malliavin Calculus, Pitman Monographs & surveys in Pure & Applied Math., 34, Longman Scientific & Technical, UK, Co-publisher John-Wiley, New York.

(15). N.Wiener,(1938), The Homogeneous Chaos, Amer. J. of Math., Vol.LV, 897-936.

(16). N.Wiener,(1958), Nonlinear Problems in Random Theory, The M.I.T Press, Cambridge.

(17). T.A.W.Dwyer, Analytic Evolution Equations in Banach Spaces, Lect. Notes.in Math. Vector Space Measures and Applications, 645, Springer-Verlag, Berlin Heidelberg New York,(1978),48-61.

(18). T.F.Chan, J.Shen and L.Vese, Variational PDE Models in Image Processing, Notices of the AMS, 50(1),2002, Joint Mathematics Meeting, San Diego.

(19). S.Angenent, E.Pichon and A. Tannenbaum, Mathematical Methods in Medical Imaging, Bulletin AMS, 43(3), (2006), 365-396.

(20). N.U.Ahmed, Optimal Control of Turbulent Flow as Measure Solutions, IJCFD, 11, (1998), 169-180.

(21). D.Nualart,(2009), Malliavin Calculus and its Applications, CBMS series no. 110, AMS, Providence, Rhode Island.

(22). S.K.Srinivasan and R. Vasudevan,(1971), Introduction to Random Differential Equations and their Applications, Elsevier, New York. London. Amsterdam.

(23). N.U.Ahmed and J.M. Skowronski, Boundary Stabilization of Nonlinear Flexible Systems, in Mechanics and Control (Ed. R.S.Gutalu), Plenum Press, New York and London, (1994), 213-222.

(24). I.I.Gihman and A.V.Skorohod,(1971), The Theory of Stochastic Processes I, Springer-Verlag, New York Heidelberg Berlin, (trans. S. Kotz).

(25). N.U.Ahmed, Linear and Nonlinear Filtering for Scientists and Engineers, (1998), World Scientific, Singapore, New Jersey, London, Hong Kong.

(26). J.Diestel and J.J.Uhl,Jr., (1977), Vector Measures, AMS, Providence, Rhode Island.

(27). A.I.Tulcea and C.I.Tulcea, (1969), Topics in the Theory of Lifting, Springer-Verlag, Berlin Heidelberg New York.

(28). N.U.Ahmed, M.Fuhrman and J.Zabczyk, On Filtering Equations in Infinite Dimensions, Journal of Functional Analysis, 143(1),(1997),180-204.

(29). A.Wulfsohn, Infinitely Divisible Stochastic Differential Equations in Space-Time, Proc. Vector Space Measures and Applications II, Lecture Notes in Mathematics, Springer-Verlag, Berlin Heidelberg, New York,(1977),199-208.

(30). D.A.Dawson, Stochastic Evolution Equations and Related Measure Processes, J. Mult. Anal. 5, (1975), 1-52.

(31). N.U.Ahmed, Evolution Equations on Fock Spaces Based on Gaussian random Fields, Dynamic Systems and Applications, 3,(1994),25-38.

(32). N.U.Ahmed, Generalized Functionals of Gaussian Random Fields and Evolution Equations Thereon, Dynamic Systems and Applications, 4,(1995),27-46.

(33). A.V.Skorohod, (1965), Studies in the Theory of Random Processes, Addison-Wesley Publishing Company, Inc. Reading, Massachusetts.

(34). G.Da Prato and J.Zabczyk, (1992), Stochastic Equations in Infinite Dimensions, Cambridge University Press.

(35). N.U.Ahmed, Semigroup Theory with Applications to Systems and Control, (1991), Pitman Research Notes in Mathematics Series, 246, Longman Scientific and Technical.

(36). N.U.Ahmed, Measure Valued Solutions of Stochastic Evolution Equations on Hilbert Space and Their Feedback Control, Discussiones Mathematicae Differential Inclusions, Control and Optimization, 25 (2005), 129-157.

(37). N.U.Ahmed, Stone-Čech Compctification with Applications to Evolution Equations on Banach Spaces, Publicationes Mathematicae, Debrechen, Tomus 59, (2001), fasc.3-4, 289-301.

(38). N.U.Ahmed, Optimal Feedback Control for Impulsive Systems on the Space of Finitely additive Measures, Publ. Math. Debrecen,70(3-4), (2007), 371-393.

(39). L.Gross, Abstract Wiener Spaces, Proc. Fifth Berkeley Symposium, Math. Stat. & Probability, Berkeley, California, 1965/66, Vol.II: Contributions to Probaility Theory, part 1, p31-42, U.C Berkeley Press, Berkeley, Cal.

(40). I.Shigekawa, (1998/2004), Stochastic Analysis, (Trans. Math. Monographs), Vol. 224, AMS, Providence, Rhode Island.

(41). P.K.Friz, An Introduction to Malliavin Calculus, Notes (2002), Courant Institute of Mathematical Sciences, New York University, www.math.nyu.edu.

(42). N.U.Ahmed, C. D. Charalambous, Optimal Measurement Strategy for Nonlinear Filtering, SIAM Journal Contr. and Optim., 45(2),(2006),519-531.

(43). N.U.Ahmed, (1988), Elements of Finite Dimensional Systems and Control Theory, Pitman Monographs and Surveys in Pure and Applied Mathematics, 37, Longman Scientific and Technical, U.K, Copublished in the United States with John-Wiley & Sons, Inc., New York.

(44). N.U.Ahmed and K.L. Teo,(1981), Optimal Control of Distributed Parameter Systems, North Holland, New York, Oxford.

(45). N.Bouleau and F.Hirsch,(1991), Dirichlet Forms and Analysis on Wiener Space, Walter de Gruyter and Co. Berlin.

(46). P.Malliavin,(1995), Integration and Probability, Springer-Verlag, New York Berlin Heidelberg London Paris Tokyo Hong Kong Barcelona Budapest.

(47). P.Malliavin, Stochastic Caculus of Variations and Hypoelliptic Operators, Proc. of the International Conference on Stochastic Differential Equations, Kyoto, Kinokunia, Tokyo, Wiely, New York, (1976), 195-263.

(48). P.Malliavin,(1997), Stochastic Analysis, Springer-Verlag, Berlin.

(49). N.Ikeda and S.Watanabe, (1989), Stochastic Differential Equations and Diffusion Processes, (2nd Edition), Kodansha/North-Holland, Tokyo/Amsterdam.

(50). S.Watanabe, (1984), Lectures on Stochastic Differential Equations and Malliavin Calculus, Tata Institute of Fundamental Research, Springer-Verlag, Berlin.

(51). N.U.Ahmed, (1988), Optimization and Identification of Systems Governed by Evolution Equations on Banach Spaces, Pitman Research Notes in Mathematics Series,184, Longman Scientific and Technical, U.K; Co-publisher: John Wiely & Sons., New York.

(52). S.Si, Hida Distributions,(in press), World Scientific Publishing Company, Singapore.

(53). N.U.Ahmed, Existence of Optimal Controls for a General Class of Impulsive systems on Banach Spaces, SIAM J, Control. Optim. 42(2), (2003),669-685.

(54). N.U.Ahmed, Optimal Control of Systems Determined by Strongly Nonlinear Operator Valued Measures, Differential Inclusions, Control and Optimization, 28, (2008), 165-189.

(55). N.U.Ahmed, Existence and Unqueness of Measure Valued Solutions for Zakai Equation, Publicationes Mathematicae, Debrechen, 49(3-4), (1996), 251-264.

(56). N.U.Ahmed, Relaxed Solutions for Stochastic Evolution Equations on Hilbert space with Polynonial Nonlinearities, Publicationes Mathematicae, Debrechen, 54(1-3), (1999), 75-101.

(57). L.Hörmander, Hypoelliptic Second Order Differential Equations, Acta Math. 119,(1967), 147-171.

(58). R.P.Feynman, Space-time Approach to Nonreativistic Quantumn Mechanics, Reviews of Modern Physics, 20, (1948), 367-387.

(59). S.Willard, (1970), General Topology, Addison Wesley Publishing Company, Reading, Massachusetts. Menlo Park, California. London. Don Mills, Ontario.

(60). N.U.Ahmed, Measure Solutions for semilinear Systems with Unbounded Nnonlinearities, Nonlinear Analysis: Theory, Methods and Applications, 35,(1999), 487-503.

(61). N.U.Ahmed, Optimal Stochastic Control of Measure Solutions on Hilbert space; Systems, Control, Modeling and Optimization, Proc. of 22nd IFIP TC7 Conference, 2005, July 18-22, Turin, Italy, Springer (2006), 1-12.

(62). R.Temam, (1988), Infinite Dimensional Dynamical Systems in Mechanics and Physics, Springer-Verlag, New York Berlin Heidelberg London Paris Tokyo.

(63). C.Foias, O.Manley, R.Rosa and R.Temam, (2001), Navier-Stokes Equations and Turbulence, Cambridge University Press.

(64). A.Sudbery, (1986), Quantum Mechanics and the Particles of Nature, Cambridge University Press,Cambridge London New York New Rochelle Melbourne Sydney.

(65). J.Yeh, (1973), Stochastic Processes and the Wiener Integral, Marcel Dekker, Inc. New Yor.

(66). Jean Leray, Essai sur le Mouvement d'un Liquide Visqueux Emplissant L'espace, Acta Math., 63(1934), 193-248.

(67). Jean Leray, Etude de Diverses équations intégrales Nonlinéaires et de Quelques Problémes que Pose L'hydrodynamique, J. Math. Pures. Appl., 12(1933),1-82.

(68). J.Mujica, Representation of Analytic Functionals by Vector Measures, Lecture Notes in Mathematics, 645, Vector Space Measures and Applications II, Proc. Dublin, 1977, (edited by R.M.Aron and S.Dineen), Springer-Verlag, Berlin Heidelberg New York.

(69). D.Nualart and B.Rozovskii, Weighted Stochastic Sobolev Spaces and Bilinear SPD's Driven by Space-time White Noise, J.Funct. Anal. 149(1997), 200-225.

(70). H.Holden, B.Øksendal, J.Ubœ, T.S. Zhang; Stochastic Partial Differential Equations: A Modeling, White Noise, Functional Approach, in "Probability and Its Applications, Birkhauser, (1996).

(71). B.Øksendal, (1997), An Introduction to Malliavin Calculus with Application to Economics, Lect. Notes, available on www.nhh.no/for/dp/1996/

(72). H. Körezlioglu and A.S.Üstünel, New Class of Distributions on the Wiener Space, in "Stochastic Analysis and Related Topics II (Conf. Proc. Silivri,1988)", Lect. Notes Math, Vol. 1444,(1990),106-121.

(73). G.Da Prato, (2007), Introduction to Stochastic Analysis and Malliavin Calculus, Vol.6 of Appunti, Scuola Normale Superiore, Pisa.

(74). P.Imkeller, Malliavin's Calculus and Applications in Stochastic Control and Finance, Vol. 1 of IMPAN Lect. Notes, Institute of Mathematics, Polish Academy of Sciences, Warsaw,(2008).

(75). D.Nualart, (2006), The Malliavin Calculus and Related Topics, Probability and Its Applications , Springer, New York,(2nd Edition).

(76). D.W.Stroock, The Malliavin Calculus and its application to second Order Parabolic Differential Operators, I,II, Mathematical System Theory, 14(1981),25-65, 141-171.

(77). D.W.Stroock, The Malliavin Calculus, Functional Analytic Approach, J. Funct. Anal., 44 (1981), 217-257.

(78). A.N.Godunov, Peano's Theorem in Infinite Dimensional Hilbert Space is false Even in a Weakened Formulation, Math. Zametki 15, (1974),467-477.

(79). H.O.Fattorini, A Remark on Existence of Solutions of Infinite Dimensional Noncompact Optimal Control Problems, SIAM J. Control and Optimization, 35(4), (1997), 1422-1433.

(80). K.R.Parthasarathy, (1967), Probability Measures on Metric Spaces, Academic Press, New York and London.

(81). N.U.Ahmed, Measure Solutions for Impulsive Evolution Equations with Measurable Vector Fields, J. Mathematical Analysis and Applications, 319 (2006), 74-93.

(82). J.K.Brooks, Weak Compactness in the Space of Vector Measures, Bulletin of the American Mathematical Society, 78(2), 1972, 284-287.

(83). J.K.Brooks and N. Dinculeanu, Strong Additivity,Absolute Continuity and Compactness in spaces of Measures, J. Math. Anal. Appl., 45 (1974),156-175.

(84). N.U.Ahmed, A General Result on Measure Solutions for Semilinear Evolution Equations, Nonlinear Analysis:TMA, 42, (2000),1335-1349.

(85). N.Dumford and J.T.Scwartz, (1958), Linear Operators, Part 1, Interscience publishers, Inc., New York.

(86). B.B.Mandelbrot and J.W.Van Ness, Fractional Brownian Motions, Fractional noises and Applications, SIAM Rev.,10(1968),422-437.

(87). N.U.Ahmed and C.D.Charalamboos, Filtering for Linear Systems Driven by Fractional Brownian Motion, SIAM J. Control Optim. 41(1),(2002), 313-330.

(88). T.E.Duncan, Y.Hu and B.P.Duncan, Stochastic Calculus for Fractional Brownian Motion I. Theory, SIAM J. Control Optim. 38(2),(2000), 582-612.

(89). N.U.Ahmed and Cheng Li, Optimum Feedback Strategy for Access Control Mechanism Modelled as Stochastic Differential Equation in Computer Network, Mathematical Problems in Engineering 2004:3(2004), 263276.

(90). N.U.Ahmed, Existence of Solutions of Nonlinear Stochastic Differential Incusions on Banach Spaces, Proc. of the 1st World Congress of Nonlinear Analysis, Tampa, Florida, Aug. 19-26, (1992), Walter de Gruyter, Berlin, New York (1996), 1699-1712.

(91). S. Hu and N.S. Papageorgiou, (1997), Handbook of Multivalued Analysis, Vol.1 (theory), Kluwer Academic Publishers.

(92). A.V. Fursikov, Stabilization for the 3-D Navier-Stokes System by Feedback Boundary Control, Discrete and Continuous Dynamical Systems 10(1& 2), (2004), 289-314.

(93). A. Pazy, Semigroups of Linear Operators and Applications to Partial Differential Equations, Springer-Verlag, New York Berlin heidelberg Tokyo.

(94). P.Mattila and R.D. Mauldin, (1997), Measure and Dimension Functions: measurability and densities, math. Proc. Camb. Phil. Soc. (1997), 121, 81-100.

Index